Technology

Excel Manual
ANNE DROUGAS
Dominican University

Minitab Manual
AUGUSTIN VUKOV
University of Toronto

TI-83 Manual
SINGH KELLY
Pensacola College

Tenth Edition

Statistics

McCLAVE • SINCICH

PEARSON
Prentice
Hall

Upper Saddle River, NJ 07458

Editor-in-Chief: Sally Yagan
Senior Acquisitions Editor: Petra Recter
Supplements Editor: Joanne Wendelken
Executive Managing Editor: Kathleen Schiaparelli
Assistant Managing Editor: Becca Richter
Production Editor: Allyson Kloss
Supplement Cover Manager: Paul Gourhan
Supplement Cover Designer: Joanne Alexandris
Manufacturing Buyer: Ilene Kahn

© 2006 Pearson Education, Inc.
Pearson Prentice Hall
Pearson Education, Inc.
Upper Saddle River, NJ 07458

Pearson Prentice Hall™ is a trademark of Pearson Education, Inc.

The author and publisher of this book have used their best efforts in preparing this book. These efforts include the development, research, and testing of the theories and programs to determine their effectiveness. The author and publisher make no warranty of any kind, expressed or implied, with regard to these programs or the documentation contained in this book. The author and publisher shall not be liable in any event for incidental or consequential damages in connection with, or arising out of, the furnishing, performance, or use of these programs.

Printed in the United States of America

10 9 8 7 6 5 4 3 2

ISBN 0-13-149826-6

Pearson Education Ltd., *London*
Pearson Education Australia Pty. Ltd., *Sydney*
Pearson Education Singapore, Pte. Ltd.
Pearson Education North Asia Ltd., *Hong Kong*
Pearson Education Canada, Inc., *Toronto*
Pearson Educación de Mexico, S.A. de C.V.
Pearson Education—Japan, *Tokyo*
Pearson Education Malaysia, Pte. Ltd.

Excel Manual

Anne Drougas
Dominican University

■

Statistics

Tenth Edition

JAMES T. McCLAVE & TERRY SINCICH

1.1   Introduction and Overview

The purpose of this manual is to provide students with a step-by-step process for solving statistical problems introduced in the textbook. By designing a "user-friendly" manual that replicates solutions to specific examples and homework problems within the text, it is hoped that the study of statistical applications may be made more palatable and enjoyable. It is important to remember that statistical software packages such as Excel and Minitab often allow the user to bypass tedious and rigorous calculations with a few clicks of the mouse and arrive at a solution to a specific problem within a few seconds. However, the derivation of the manual calculations is only half the battle. It is secondary to understanding the philosophy and ideology of the statistical concepts and theorems presented in the textbook. Spending less time on computation and more time on developing the analytical skills necessary to interpret output is the primary goal of this manual.

By design, each chapter in this manual follows a three-step process. First, a brief introduction of the statistical concepts that need to be reviewed prior to using Excel software is presented. Remember that it will be difficult to perform calculations in Excel unless the student has a working knowledge of the concepts in the textbook. Second, a step-by-step process on how to find the relevant statistical functions is provided using pictures of where each function is located on the Excel menu. Third, examples and/or homework problems from the text are used in order to illustrate how to use Excel to derive solutions

For those that have worked with spreadsheets in Excel, Minitab, SPSS, SAS, Lotus 1-2-3, or Quattro, most of this introductory chapter may be skipped. It is important to note that all spreadsheets differ slightly in terms of presentation of menus and terminology in describing statistical functions. If new to Excel or new to spreadsheets, this chapter is a must read.

1. 2   The Basics of Spreadsheets: Using the Mouse

Examine the mouse that accompanies your laptop or desktop computer. Each mouse should have at least one, most two, buttons that may be used to click once, twice, or hold in order to perform various operations. If using a laptop, these buttons may not be contingent to the mouse but located at the lower end of the keyboard. There are four primary functions for the buttons on the mouse: point, click, double-click, and hold to drag.

1. **Point**: Pointing to objects within Excel or on a screen simply require sliding the mouse across the desktop toward the direction of the icon or cell that requires attention. Sliding the mouse to the right will move the screen arrow towards the right. Sliding the mouse to the left will move the screen arrow towards the left. When in Excel, the screen pointer may not appear to be in the shape of an arrow. Within the spreadsheet, the pointer will commonly change to an outlined arrow (in the shape of a cross) or a plus sign when pointing at the edge of a cell.

2. **Right-Click**: The term "click" means to press and then release the appropriate mouse button in order to execute a command or highlight a cell in Excel. When you right-click (press and then release the right mouse button on the mouse), a shortcut menu appears in most programs.

3. **Insert a clean worksheet**. Right click and a menu will appear (cut, copy, paste, paste special, etc.) Paste the Excel worksheet with the menu in Excel.

4. **Left-Click**: If you point to a cell within the Excel spreadsheet and left-click (press and release the left-hand button on the mouse), the cell will be activated in Excel. Visually, the cell will be highlighted on all four sides in bold and implies that the cell is ready to receive qualitative or quantitative information. If you point at a command within Excel (such as File, Edit, View, etc.) and left-click, the command is delivered

5. **Changing the Mouse Buttons**: The following discussion about the functions of the mouse assumes that the settings for the mouse are set at default. This usually implies that the mouse is designed to be used by the right hand for a right-handed individual. If left-handed, the description of the functions still applies; however, the movements of the mouse may be slightly uncomfortable. If you want to reverse the mouse buttons in Windows 98, left-click on the Start button and then left-click on Help. Find the tab option Index. Left-click on Index. Left-click within the space to type in your query and type in the word mouse. Highlight the option that specifies reversing directions and follow the directions. Notice that you may change various specifications including the clicking speed of the mouse, the appearance of the pointer, etc.

6. **Double-Left-Click**: The double-click feature is used to rapidly open a program or to execute a command quickly in order to bypass the OK option after a single click. To double click to activate program, press and release the left-hand mouse button twice quickly. Practice by placing the mouse over any icon on the desktop such as the Recycle Bin icon or the My Computer icon. Notice that a single left-click highlights the icon. A double left-click activates the icon and allows you to enter into the program.

7. **Left-Click and Drag**: By holding down the left-click button and moving the mouse at the same time, you may move objects across the screen or across cells. To drag at item to a different location, first place the mouse pointer on the object you would like to move. Once the pointer is on the object, press the left-click button down and hold it. While holding the button, slide the mouse to the place where you would like to move and drag the object. Notice that as the mouse is moving, you can visually see the object moving as well. When the destination is reached, release the mouse button and the object will appear in its new location.

1. 3   Starting and Exiting Excel

To start Excel in Windows 98/95, left-click on Start on the lower left hand side of the screen. Refer to the following steps.

Step 1: Left-click on Programs and move through the options until Microsoft Excel icon appears.

Step 2: Then left-click on that icon and begin. When Excel opens, a new worksheet should automatically appear.

Step 3: If not, left-click on File and then New Worksheet.

Step 4: When a new worksheet is opened, the Excel document should resemble the one below with cell A1 highlighted and ready for data entry.

Figure 1.1

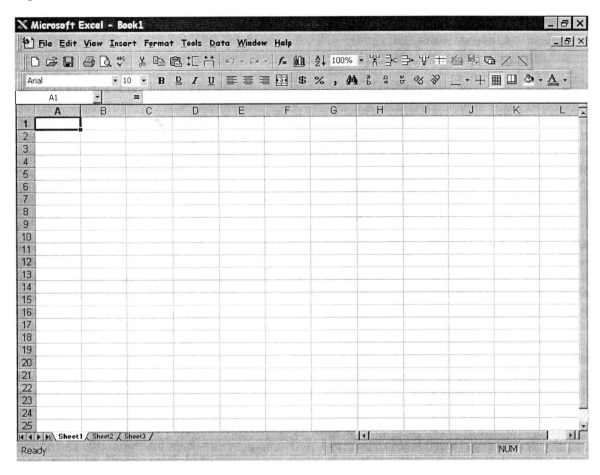

To exit Excel in Windows 98/95,

Step 1: Left-click on the upper "X" (close button) located on the top right-hand corner of the screen.

Step 2: Notice that there are two sets of duplicate buttons. The first set of buttons, a minus sign (-), a square, and the "X" button, refer to the entire Excel application. The second set of identical buttons directly below the first set, refer to the individual Excel spreadsheet being modified.

Step 3: Left-click on the upper "X" button (close button) to exit Excel entirely. At this time, Excel should prompt you to save your work before exiting. Left-click on the lower "X" to close the Excel worksheet you are currently using. In this case, you will not exit the Excel application but the current worksheet displaying your data.

Figure 1.2

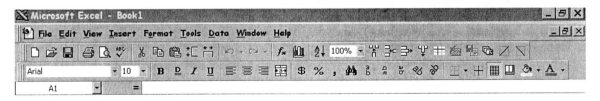

3

1.4   Starting and Exiting PHStat

To start PHStat using Windows 95/98, click on Start.

Step 1: Find Programs from the list of options.

Step 2: Move the mouse through the menus and find the PHStat icon. Click to begin.

Step 3: Once PHStat is opened, click on Enable Macros and Continue to open the program.

If an icon currently appears on the desktop, double click on the PHStat icon. Once PHStat is opened, click on Enable Macros and Continue to open the program.

The procedure for exiting PHStat is identical to that for exiting Excel. To exit PHStat in Windows 98/95, left-click on the upper "X" (close button) located on the top right-hand corner of the screen.

Step 1: Notice that there may be two sets of duplicate buttons. The first set of buttons, a minus sign (-), a square, and the "X" button, refer to the entire PHStat application. The second set of identical buttons directly below the first set, refer to the individual PHStat spreadsheet being modified.

Step 2: Left-click on the upper "X" button (close button) to exit PHStat entirely. At this time, PHStat should prompt you to save your work before exiting. Left-click on the lower "X" to close the PHStat worksheet you are currently using. In this case, you will not exit the PHStat application but the current worksheet displaying your data.

1. 5   Worksheet and Workbook Displays in PHStat

When PHStat for Excel 97 and 2000 is opened, the screen should be similar to the following figure. Slight differences in presentation of program title bars and icons may be attributed to an earlier version of Excel or PHStat or perhaps different default settings for the displays within Excel itself.

All items currently displayed on the screen will be explained. Please be sure to have a good working knowledge of the title bars, toolbars, and icons before attempting to perform any of the statistical applications presented in the textbook.

4

Figure 1.3

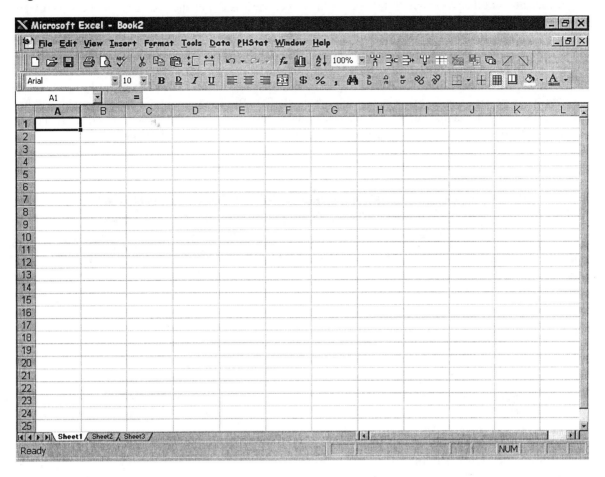

1. **Program Title Bar**: "Microsoft Excel - Book 1" At the very top of the screen, a program title bar labeled "Microsoft Excel - Book 1" should appear by default. This indicates the name of the current workbook being used and the name of this workbook may be changed when the option to save is presented. Also noted that when the user is prompted to save work, all worksheets will be saved together in a workbook with a name created by the user.

2. **Creating Worksheets within a Workbook** Each workbook contains by default three sheets stored within the book. To create more worksheets, simply right-click over the tab labeled "sheet 1", and left-click "insert". Then click "OK". A new worksheet will appear. Notice that right clicking over "sheet 1" provides the user with the option of changing the title of the sheet, moving the sheet, etc.

3. **Program Icons** Previously discussed, the sizing buttons in the upper right hand corner of the workbook allow the user to minimize, maximize, and close the entire workbook.

4. **Insert minimize button here** The minimize button reduces the current workbook with its collection of worksheets to a button at the very bottom of the screen. Notice that the sheet is not closed but is dormant until the user reactivates the workbook. Why minimize a workbook? If working with several applications (for example, Word and Excel), it may be faster to reopen a minimized spreadsheet for reference than to close one application and reopen the spreadsheet.

5

5. **Insert maximize button here** The maximize button allows the user to increase the screen presentation of the workbook to full-size if the screen view is diminished or decrease the screen view of the workbook if the screen is displayed in full.

6. **Insert close button here** The close button will close the Excel application. At this time, the user will be prompted to save work.

7. **Main Menu bar**: The first row beneath the Program Title Bar is commonly called the main menu bar and displays commands that are grouped into categories called File, Edit, View, Insert, Format, etc. Left-click on "File" and notice that a group of subcommands falls from the file option.

Figure 1.4

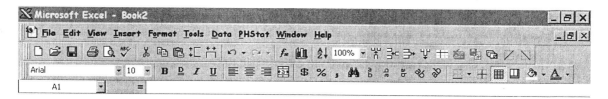

8. **Toolbar**: The row of icons (pictures) below the main menu bar is referred to as the toolbar. It consists of a row of pictures that allow the user to open a new workbook, save onto disk, print, check for spelling errors, etc. without having to hunt through the options displayed within the main menu bar. If the mouse is moved over the icon, a one or two word explanation of its function will appear. At the end of the row, the help icon (a question mark) may be used to find out more information about any information on the toolbar. A single click on the question mark will open a help menu that will ask the user to type a question into the appropriate space. After the question is typed, Excel will display a variety of topics that cover the question asked.

9. **Font Size and Justifications:** The row below the toolbar contains specific options for adjusting the font, size, and appearance of the data being entered into Excel. Notice that the font, by default, is Arial with a 10pt size. The user may scroll down the options to select the font and font size that is appropriate. In addition, there are options to change the appearance of the font to bold and italics and to change the justification to right, left, or center. The remaining options control the type of currency being entered, percentage, use of commas, and numbers of decimals displayed.

10. **Formula Bar**: The last row displays two primary windows. The first window shows the cell that is currently active within the worksheet. When opening a new worksheet for the first time, the active cell by default is A1. In the second window, the user may begin to input qualitative or quantitative data.

In the second window, next to the equal sign, type in **"PHStat 1"**. Notice that the formula bar now displays three more buttons.

6

Figure 1.5

The red "X" may be clicked when it is necessary to delete the information in the active cell.

The green check mark may be clicked when the data is entered. The data will then remain in the cell.

If the "=" sign is clicked, prior to typing in data, the formula bar will prompt the user to enter a formula. The f_x icon will provide a variety of equations from finance to statistics to use to evaluate data.

11. **Worksheet Area**: The worksheet area generally consists of columns labeled as letters and rows as numbers. There are roughly 256 columns and 65,536 cells. Press CTRL and DOWN ARROW at the same time to move to the very last row and CTRL and RIGHT ARROW to move to the very last column.

Figure 1.6

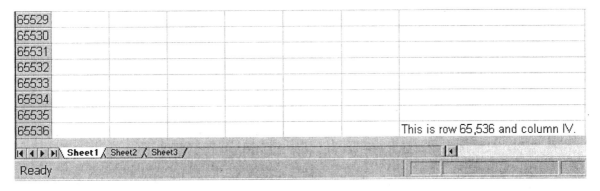

12. **Scroll Bars**: There are two sets of scroll bars located on the far right and at the bottom of the worksheet. The arrows at each end allow the user to scroll up and down the worksheet and right and left within the worksheet.

13. **Worksheet Tabs**: Already discussed, each workbook contains by default three sheets stored within the book. To create more worksheets, simply right-click over the tab labeled "sheet 1", and left-click "insert". Then click "OK". A new worksheet will appear. Notice that right clicking over "sheet 1" provides the user with the option of changing the title of the sheet, moving the sheet, etc.

1.6   Maneuvering through Menus and Help Tools in Excel

When in Excel, the first row of commands is referred to as the **Main Menu Bar**.

Step 1: Click on the word **File** and notice that a variety of subcommands appear in a column below the File command. Notice at the end of that column, a double arrow pointing downwards appears in Excel 2000.

7

Step 2: Click on that arrow and the full array of subcommands under File will appear. Notice that **New** will open a new workbook and **Save** will save the current workbook to a user's specification.

Step 3: Clicking within any cell in the existing spreadsheet will close the subcommand.

Step 4: **Page Setup** will allow the user to adjust margins, headers and footers, etc.

Figure 1.7

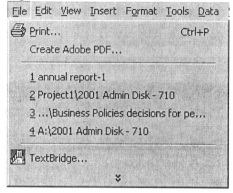

Figure 1.8

8

1.7   Help Function

In order to quickly find out where to locate a function or command,

Step 1: Click on the last icon at the extreme right within Excel, the question mark icon. This icon, referred to as the **Help Tool**, allows the user to find the appropriate series of commands needed to perform a task.

Step 2: Click on the Help Tool and notice a box appears containing three tabs: **Contents, Answer Wizard,** and **Index**.

Step 3: If the tab **Answer Wizard** is activated, the user may type in a specific question and the help key will formulate a solution to the question generated.

Figure 1.9

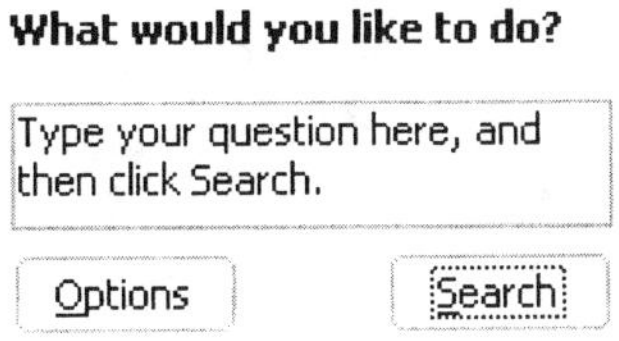

What would you like to do?

Type your question here, and
then click Search.

| Options | | Search |

For example, suppose more specific questions regarding printing a document is needed. Often, a statistician may want to change the page orientation from **Portrait** (a standard 8 1/2 by 11 print) to **Landscape** (11 by 8 1/2 print).

Step 1: Click on the **Help Tool** button at the far right of the Excel menu.

Step 2: Within the **Answer Wizard**, type in **Print: Portrait versus Landscape**. Then click **Search**.

Step 3: Notice that various topics are displayed. Highlight **Change the Page Orientation** and step-by-step directions for changing from Portrait to Landscape and from Landscape to Portrait are presented.

Figure 1.10

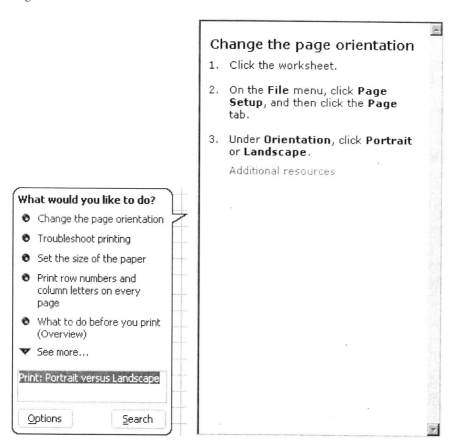

In addition, suppose more specific questions regarding the Binomial Distribution (covered in Chapter 3 of the textbook) are needed.

Step 1: Click on the **Help Tool** button at the far right of the Excel menu.

Step 2: Within the **Answer Wizard**, type in **Binomial Distribution**. Then click **Search**.

Step 3: Click on **BINOMDIST Worksheet Function** and a description of the Excel commands for formulating Binomial Distributions will appear.

Figure 1.11

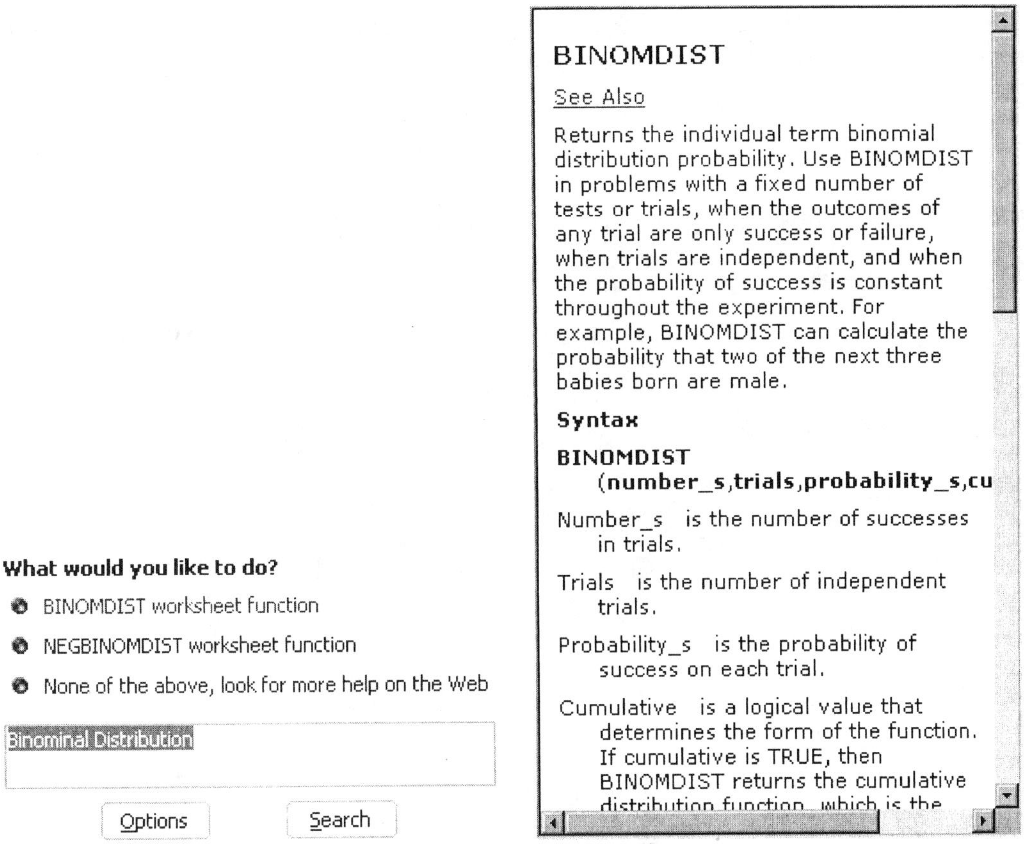

BINOMDIST

See Also

Returns the individual term binomial distribution probability. Use BINOMDIST in problems with a fixed number of tests or trials, when the outcomes of any trial are only success or failure, when trials are independent, and when the probability of success is constant throughout the experiment. For example, BINOMDIST can calculate the probability that two of the next three babies born are male.

Syntax

BINOMDIST (number_s,trials,probability_s,cu

Number_s   is the number of successes in trials.

Trials   is the number of independent trials.

Probability_s   is the probability of success on each trial.

Cumulative   is a logical value that determines the form of the function. If cumulative is TRUE, then BINOMDIST returns the cumulative distribution function, which is the

What would you like to do?

● BINOMDIST worksheet function

● NEGBINOMDIST worksheet function

● None of the above, look for more help on the Web

Binominal Distribution

Options   Search

1.8   Status Bar

At the very bottom of the Excel screen, the **Status Bar** indicates what function is currently being undertaken by the user. Notice that **Ready** is displayed at the left of the status bar when a cell is idle or inactive. When editing a cell, the display changes from **Ready** to **Edit**. To indicate the difference, highlight cell A1 and type in "123456". Click the green check button to accept the number sequence. Notice that once the green check button is clicked, the number sequence "123456" is displayed in A1 and in the Formula Bar. Notice how the three icons, including the green check mark, disappear. Now, edit the number sequence by changing the number "4" to "9". Move the pointer to the numbers in the formula bar and notice that the pointer changes to an **I-beam**. Place the pointer to the right of the number "4" and then click in that location. Now the three icons, including the green check mark, reappear. Press the backspace key and the number "4" disappears. Manually enter the number "9" in its place and click on the green check mark to indicate acceptance.

The right hand side of the Status Bar indicates whether the **Number Lock**, **Scroll Lock**, and **Caps Lock** are turned off or on. If the **Number Lock** is turned on, the word **Num Lock** will appear on the status bar. This indicates that the keypad can be used to enter numbers. The Num Lock key may be turned off by pressing Num Lock on the keypad. The same analogy applies for the Caps Lock and Scroll Lock keys.

11

1.9   Dialog Box

A **Dialog Box** allows the user to select an option from a number of alternatives. For example, click **File** from the menu option and then select **Print**. A dialog box should appear similar to the one in the following figure.

Figure 1.12

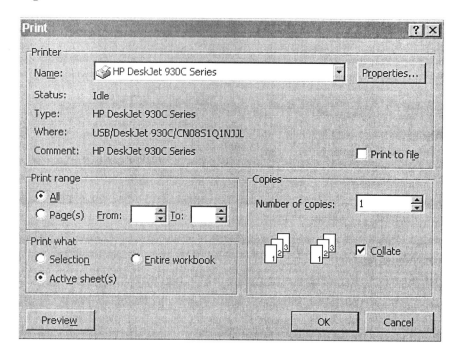

At the top of the dialog box, the user may select the type of printer to be used in order to print out the Excel document. Usually, by default, the printer currently displayed is the one connected to the computer. The **Print Range** allows the user to select the number of pages to be printed and the number of copies of each page to be made. In addition, the option to print a selection of a page, the active sheet, or the entire workbook is provided.

Step 1: Click on **Properties** and a series of tab options providing specifications for features and color appear.

Step 2: Select **Features** and notice that the page orientation may be changed from portrait to landscape. Click **OK** to keep the additions made for this print job only.

Step 3: Click **Default** to change any additions made to all print jobs in the future.

Figure 1.13

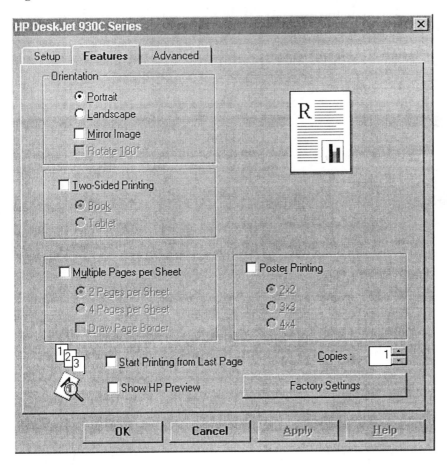

1.10 Help Icons: Finding Help on Specific Tasks

The most important command needed to navigate through Excel is the Help command, which is designed to solve specific problems defined by the user. Click on the question mark icon located in the upper right hand corner of the Excel 2000 worksheet. The menu is subdivided into three categories or tab options: **Contents**, **Answer Wizard**, and **Index**. The names of the options may differ slightly in Excel 97 or 95.

1. **Contents** provide a broad overview of main categories for troubleshooting such as formatting worksheets.
2. After clicking on a broad category, a list of more defined categories will appear. **Answer Wizard** allows the user to type in a specific question upon which a series of categories that match that description will appear in the empty dialog box below.
3. **Index** presents an alphabetized list of topics similar to an index at the back of a textbook.

13

Figure 1.14

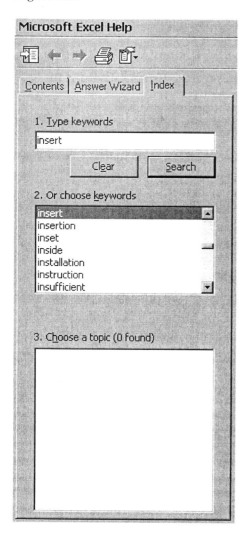

For example, suppose that more information on inserting a worksheet is needed.

Step 1: Type in the keyword **Insert** as directed and click **Search**. Notice that 380 topics on Insert are found.

Step 2: Click on the topic **"Insert a New Worksheet"** and notice that to the right of the screen, the directions for inserting a new worksheet are provided in stepwise form.

Figure1.15

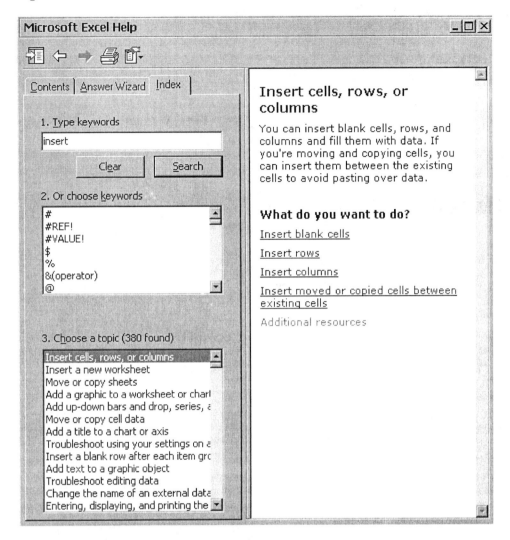

1.11 Changing the Name of a Worksheet and Inserting/Deleting Worksheets

If Excel is started by clicking on an Excel icon, Excel will automatically open to a new blank worksheet. By default, the title at the top is called **Book 1** and the three worksheets accompanying it are titled **Sheet 1, Sheet 2,** and **Sheet 3**. All three sheets are stored in a unit called a **workbook**, which is simply a collection of worksheets.

Step 1: By default, each worksheet is titled Sheet1, Sheet 2, etc. Right click over the title Sheet 1 and notice that a series of subcommands (**Insert, Delete, Rename**, **Move or Copy**, etc.) appear within the current worksheet.

Step 2: Click **Rename** and title it with a descriptive name representative of the data set being analyzed. Click **Insert** and the user is prompted to add a new worksheet to the workbook.

Step 3: Click **Delete** and the user is prompted to delete the existing worksheet from the workbook.

15

Figure 1.16

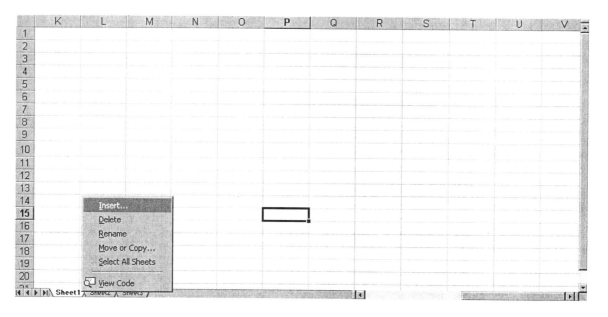

Figure 1.17

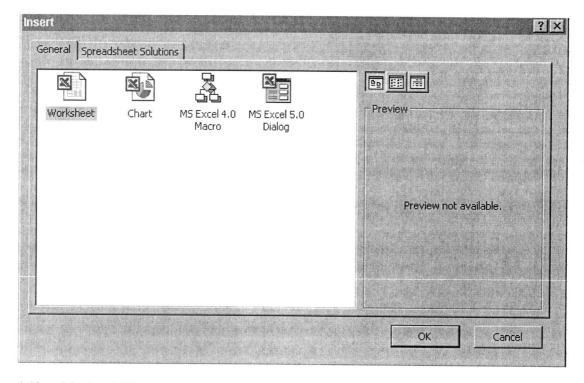

1.12   Moving A Worksheet

Step 1: There is also the option to change the position of the worksheet in its array. Suppose that a researcher prefers that Sheet 2 appear before Sheet 1 in an array of worksheets.

16

Step 2: Right click over Sheet 1 and click **Move or Copy**. The user may move the worksheet, Sheet 1, to another workbook altogether or move Sheet 1 **before sheet** 3. This would place sheet 1 in position behind Sheet 2 as preferred by the user.

1.13   Activating a Cell or a Range of Cells

Each **cell** is defined by a combination of letters and numbers located at the top and left side of the spreadsheet. The letters refer to the column location and the numbers refer to the row location. For example, the cell indicating the very first row and column of the spreadsheet is referred to as cell A1. A1 is considered the **address** of that cell, much like the address of any individual residence. There are roughly 256 columns and 65,536 cells. Press CTRL and DOWN ARROW at the same time to move to the very last row and CTRL and RIGHT ARROW to move to the very last column. The last cell is located in row 65,536 and column IV.

Notice that when a new worksheet is opened, cell A1 is automatically the active cell. A cell is "active" when its address (in this case, A1) appears displayed in the Name Box, next to the formula bar. It is also active when a cell is highlighted, in bold, on all four sides as indicated. For practice, click on cell B5. Notice that B5 appears in the Name Box and is highlighted in bold on all four sides.

Figure 1.18

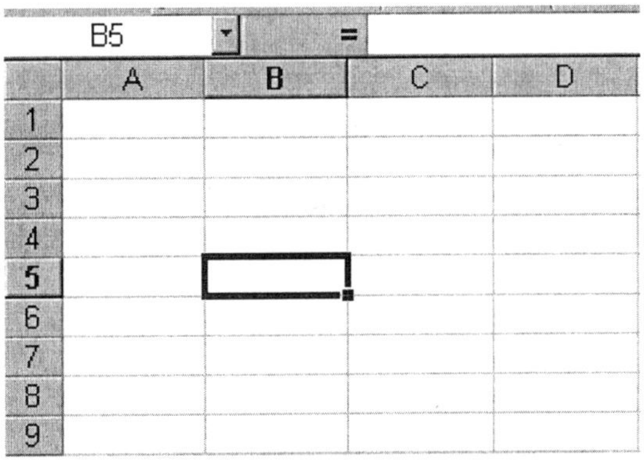

A **range** of cells is a group of cells. Now that B5 is active, click on B5, hold the mouse button down and drag to B8. When the mouse button is released, cell B6 through B8 is highlighted. The address of this range is now called B5:B8 where a colon separates the top cell from the bottom cell. Notice how the address within the name box does not change.

To practice, activate cell B5. Once active, click on B5, hold the mouse button down and drag to C8. When the mouse button is released, cell B6 through C8 is highlighted in bold. The address of this range is now called B5:C8 where B5 refers to the upper left cell and C8 refers to the lower right cell.

Figure 1.19

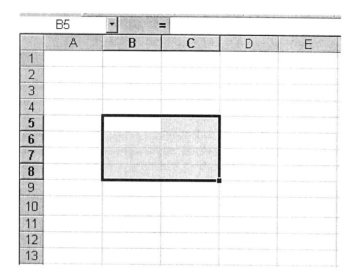

Continue to practice activating many cells by placing the cursor in the upper left hand cell and dragging to the lower right. Or place the cursor in the lower right hand cell and drag to the upper left.

1.14   Cutting and Copying Information into Cells

In order to remove information from one cell and insert that information into a new cell or to copy information into a new cell, use the **cut-copy** commands located within Excel. Notice that cutting a cell means to remove the original cell contents from that cell with the intent of copying it elsewhere. This implies that the cell is empty after the data is "cut" from it. Copying a cell implies that the data will exist in two locations after the command is executed.

Let's use the same number sequence, 13579, to illustrate the technique of copying. Enter the number sequence into cell A1 on the worksheet.

Step 1: To copy the number sequence into cell B1, first activate A1 by clicking on it. Notice that the cell is highlighted.

Step 2: Click **Edit** from the menu option and select **Copy** from the list of subcommands. Notice how the cell is no longer highlighted but is illustrated by a dotted line surrounding the cell that blinks.

Step 3: Click on the cell where the information is to be copied, B1. This cell should now be activated.

Step 4: Click **Edit** from the menu option and select **Paste**. Now the number sequence, 13579, appears in two places.

Step 5: Activate cell C1, click **Edit** and then **Paste**. Notice the number sequence is copied again.

Another way to copy information is to use the copy icon on the standard toolbar. For practice, activate cell C1. Click on the copy icon, which looks like two pieces of paper on top of each other, located next to the scissors icon. Activate cell D1 and then click on the paste icon, which is depicted as a clipboard with a paper attached to it. Now the number sequence, 13579, also appears in cell D1.

18

1.15   Formatting Numbers within a Cell or within a Range

Step 1: To format numbers within Excel, click **Format** and then **Cell**. Notice that a Dialog Box appears with many options for formatting. By default, all cells are specified as **general**. However, once a cell or range of cells is highlighted, the user may specify currency, decimal, percentage, etc. This dialog box presents all possible formatting options; however, icons on the toolbar for the more popular options (percentages, dollars, etc.) can be easily accessed for conversion.

Figure 1.20

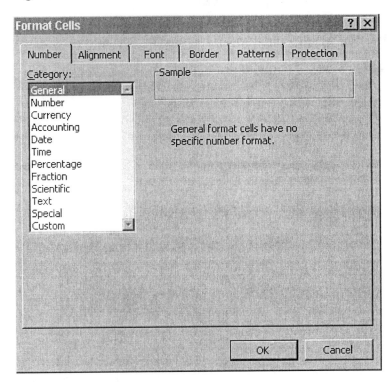

Step 2: Please open to a clean spreadsheet or delete the contents of the spreadsheet within Excel. (Press **CTRL+A and then Delete** to delete the entire spreadsheet.) Please enter the following numbers into the Excel spreadsheet as shown below in cells A1 through A5. Then copy those numbers into columns B, C, D, and E.

19

Figure 1.21

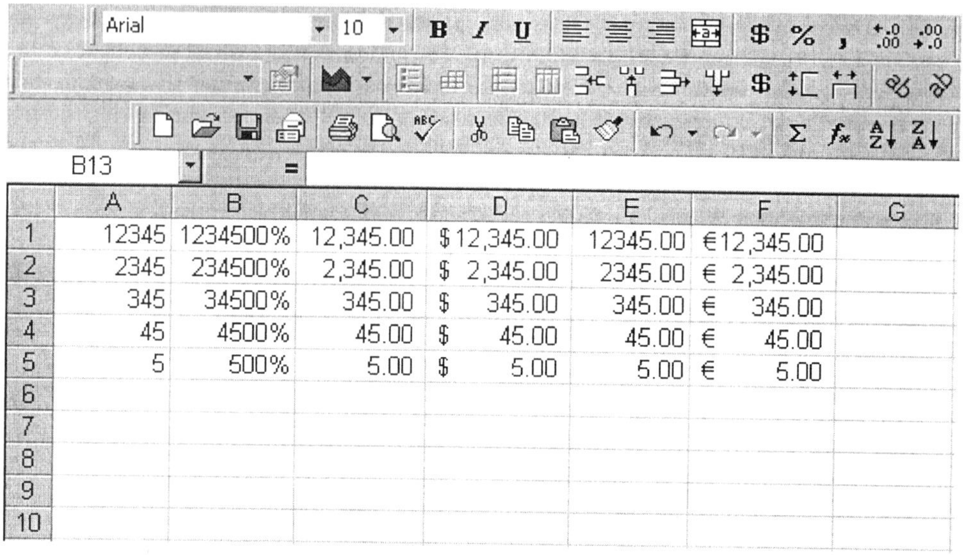

	A	B	C	D	E	F	G
1	12345	1234500%	12,345.00	$ 12,345.00	12345.00	€12,345.00	
2	2345	234500%	2,345.00	$ 2,345.00	2345.00	€ 2,345.00	
3	345	34500%	345.00	$ 345.00	345.00	€ 345.00	
4	45	4500%	45.00	$ 45.00	45.00	€ 45.00	
5	5	500%	5.00	$ 5.00	5.00	€ 5.00	
6							
7							
8							
9							
10							

Step 3: Percents: To change the format of the cell contents to a percentage, click on the "B" at the top of the second column. Notice how the entire column is highlighted. Locate the icon, **%**, and click on it. Notice how the numbers expressed in column A are changed into percentages. An alternative method is to highlight column B as above and click **Format** and select **Cells**. A dialog box appears that allows the user to change the cell format to **percentage**. Click **OK** and now the numbers are displayed in percentage form.

Step 4: Comma: To change the format of the cell contents to a comma, click on the "C" at the top of the third column. Locate the "**,**" icon on the toolbar and click it. Now the data in the column is expressed with a comma.

Step 5: Currency: Click on the "D" at the top of the fourth column and locate the "**$**" icon on the toolbar. **Click** the **$ icon** and notice how column D is now expressed in dollars. An alternative method is to highlight column "D" and click **Format-Cells-Currency** or **Format-Cells-Accounting**. Notice the difference between the two commands. The accounting format will identify the difference between negative and positive dollar amounts. By convention, negative dollars will be marked in red or in parenthesis while positive dollar amounts will be expressed as normal. Scroll through the options under **Format-Cells-Accounting** and be familiar with those options.

Step 6: Changing the Decimal: Click on the "E" at the top of the fifth column and locate the two decimal icons to increase or decrease the decimal point. **Click Increase Decimal** twice and notice that two decimal places appear. **Click Decrease Decimal** twice and notice that the two decimal places disappear. An alternative method is to highlight column E, and click **Format-Cells-Number** and set the number of decimals to the option desired.

Step 7: Euro-currency: In Excel 2000, the user has the option of converting numbers in **Euros**. Click on the "F" at the top of the sixth column and locate **Euro** icon on the toolbar. Click it and notice that the numbers are displayed in Euros.

20

1.16 Formatting Numbers that Exceed Seven or Eight Digits

Step 1: In an empty cell (A1), type in the number sequence 123456789123456789. Notice that when enter is pressed, the number displayed in cell A1 is 1.23457E+17 (This display may be different given the default settings for Excel). One way to display the entire number sequence is as follows

Step 2: Highlight cell A1

Step 3: One the main menu bar, click **Format-Cell**.

Step 4: Open the number tab from the list of options and select **NUMBER**. Click **OK**.

Step 5: Notice how cell A1 displays #######. This symbol indicates that the number is too large to be placed in cell A1 given its existing width. Place the cursor between "A" and "B" at the top of the column.

Step 6: Double left-click between "A" and "B" and the entire number sequence will be displayed.

Figure 1.22

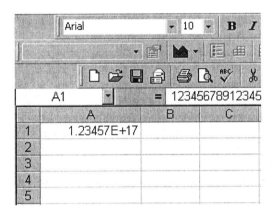

Figure 1.23

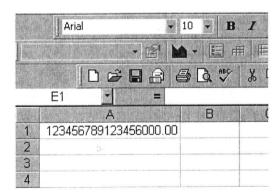

1.17   Inserting or Deleting Rows and Columns

Step 1: Use the spreadsheet from the preceding section, which contains data in columns A through F. Suppose that a column needs to be inserted between A, and B. To do this, click on the letter "B" at the top of the column to highlight it.

Step 2: On the main menu bar, select **Insert-Columns**

Step 3: Insert more columns, as desired, by continuing to click the mouse.

Step 4: Suppose that two columns need to be inserted between A and B. To do this, highlight columns B and C simultaneously

Step 5: On the main menu bar, select **Insert-Columns**

Step 6: To delete these two new columns, highlight them and then click **Edit-Delete**

The same procedure is followed for inserting or deleting rows. Suppose that a row needs to be inserted between row 1 and 2. To do this,

Step 1: Click on the number "2" at the top of the row to highlight it.

Step 2: On the main menu bar, select, **Insert-Rows**

1.18   Aligning Information

Within a clean worksheet, please enter a series of unaligned data such as the following. As the last row is entered, 0.765432 is displayed instead of 0.7654321. Because Excel has been set to display numbers to six decimals by default, it rounds off any values that exceed that size. Notice in cell A8 the rounded number is displayed; however, the unrounded number is displayed in the edit window. Mark cell A8 and notice the difference.

Figure 1.24

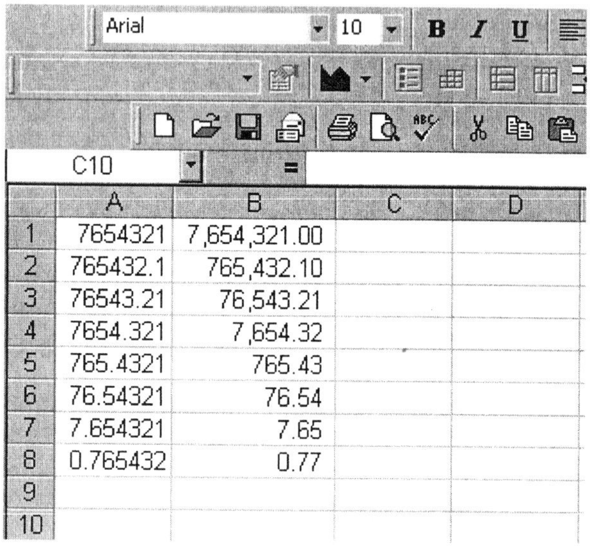

22

There is a simple way to align the data.

Step 1: Highlight the column and click on the comma,, , icon. This will display numbers to two decimal places.

Step 2: Increase or decrease the number of decimals by using the appropriate icons.

1.19   Creating a Series: Using the Series Command

Use the series command within Excel to generate a series that is not readily identifiable by Excel or to create a specific identifiable series in Excel. For example, days of the week and months of the year are special series that have a universal, agreed upon sequence. This is recommended for use when a user wants to specify specific options. Try this example.

Step 1: Enter the number "6" in cell A1, "8" in cell A2, and "10" in cell A3

Step 2: Activate the range A1:A12

Step 3: Click **Edit-Fill-Series** and a dialog box is displayed that prompts the user to provide information about the series. By default, Excel's interpretation of the series is provided. Notice how the series will be stored in columns, it is a linear series and the step value is "2".

Step 4: Click **OK** and the series will be produced.

Figure 1.25

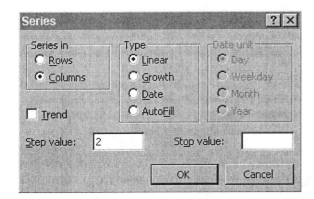

1.20   Sorting Data

Sorting is useful if the data needs to be rearranged from the original format of entry. For example, a user can alphabetize the names in a spreadsheet if presented in an unalphabetized manner. Using a new spreadsheet, enter the following data.

Suppose that an address book with names and telephone numbers are entered into Excel and they need to be alphabetized.

Step 1: Highlight the data.

Step 2: Find the menu option **Data-Sort**

Step 3: The upper left cell is listed as the default variable to use. Notice that if cell A1 is nonnumeric (a label to identify the column) click on **My list has a header row**.

Step 4: Notice that **Sort by** Name should appear in the first entry prompted by the dialog box.

Step 5: Choose whether to sort the data in ascending order or descending order. In this case, ascending order would alphabetize the list from A to Z. Then click OK.

Notice that the names are now arranged in alphabetical order and the corresponding data is located in the rows next to them.

Figure 1.26

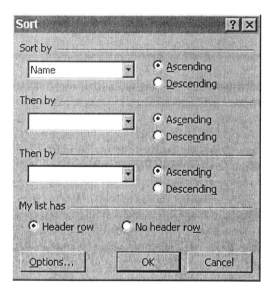

Figure 1.27

	A	B	C	D
1	Name	Home Phone	Work Phone	
2	Stephen	555-2361	555-2112	
3	Catherine	555-1938	555-1139	
4	Joseph	555-2963	555-9998	
5	Peter	555-1212	555-3456	
6	Anne	555-1313	555-8455	
7				
8				

There are three options for sorting and in this example, each name used is unique. This may not always be the case. For example, when dealing with a large data set, it might be required to sort information by zip code or area code. In this case, there may be many individuals with the same zip or area code. If this occurs, it would be necessary to sort first by area code and then by name, for example.

24

1.21   Saving Information

It is always recommended that work be saved regularly in case of accidents or power outages.

Step 1: Within a new worksheet, click on **File** within the Main Menu and click on **Save**.

Step 2: The program will prompt the user to provide a descriptive file name to the collection of worksheets.

Step 3: Clicking **OK** confirms that the file is saved however it will be saved in the last location used. This could be on the hard drive or on the network for example. Some Excel programs will save work into a folder called **My Documents**.

It is recommended that when work is saved,

Step 1: Click on **File-Save As**. A dialog box will appear. At the top of the dialog box, the program prompts the user to save work in a file or folder.

Step 2: Click on the arrow key and select where the information should be saved, such as on a floppy disk (a:) or into the **My Document** folder on the hard drive. Be sure to provide the file with a specific name that will be remembered.

Step 3: Save as Type should be labeled as **Microsoft Excel Workbook** or a specific Excel file indicated by default within Excel. Click on the arrow key and scroll through the options listed.

Figure 1.28

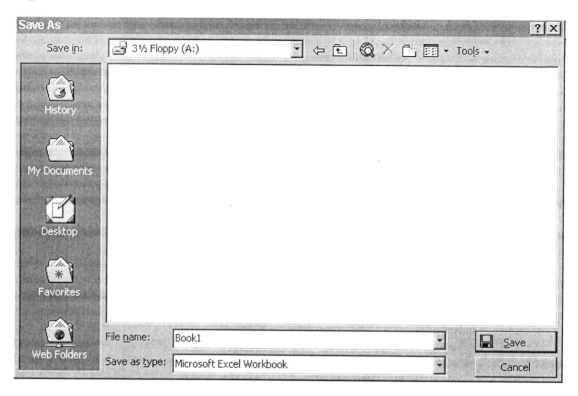

1.22   Printing: Page Setup

Using **File-Page Setup**, it is easy to determine how the Excel document will look once it is printed. Notice that the Page Setup dialog box displays four tabs - **Page**, **Margins**, **Header/Footer**, and **Sheet**.

1. **Page** allows the user to change the orientation of the print from **portrait** (8 1/2 by 11) to **Landscape** (11 by 8 1/2). **Scaling** allows the image to be reduced on the printed page or enlargened on the printed page. **Fit to** commands the program to automatically adjust the size to fit any number of pages.

2. **Paper Size** and **Print Quality** allows the user to adjust the letter size use and the print quality. Remember that changing the print quality to a higher dpi value may increase the printing time as well as the amount of ink used.

3. **First Page Number** is usually set to **Auto** as default. If number is required to start at a specific page, highlight auto and replace it with the required number.

Figure 1.29

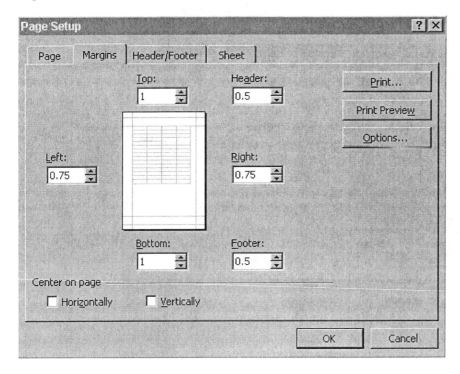

4. Using the Margins tab, the margins may be adjusted on all four sides of the document. It is recommended that **Print Preview** be always used to view changes prior to printing. At the bottom of the dialog box, the option of centering output generated horizontally or vertically is also provided.

Figure 1.30

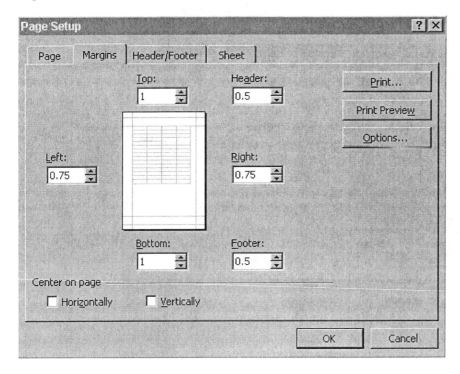

27

1.23 Printing: Header/Footer

Headers contain descriptive information printed across the top of a page whereas **footers** contain information at the bottom of the page. Why use headers/footers? Sometimes it is necessary to print a description across each page, such as a chapter title, page number, name, etc.

Step 1: Click **File-Page Setup** and select the tab for **Header/Footer**. Notice how the default setting indicates no headers or footers.

Step 2: Under **Header**, click on the arrow at the right and a display of previously used headers, if any, will be listed. If there is a header that is appropriate, select it and edit it.

Figure 1.31

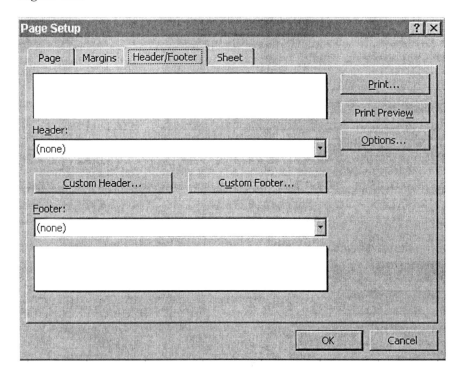

Step 3: It is also possible to create a header. Click **Options** and notice that the following dialog box appears.

Step 4: To format the text, the header may be placed in the left, center, or right of the printed page. Note that this refers to the top of the document. The icon, **A**, allows the user to write in text and specify the font and size of the font for the header. The remaining icons provide information regarding page numbering, date and time, file name, and sheet name.

Step 5: Click on each icon and become familiar with customizing the header. Note that the same dialog boxes will appear for footers except the placement of footer is at the bottom of a page.

28

Figure 1.32

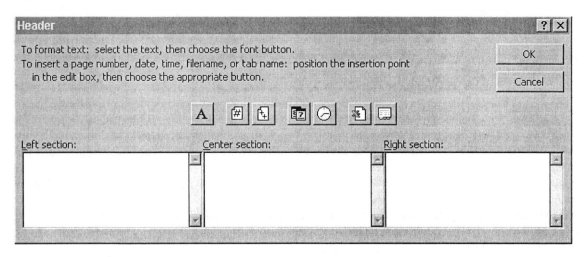

1.24 Printing: Sheet

The last tab found by using **File-Page Setup-Page** allows the user to choose the area of the spreadsheet for printing, the quality of the print, etc. Notice that when the Sheet tab is activated, **Print Area** allows the user to select the range to print; this may include specific columns or specific rows or a combination depending upon the user's specification. For clarification regarding each option, click on the question mark and point to each area for specific instructions. It is ALWAYS recommended that **Print Preview** be selected prior to printing. This will save time as well as paper while adjustments to the print job are being made.

Figure 1.33

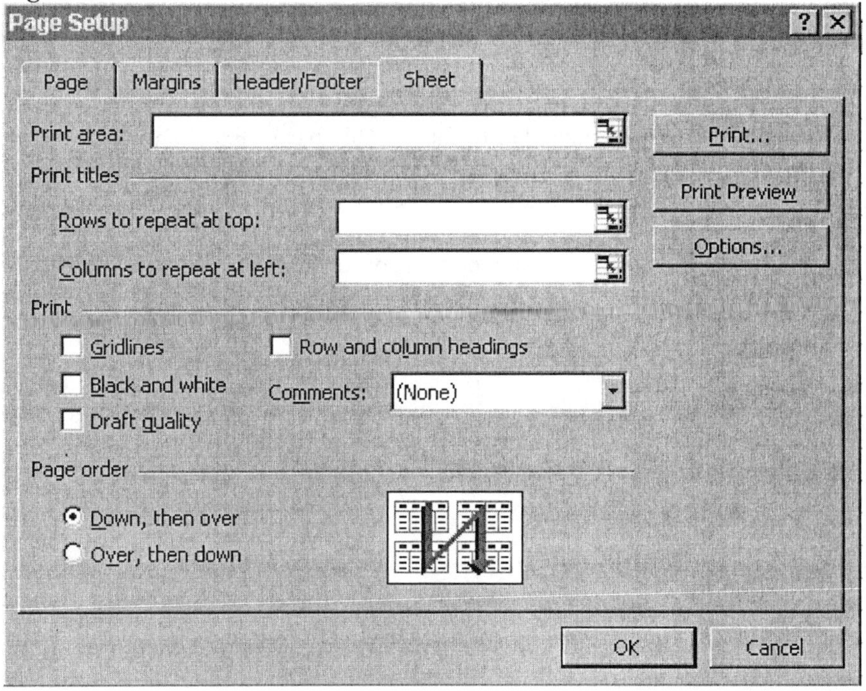

1.25   Algebraic Symbols in Excel

When creating formulas in Excel, it is necessary to remember the basic symbols used within Excel. All formulas are created in the formula bar within Excel. It is important to remember that the equal sign, =, must be clicked prior to starting a calculation. Before creating formulas in Excel, let's introduce the basic symbols needed for formulas, which may vary from the traditional symbols used in a basic mathematics class.

1. **Addition**: The symbol for addition is the plus sign, +. To write a formula to find the sum of 2 and 3, simply write 2+3.
2. **Multiplication**: The symbol for addition is the asterisk, . To write a formula to find 2 multiplied by 3, write 2*3.
3. **Subtraction**: The symbol for subtraction is the minus sign, -. To write a formula subtracting 2 from 3, write 3-2.
4. **Division**: The symbol for division is the backward slash, /. To divide 3 by 2, write 3/2 in Excel
5. **Exponentiation**: The symbol for raising a number to a power, or exponent, is the carat sign, ^. To raise 2 to the third power, write 2^3 in Excel.
6. **Percent**: The symbol for expressing a number in percentage terms is %. To express 2 as a percent, write 2% in Excel and the number will be converted to 0.02.
7. **Negation**: The symbol for negation is the minus sign, -. To indicate 2 as a negative number, write -2 in Excel.

1.26   Order of Operations in Excel

By convention, Excel will calculate expressions in parentheses first and then use results to continue on with the rest of the formula. Without parentheses, the order of operations is
1. Negation (-2 or -6)
2. Percent
3. Exponentiation
4. Multiplication/Division
5. Addition/Subtraction

Using an example, calculate the following expressions. Go to the formula bar in Excel and click on the equal sign. Now that the formula bar is activated, proceed with the following calculations.
1. =3 + 4*5   produces 23 because multiplication occurs before addition
2. =3 +(4*5)   produces 23 because parenthetical operations occur first
3. =(3+4)*5   produces 35 because parenthetical operations occur first
4. =3 + 4*5^2

2.1   Introduction

Chapter 1 was designed to introduce many of the basic definitions and concepts needed to interpret various types of datasets. Within this chapter, two primary areas of statistics are introduced: descriptive statistics and inferential statistics. Chapter 2 of the text sets the foundation for understanding and interpreting descriptive statistics by employing techniques to summarize and categorize data numerically as well as pictorially through various types of graphs.

Within Excel 2000/2003, there are a variety of options that can be utilized to graph qualitative as well as quantitative data. For qualitative data (nonnumeric in nature), **pie charts** and **bar charts** can be customized to the statistician's preference in terms of presentation. For quantitative data, **box plots, histograms, scatter plots,** and **stem-and-leaf displays** can all be used to create a visible picture of numeric data for presentation purposes.

In addition to various graphing options presented in Excel, the three primary measures of central tendency (**mean, median,** and **mode**) and dispersion (**range, variance,** and **standard deviation**) can all be calculated using the Data Analysis-Descriptive Statistics function within Excel.

The following examples will be used to illustrate techniques in PHStat and Excel.

> Discussion from Section 2.1 (Table 2.1), page 30 (Data on 22 Adult Aphasiacs)
> Example 2.2, (Table 2.4), page 45 (Percentage of Student Loans (per state) in Default)
> Example 2.4, page 56 and (Table 2.2), page 41 (EPA Mileage Rating on 100 Cars)
> Sample Exercise #1: Problem 2.163 (Gallup Poll on Smoking), page 107
> Sample Exercise #2: Problem 2.166 (Competition for bird next holes), page 107
> Sample Exercise #3: Problem 2.83 (Ammonia in car exhaust), page 72

2.2 Graphical Techniques in Excel: Pie Charts in Excel

Section 2.2 of the text presents various graphical techniques for describing quantitative data. Excel 2000/2003 allows a statistician to present and customize bar graphs, histograms, stem-and-leaf displays, pie charts, and scatter plots for a variety of datasets. In most instances, graphics present a more easily readable presentation to a viewer and allows that viewer to analyze trends or anomalies in data quickly. Please refer page 30 of the text (Table 2.1) which contains summary data on the proportion and number of aphasia types. The steps to constructing Pie Charts are as follows.

Step 1: Open a new workbook and place the cursor in the upper left-hand corner of the spreadsheet.

Step 2: Open the database or input the data manually.

Step 3: Select f$_x$ from the menu options on the toolbar.

Step 4: Select Pie Chart from the choices available. Notice how several different pie charts are presented. Select the first pie chart highlighted in the upper left-hand corner by default. This is referred to as **Step 1 of 4 in Chart Wizard. (Figure 2.1a)**

Step 5: In **Step 2 of 4**, **Highlight the data**. Be sure that the data is grouped either by **columns or rows** given the presentation of the data. **(Figure 2.1b)**

Step 6: In **Step 3 of 4**, an output title may be assigned to the data if desired. **(Figure 2.1c)**

Step 7: In **Step 4 of 4**, place the chart either within the existing sheet or within a new workbook. **(Figure 2.1d)**

Step 8: Click OK to finish.

Figure 2.1a

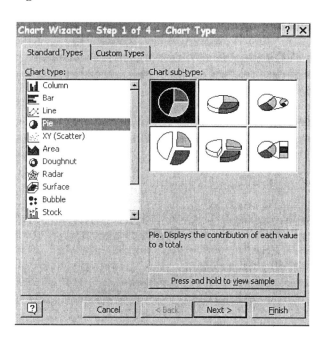

Figure 2.1b

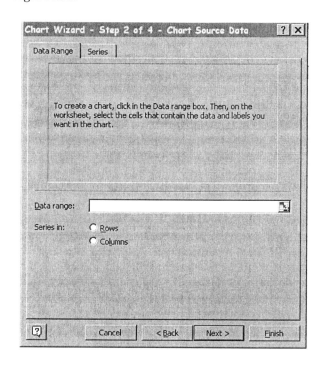

Figure 2.1c

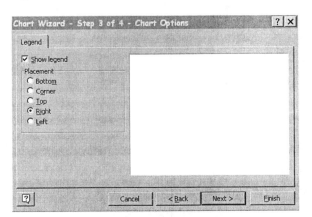

Figure 2.1d

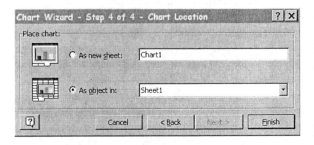

The following example will be used from the text to illustrate how to construct a pie chart. Since none of the examples in the text specifically requires the construction of pie charts, the data from Table 2.1 of the text will be used.

Example 2.1: Please refer to Section 2.1 and the data listed in Table 2.1 (page 30) of the text.

Consider a study of aphasia published in the *Journal of Communication Disorders* (Mar. 1995). In Table 2.1 of the text, five aphasiacs in the study were diagnosed as suffering from Broca's aphasia, 7 from conduction aphasia, and 10 from anomic aphasia. Construct a pie chart showing the disparity within the three types of aphasia. In order to answer this question, we must first organize the data into categories using the COUNTIF function. The steps to using COUNTIF are as follows:

Step 1: Click the fx icon on the toolbar. This is the insert function. Find COUNTIF and click OK.
Figure 2.1e

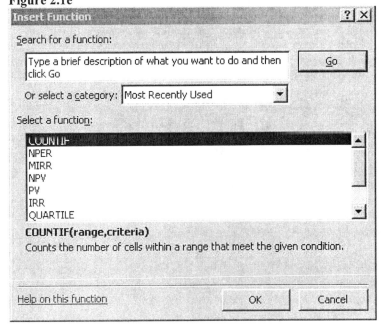

Step 2: Select the range, which in this case is the qualitative data for aphasia, B2:B23, and then select criteria, such as Broca's first.

Figure 2.1f

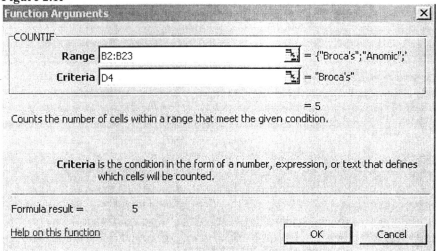

Step 3: Click Ok and the number "5" will appear in the column.

Table 2.1

	File	Edit	View	Insert	Format	Tools	Data	Window	Help

E4			f_x =COUNTIF($B$2:$B$23,D4)		

	A	B	C	D	E
1	Subject	Types of Aphasia			
2	1	Broca's			
3	2	Anomic		Type of Aphasia	Number of Subjects
4	3	Anomic		Broca's	5
5	4	Conduction		Conduction	7
6	5	Broca's		Anomic	10
7	6	Conduction			22
8	7	Conduction			
9	8	Anomic			
10	9	Conduction			
11	10	Anomic			
12	11	Conduction			
13	12	Broca's			
14	13	Anomic			
15	14	Broca's			
16	15	Anomic			
17	16	Anomic			
18	17	Anomic			
19	18	Conduction			
20	19	Broca's			
21	20	Anomic			
22	21	Conduction			
23	22	Anomic			
24					

Sheet1 / Sheet2 / Sheet3 /

Solution:

Please refer to the following steps to answer this question.

Step 1: Open a new workbook and place the cursor in the upper left-hand corner of the spreadsheet.

Step 2: Open the database or input the data manually. Notice that the data is located in this example in cells as follows. Notice also that both the number experiencing each aphasia as well as the relative proportion of those experiencing aphasia are provided. (Table 2.1a)

Table 2.1a

	A	B	C
1	Class	Frequency	Relative Frequency
2	Type of Aphasia	Number of Subjects	Proportion
3	Broca's	5	0.227
4	Conduction	7	0.318
5	Anomic	10	0.455

35

Step 3: Select f_x from the menu options on the toolbar.

Step 4: Select Pie Chart from the choices available. Notice how several different pie charts are presented. Select the first pie chart highlighted in the upper left-hand corner by default. This is referred to as **Step 1 of 4 in Chart Wizard. (Figure 2.2a)**

Step 5: In **Step 2 of 4**, **highlight the data** be sure that the data is grouped either by **columns. (Figure 2.2b)**

Step 6: In **Step 3 of 4**, an output title may be assigned to the data if desired. In this case, "Type of Aphasia" is used as the title. **(Figure 2.2c)**

Step 7: In **Step 4 of 4**, place the chart either within the existing sheet or within a new workbook. In this case, it is placed **as an object in** the existing worksheet. **(Figure 2.2d)**

Step 8: Click OK to finish. **(Figure 2.2e)**

36

Figure 2.2a

Figure 2.2b

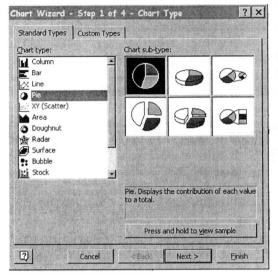

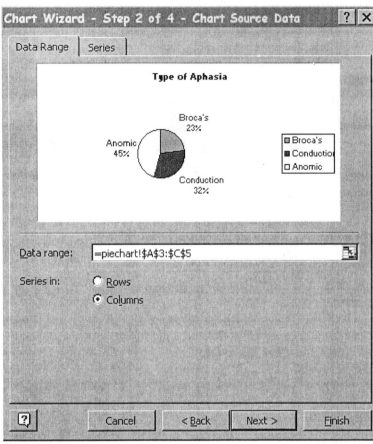

Figure 2.2c

Figure 2.2d

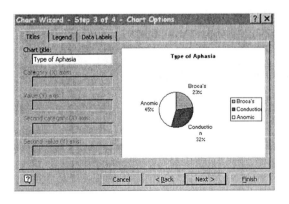

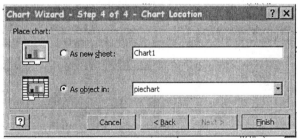

37

Notice how the pie chart is displayed. If the actual proportions of those experiencing each type of aphasia are displayed, the data can be formatted. Please refer to the following steps.

Figure 2.2e

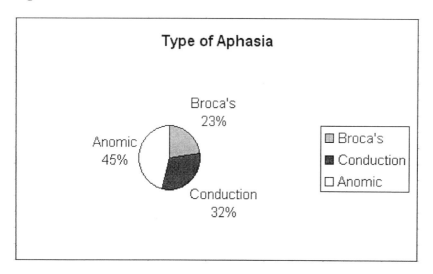

Step 1: Right click the chart itself.

Step 2: Select Format Data Series.

Step 3: Select the tab **Data Labels** and check the box corresponding to **Show Label and Percent**.

Step 4: Click OK to finish.

Once these steps are completed, the percentage of each type of aphasia is presented along side the corresponding labels.

2.3 Graphical Methods for Describing Data: Bar Graphs

For consistency in presentation, the aphasia data in the textbook will now be used to illustrate bar graphs. To create bar graphs, please refer to the following steps.

Step 1: Open a clean workbook and place the cursor in the upper left-hand cell of the worksheet.

Step 2: Select Insert - Chart from the menu options at the top of the screen.

Step 3: In Step 1 of 4, select Column chart. Leave the setting at default. **Click next. (Figure 2.3a)**

Step 4: In Step 2 of 4, highlight the **data range**. Be sure to indicate if the data are grouped by columns or rows. **(Figure 2.3 b)**

Step 5: In Step 3 of 4, enter options to change the location of the legend, width of the axes, etc. **(Figure 2.3c)**

Step 6: In Step 4 of 4, place the chart in a new worksheet or as an object in an existing worksheet. **(Figure 2.3d)**

Step 7: Click OK to finish.

Figure 2.3a

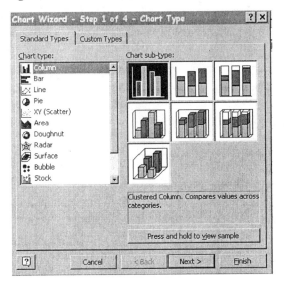

Figure 2.3b

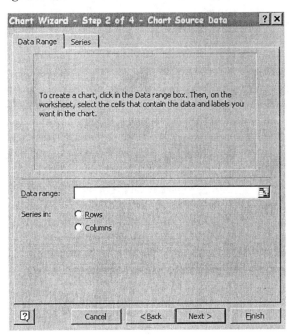

Figure 2.3c

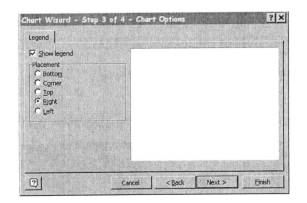

Figure 2.3d

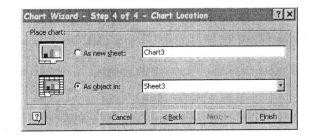

Example 2.2 : Please refer to Section 2.1 and the data listed in Table 2.1 (page 30) of the text to create a bar graph.

Consider a study of aphasia published in the *Journal of Communication Disorders* (Mar. 1995). In Table 2.1 of the text, five aphasiacs in the study were diagnosed as suffering from Broca's aphasia, 7 from conduction aphasia, and 10 from anomic aphasia. Construct a bar graph showing the disparity within the three types of aphasia.

39

Table 2.2

Class	Frequency	Relative Frequency
Type of Aphasia	Number of Subjects	Proportion
Broca's	5	0.227
Conduction	7	0.318
Anomic	10	0.455
Totals	22	1.000

Solution:

Please refer to the following steps to answer this question.

Step 1: Open a new workbook and place the cursor in the upper left-hand corner of the spreadsheet.

Step 2: Select Insert -Chart

Step 3: In Step 1 of 4, select **Column** and leave the chart subtype setting on default **(Figure 2.4a)**. **Click Next.**

Step 4: In Step 2 of 4, highlight the data range as A3:B5. The bar chart should appear. **Click next. (Figure 2.4b)**

Step 5: In Step 3 of 4, adjust the legend and axes as desired. **Click next. (Figure 2.4c)**

Step 6: In Step 4 of 4, place the chart as an object in the current worksheet or a new worksheet. **Click finish. (Figure 2.4d)**

Figure 2.4a

Figure 2.4b

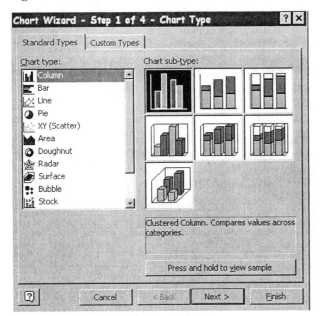

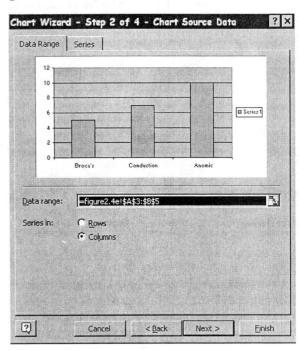

Figure 2.4c
Figure 2.4d

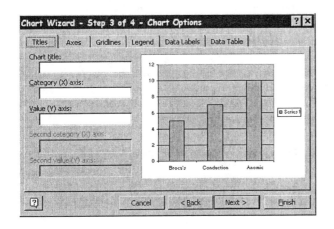

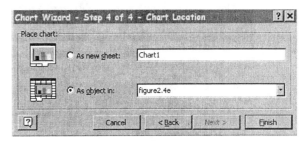

The bar graph generated is presented in **Figure 2.4e** below. Notice the differences in presentation in comparison to the text.

Figure 2.4 e

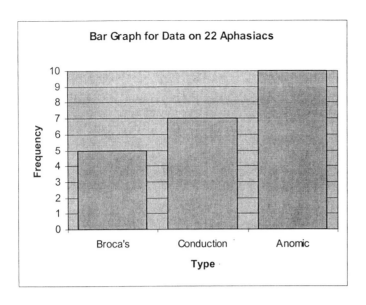

Bar Graph for Data on 22 Aphasiacs

2.4   Graphical Methods for Describing Data: Histograms

To create histograms (a special type of bar graph) in Excel, please refer to the following steps.

Step 1: Open a clean workbook and place the cursor in the upper left-hand cell of the worksheet.

Step 2: Select Tools-Data Analysis from the menu options at the top of the screen.

Step 3: Select Histogram from the options listed.

The following example will be used to illustrate how to construct histograms and bar charts. Stem-and-leaf displays will be presented in the following section.

Example 2.3 : Please refer to Example 2.2 from the textbook (page 45)

The data present, by state, the percentages of the total number of college or university loans that are in default.
   a.   Use a statistical software package to create a histogram for these data.
   b.   Use a statistical software package to create a stem-and-leaf display for these data.

Table 2.3

Percentage of Student Loans (per State) in Default							
State	%	State	%	State	%	State	%
Ala.	12.0	Ill.	9.3	Mont.	6.4	R.I.	8.8
Alaska	19.7	Ind.	6.7	Nebr.	4.9	S.C.	14.1
Ariz.	12.1	Iowa	6.2	Nev.	10.1	S.Dak.	5.5
Ark.	12.9	Kans.	5.7	N.H.	7.9	Tenn.	12.3
Calif.	11.4	Ky.	10.3	N.J.	12.0	Tex.	15.2
Colo.	9.5	La.	13.5	N.Mex.	7.5	Utah	6.0
Conn.	8.8	Maine	9.7	N.Y.	11.3	Vt.	8.3
Del.	10.9	Md.	16.6	N.C.	15.5	Va.	14.4

42

D.C.	14.7	Mass.	8.3	N.Dak.	4.8	Wash.	8.4
Fla.	11.8	Mich.	11.4	Ohio	10.4	WVa.	9.5
Ga.	14.8	Minn.	6.6	Okla.	11.2	Wis.	9.0
Hawaii	12.8	Miss.	15.6	Oreg.	7.9	Wyo.	2.7
Idaho	7.1	Mo.	8.8	Pa.	8.7		

Solution:

In order to answer part a., we must first access the data from the disk accompanying this text or manually type the data into Excel in a column. Please refer to the following steps to answer this question.

Step 1: Open a new workbook and place the cursor in the upper left-hand corner of the spreadsheet.

Step 2: Open the file where the data are found or input the data manually from the example. Be sure that the data are inputted as in Table 2.3a below. Table 2.4 (page 45 of the text) is reformatted into a single column displayed in Table 2.3a below

43

Table 2.3a

	A	B
1	Percentage of Student L	
2		
3	State	Percentage
4	Ala.	12.0
5	Alaska	19.7
6	Ariz.	12.1
7	Ark.	12.9
8	Calif.	11.4
9	Colo.	9.5
10	Conn.	8.8
11	Del.	10.9
12	D.c.	14.7
13	Fla.	11.8
14	Ga.	14.8

Step 3: Select Tools from the toolbar at the top of Excel.

Step 4: Select Data Analysis and then choose **Histogram** from the list of options provided.

Step 5: When the histogram menu appears, first enter the rows and columns where the data is located in the **Input Range** of the histogram menu. In this example, the input range is **B4:B54**. This may be done by typing the input range in manually or by **clicking and dragging** the appropriate range of data cells.

Step 6: Specify the **Output Range** where the Histogram may be placed within the current worksheet. We have selected F1. Remember when selecting a range to place output, be sure to check that the cell(s) do not contain any data! (See **Figure 2.5**)

Step 7: Check the Chart Output option in the menu so that the histogram may be generated.
Figure 2.5

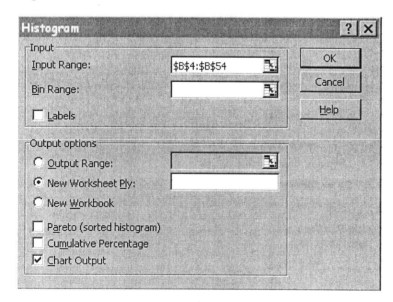

Step 8: When this is complete, **click OK** to finish. Two pieces of output are generated:

Once the output has been generated, notice that two pieces of information have been generated. (See **Table 2.4 below**) First, a bin and frequency table provides the viewer with the upper bound or endpoint of each category of the histogram (**bin**) and the number of observations that appear in each corresponding category (**frequency**).

Table 2.4

Bin	Frequency
2.7	1
5.128571429	2
7.557142857	9
9.985714286	14
12.41428571	13
14.84285714	7
17.27142857	4
More	1

The histogram, presented below (**Figure 2.6**), can be altered to make the chart easier to read. Clicking inside the histogram and moving the "squares" on its outline can make the graph larger or smaller. Notice how the histogram produced in Excel differs from the one presented in the text. First, the bars displayed in the text between categories are contiguous; the bar of one class touches the bar of the next class. In Excel, the bars of each category are separated. Second, the endpoints of each class or category differ from the ones generated using Excel.

Figure 2.6

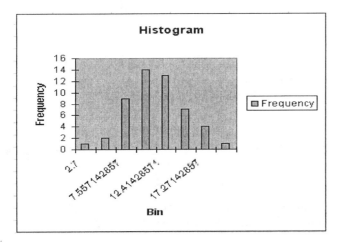

In order to create a histogram where the bars for each subsequent class or category touch, please refer to the following steps.

Step 1: Left click on any one of the bars and then **right click** and then select **Format Data Series**.

Step 2: Select the **Options tab** within **Format Data Series** and find and select **Gap Width**.

45

Step 3: If **Gap Width** is set to zero, the spaces between each bars of each subsequent class or category will be removed. (See **Figure 2.7**) Figure 2.7a presents the histogram with the bar lines connected without spaces between classes or categories.

Figure 2.7

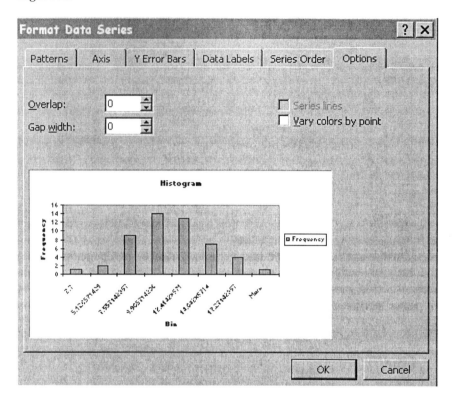

In order to create a histogram where the upper endpoints of each class or category are specified in even intervals (5, 10, 15, 20, etc.), please refer to the following steps.

Step 1: Open a new workbook and place the cursor in the upper left-hand corner of the spreadsheet.

Step 2: Open the database or input the data manually.

Step 3: Create a separate column labeled **Bin** next to the data.

Step 4: Within the **Bin** column, enter the numbers: 3, 5, 7, 9, 11, 13, 15, 17, 19, 21, and 23. In this example, the Bin column was placed in column C of the spreadsheet. (See **Table 2.5**)

Table 2.5

	A	B	C
1	Percentage of Student Loans (per		
2			
3	State	Percentage	Bin
4	Ala.	12.0	3
5	Alaska	19.7	5
6	Ariz.	12.1	7
7	Ark.	12.9	9
8	Calif.	11.4	11
9	Colo.	9.5	13
10	Conn.	8.8	15
11	Del.	10.9	17
12	D.c.	14.7	19
13	Fla.	11.8	21
14	Ga.	14.8	

Step 5: Select Insert-Chart.

Step 6: Enter Input Range and Bin Range created in the space provided.

Step 7: Click Labels in First Row if the first row highlighted contains nonnumeric descriptive headings.

Figure 2.8

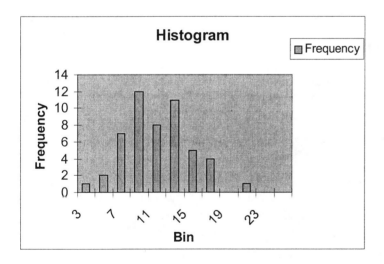

Step 8: Select the Output Range (either in a new location, a new worksheet, or the same worksheet). If selecting an output range within the existing spreadsheet, be sure that the output range selected does NOT overwrite data.

Step 9: Select Chart Output

Step 10: Click OK to finish.

The histogram should appear as follows. Once the changes are complete, notice how the new histogram mirrors the one presented in the textbook.

Figure 2.8a

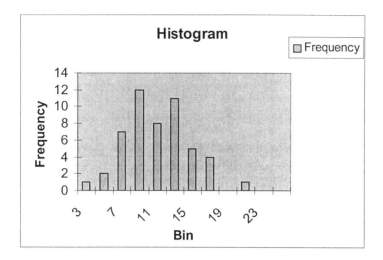

2.5   Graphical Methods for Describing Quantitative Data: Stem-and-Leaf Displays

Example 2.2 on page 45 of the text will now be used to illustrate stem-and-leaf diagrams. To create stem-and-leaf displays in Excel, please refer to the following steps.

Step 1: Open a clean workbook and place the cursor in the upper left-hand cell of the worksheet.

Step 2: Select PHStat-Stem-and-Leaf Display from the options listed.

Step 3: Enter the variable cell range and stem unit options.

Figure 2.8b

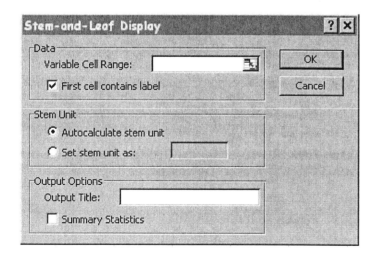

Step 4: Click OK to finish.

Example 2.4 : Please refer to Example 2.2 from the textbook (page 45)

The data present, by state, the percentages of the total number of college or university loans that are in default.

    a. Use a statistical software package to create a histogram for these data.

    b. Use a statistical software package to create a stem-and-leaf display for these data.

We will now use PHStat to generate a stem-and-leaf display. Please refer to the following steps.

Step 1: Open a new workbook and enter the data in a column that appears as in the following table.

Step 2: Select PHStat-Stem-and-Leaf Display
Table 2.6

Percentage of Student Loans (per State) in Default							
State	%	State	%	State	%	State	%
Ala.	12.0	Ill.	9.3	Mont.	6.4	R.I.	8.8
Alaska	19.7	Ind.	6.7	Nebr.	4.9	S.C.	14.1
Ariz.	12.1	Iowa	6.2	Nev.	10.1	S.Dak.	5.5
Ark.	12.9	Kans.	5.7	N.H.	7.9	Tenn.	12.3
Calif.	11.4	Ky.	10.3	N.J.	12.0	Tex.	15.2
Colo.	9.5	La.	13.5	N.Mex.	7.5	Utah	6.0
Conn.	8.8	Maine	9.7	N.Y.	11.3	Vt.	8.3
Del.	10.9	Md.	16.6	N.C.	15.5	Va.	14.4
D.C.	14.7	Mass.	8.3	N.Dak.	4.8	Wash.	8.4
Fla.	11.8	Mich.	11.4	Ohio	10.4	WVa.	9.5
Ga.	14.8	Minn.	6.6	Okla.	11.2	Wis.	9.0
Hawaii	12.8	Miss.	15.6	Oreg.	7.9	Wyo.	2.7
Idaho	7.1	Mo.	8.8	Pa.	8.7		

Step 3: Enter Variable Cell Range as B3:B54

Step 4: Click first cell contains label.

Step 5: Select autocalculate stem unit

Step 6: Enter an **output title** if desired.

Step 7: Click summary statistics. (See Figure 2.10)

49

Figure 2.10

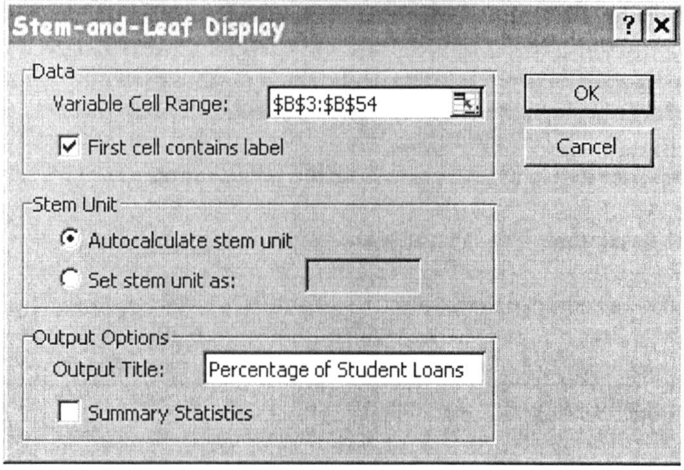

Step 8: Click OK to finish.

Table 2.7

Statistics		Stem-and-Leaf Display for % Stem unit: 1	
		2	7
Sample Size	51	3	
Mean	10.19608	4	8 9
Median	9.7	5	5 7
Std. Deviation	3.504395	6	0 2 4 6 7
Minimum	2.7	7	1 5 9 9
Maximum	19.7	8	3 3 4 7 8 8 8
		9	0 3 5 5 7
		10	1 3 4 9
		11	2 3 4 4 8
		12	0 0 1 3 8 9
		13	5
		14	1 4 7 8
		15	2 5 6
		16	6
		17	
		18	
		19	7

2.6 Numerical Techniques in Excel: Measures of Central Tendency

The following steps may be used to calculate the arithmetic mean, median, and modal values of a dataset.

Step 1: Open the database or input the data manually.

50

Step 2: Select Tools from the toolbar at the top of the screen.

Step 3: Select Data Analysis.

Step 4: Select Descriptive Statistics. (Figure 2.11)

Step 5: Specify the **Input Range**, indicating whether the data are grouped by rows or columns.

Step 6: If a descriptive, nonnumeric title is located within the first cell of the dataset, please check the **Labels in First Row** option.

Step 7: The **Output Range** may be placed either in a new worksheet, a new workbook, or within the existing spreadsheet. If the output is placed within the existing worksheet, please be sure that when specifying the cells where the output is to be placed, that no old data is overwritten.

Step 8: Select Summary Statistics and specify a confidence level if desired.

Step 9: Click OK to finish.

(Figure 2.11)

The following example from the text will be used to illustrate how to calculate measures of central tendency.

Example 2.5 : Please refer to Example 2.4 (page 56) , Table 2.2(page 41) of the textbook.

The Environmental Protection Agency performs extensive tests on all new car models to determine their mileage ratings. Suppose that the 100 measurements represent the results of such tests on a certain new car model. First find the measures of central tendency (mean, median, and mode).

Solution:

Notice how the data are presented within Excel. Column A refers to the each specific car number tested by the EPA (from 1 to 100) and Column B refers to the Ratings of those cars.

(Table 2.8)

	A	B
1	**Number of Ratings**	**Ratings**
2	1	36.3
3	2	32.7
4	3	40.5
5	4	36.2
6	5	38.5
7	6	36.3
8	7	41
9	8	37
10	9	37.1
11	10	39.9

Please refer to the following steps in order to solve this problem.

Step 1: Open the database or input the data manually.

Step 2: Select Tools-Data Analysis - Descriptive Statistics.

Step 3: Specify the **Input Range**. In this case the data are grouped by columns (**B1:B101**).

Step 4: Check the **Labels in First Row** option since B1 contains the title "Ratings of Cars".

Step 5: The **Output Range** may be placed either in a new worksheet, a new workbook, or within the existing spreadsheet. In this case, **New Worksheet Ply** is selected.

Step 6: Select Summary Statistics.

Step 7: Click OK to finish.

Figure 2.12

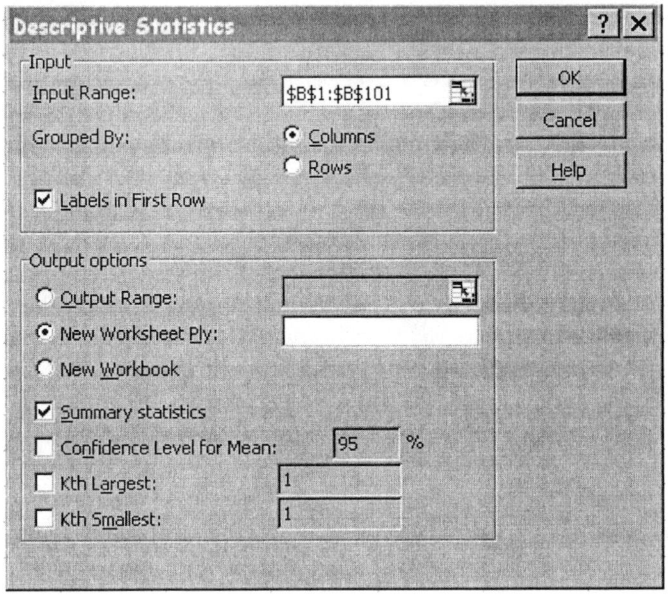

The output table reveals that the mean mileage rating of 100 cars tested by the EPA is 36.9940 with a median and modal value of 37 miles per gallon.

Table 2.9

Column1	
Mean	36.994
Standard Error	0.241789707
Median	37
Mode	37
Standard Deviation	2.417897074
Sample Variance	5.846226263
Kurtosis	0.769922694
Skewness	0.050908779
Range	14.9
Minimum	30
Maximum	44.9
Sum	3699.4
Count	100

2.7 Measures of Central Tendency: Measures of Spread or Dispersion

In order to calculate the measures of dispersion (range, variance, and standard deviation) of a dataset, follow the exact same steps as those for finding the measures of central tendency: **Tools-Data Analysis-Descriptive Statistics**. Notice when summary statistics were calculated for the EPA Mileage data in the preceding example, the measures of dispersion were automatically calculated. The sample standard

53

deviation for the 100 EPA cars is approximately 2.4179 and the sample variance (the square of the sample standard deviation) is approximately 5.8462.

2.8   Sample Exercises

The following exercises were selected from the textbook to test the reader's understanding of Excel 97/2000 and its usage within the chapter. Try to replicate the output for the following problems.

Problem 2.163 (Gallup Poll on Smoking), page 107

USA Today (December 31, 1999) reported on a Gallup Survey which asked: "Do you think cigarette smoking is one of the causes of lung cancer?" The results of the 1999 survey are compared to the results of a similar survey conducted 25 years earlier in the accompanying table. Construct a pie chart for the 1954 survey results and the 1999 survey results separately.

Is Smoking a Cause of Lung Cancer	1954 Survey	1999 Survey
Yes	41%	92%
No	31%	6%
No Opinion	28%	2%

Solution to Problem 2.163

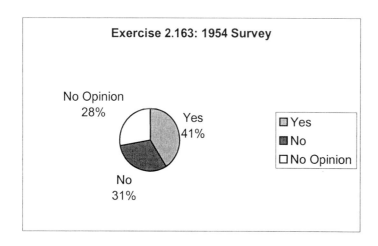

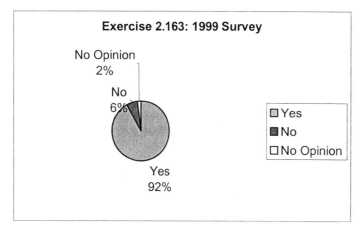

Problem 2.166 (Competition for bird nest holes), page 107

The Condor (May 1995) published a study of competition for nest holes among collared flycatchers, a bird species. The authors collected the data for the study by periodically inspecting nest boxes located on the island of Gotland in Sweden. The nest boxes were grouped into 14 discrete locations (called plots). The table presents the number of flycatchers killed and the number of flycatchers breeding at each plot. Calculate the mean, median, and mode for the number of flycatchers killed at the 14 study plots.

Plot Number	Number Killed	Number of Breeders
1	5	30
2	4	28
3	3	38
4	2	34
5	2	26
6	1	124
7	1	68
8	1	86
9	1	32
10	0	30
11	0	46
12	0	132
13	0	100
14	0	6

Solution to Problem 2.166

Number Killed		Number of Breeders	
Mean	1.428571429	Mean	55.71428571
Standard Error	0.428571429	Standard Error	10.58686132
Median	1	Median	36
Mode	0	Mode	30
Standard Deviation	1.603567451	Standard Deviation	39.61240787
Sample Variance	2.571428571	Sample Variance	1569.142857
Kurtosis	0.484071484	Kurtosis	-0.395617663
Skewness	1.129959211	Skewness	0.930386204
Range	5	Range	126
Minimum	0	Minimum	6
Maximum	5	Maximum	132
Sum	20	Sum	780
Count	14	Count	14

Problem 2.83 (Ammonia in car exhaust), page 72

Refer to the *Environmental Science & Technology* (Sep. 1, 2000) study on the ammonia levels near the exit ramp of a San Francisco highway tunnel. The data (in parts per million) for 8 days during afternoon drive-time are reproduced here in the table. Find the range, variance, and standard deviation of the ammonia levels.

1.53	1.50	1.37	1.51	1.55	1.42	1.41	1.48

Solution to Problem 2.83

Row1	
Mean	1.47125
Standard Error	0.022633532
Median	1.49
Mode	#N/A
Standard Deviation	0.064017297
Sample Variance	0.004098214
Kurtosis	-1.23809065
Skewness	-0.45637176
Range	0.18
Minimum	1.37
Maximum	1.55
Sum	11.77
Count	8

3.1   Introduction

Chapter 3 in the text discusses how probability theory provides the basis of analysis that allows that statistician to make inferences about a population using sample statistics. Although it is assumed in this chapter that the population from which samples is taken is known, the purpose of probability theory is to calculate the chance of selecting all possible samples of size, n, from that population in order to infer information about that population. If the population is known, why does a statistician need to take samples? Sometimes, it is too costly or time consuming for a researcher to review all population data or a researcher may not have access to all data in the population, even if it is known.

The following examples will be used to demonstrate techniques in PHStat and Excel.

> Example 3.15, page 146, (Probabilities of Smoking & Developing Cancer)
> Random Sampling
> Sample Exercise #1: Problem 3.158, page 181 (Study of aggressiveness and birth order)
> Sample Exercise #2: Problem 3.102, page 164 (Auditing an accounting system)

3.2   Probabilities using PHStat: 2x2 Contingency Tables

PHStat is a supplementary addition to Excel that provides the statistician with an alternative technique to calculating probabilities and selecting samples from a population. However, PHStat will only calculate the probabilities of unions and intersections if the data are presented in a 2x2 contingency table.
To calculate conditional and joint probabilities, please complete the following steps.

Step 1: Open a new worksheet in Excel and place the cursor in the upper left-hand cell.

Step 2: Click on the **PHStat** menu at the top of the screen.

Step 3: Select Data Preparation from the list of options and **Select Probabilities**. **(Figure 3.1)**

Step 4: Table 3.1 presents the output generated by the Probability function within PHStat. Notice that the default settings calculate probabilities given $P(A1) = P(A2) = P(B1) = P(B2) = 0.50$. This template may be changed to the specification of the problem. The following example will illustrate how to calculate these probabilities using PHStat.

Figure 3.1

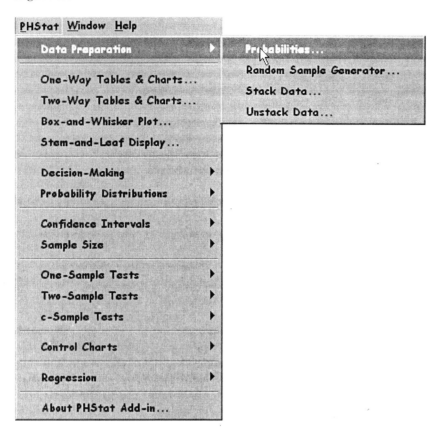

Table 3.1

Probabilities Calculations

Sample Space

		Event B		
		B1	B2	Totals
Event A	A1	1	1	2
	A2	1	1	2
	Totals	2	2	4

Simple Probabilities

P(A1)	0.50
P(A2)	0.50
P(B1)	0.50
P(B2)	0.50

Joint Probabilities

P(A1 and B1)	0.25
P(A1 and B2)	0.25
P(A2 and B1)	0.25
P(A2 and B2)	0.25

Addition Rule

P(A1 or B1)	0.75
P(A1 or B2)	0.75
P(A2 or B1)	0.75
P(A2 or B2)	0.75

Example 3.1 Please refer to Example 3.15 (page 146) in the textbook.

Many medical researchers have conducted experiments to examine the relationship between cigarette smoking and cancer. Consider an individual randomly selected from the adult male population. Let A represent the event that the individual smokes, and let A^c denote the complement of A (the event that the individual does not smoke). Similarly, let B represent the event that the individual develops cancer and let B^c be the complement of that event. Use the data presented in the table to find the following.

    a. Find P(A), the probability that an adult male smokes
    b. Find P(B|A), the probability that an adult male smoker develops cancer.
    c. Find $P(B^c|A)$, the probability that an adult male smoker does not develop cancer.
    d. Find $P(B|A^c)$, the probability that an adult male non-smoker develops cancer.
    e. Find $P(B^c|A^c)$, the probability that an adult male nonsmoker does not develop cancer.

Probabilities of Smoking and Developing Cancer		
	Develops Cancer	
	Yes, B (Cancer)	No, B^c (Non cancer)
Yes, A (Smoker)	0.05	0.20
No, A^c (Nonsmoker)	0.03	0.72

Solution

Please complete the following steps in order to reconstruct the template provided in the table.

Step 1: Open a new worksheet and be sure the cursor is located in the left-hand corner of the spreadsheet.

Step 2: Select PHStat-Data Preparation-Probabilities

Step 3: The worksheet that appears is actually a template that the user can manipulate in order to derive several types of probabilities. **(Table 3.2)**

Table 3.2

Probabilities Calculations					
Sample Space			**Developing Cancer**		
			Cancer	Non Cancer	Totals
Smoker	Smoker		0.05	0.20	0.25
	Nonsmoker		0.03	0.72	0.75
	Totals		0.08	0.92	1
Simple Probabilities					
P(Smoker)	0.25				
P(Nonsmoker)	0.75				
P(Cancer)	0.08				
P(Non Cancer)	0.92				
Joint Probabilities					
P(Smoker and Cancer)	0.05				
P(Smoker and Non Cancer)	0.20				
P(Nonsmoker and Cancer)	0.03				
P(Nonsmoker and Non Cancer)	0.72				
Addition Rule					
P(Smoker or Cancer)	0.28				
P(Smoker or Non Cancer)	0.97				
P(Nonsmoker or Cancer)	0.80				
P(Nonsmoker or Non Cancer)	0.95				

Step 4: Replace **Event A** and **Event B** with the variable names, **Developing Cancer** and **Smoker**.

Step 5: Next, replace the outcomes that are listed in the table as A1, A2, B1, and B2 with (**smoker (yes), nonsmoker (no), cancer (yes), not cancer (no)**).

Step 6: Then change the numbers within the template to reflect the probabilities associated with the example in the text. The new worksheet should with probabilities required for this problem should appear as in Table 3.2.

Now the revised template can be used to answer the questions from Example 3.15 (page 147) in the text. Before the solutions are presented, remember the symbolism used in the text and how that symbolism is correlated with the template. Remember that "and" probabilities reflect multiplication rules whereas "or" probabilities reflect addition rules.

$P(A \cap B) = P(A$ and $B)$

$P(A \cup B) = P(A$ or $B)$

$P(B|A) = P(A \cap B)/P(A) = P(A$ and $B)/P(A)$

Step 7: Solution to part a: Find P(A), the probability that an adult male smokes. From the template, P(Smoker) is 0.25 or 25%. (**Table 3.2a**)

Table 3.2a

| | Developing Cancer | | |
	Cancer	Non Cancer	Totals
Smoker	0.05	0.20	0.25
Nonsmoker	0.03	0.72	0.75
Totals	0.08	0.92	1

Step 8: Solution to part b: Find P(B|A), the probability that an adult male smoker develops cancer. To calculate this probability, we must remember that the $P(B|A) = P(A \cap B)/P(A)$. So, P(Smoker and Cancer)/ P(Smoker) = 0.05/.25 = 0.20. (**Table 3.2b**)

Table 3.2b

| | Developing Cancer | | |
	Cancer	Non Cancer	Totals
Smoker	0.05	0.20	0.25
Nonsmoker	0.03	0.72	0.75
Totals	0.08	0.92	1

Step 9: Solution to part c: Find $P(B^c|A)$, the probability that an adult male smoker does not develop cancer. To calculate this probability, remember that $P(B^c|A) = P(A \cap B^c)/P(A)$. So, P(Smoker and No Cancer)/P(Smoker) – 0.20/0.25 – 0.80. (**Table 3.2c**)

Table 3.2c

| | Developing Cancer | | |
	Cancer	Non Cancer	Totals
Smoker	0.05	0.20	0.25
Nonsmoker	0.03	0.72	0.75
Totals	0.08	0.92	1

Step 10: Solution to part d: Find $P(B|A^c)$, the probability that an adult male non-smoker develops cancer. To calculate this probability, remember that $P(B|A^c) = P(A^c \cap B)/P(A^c)$. So, P(Nonsmoker and Cancer)/P(Nonsmoker) = 0.03/0.75 = 0.04. (**Table 3.2d**)

Table 3.2d

| | Developing Cancer | | |
	Cancer	Non Cancer	Totals
Smoker	0.05	0.20	0.25
Nonsmoker	0.03	0.72	0.75
Totals	0.08	0.92	1

Step 11: Solution to part e: Find $P(B^c|A^c)$, the probability that an adult male nonsmoker does not develop cancer. To calculate this probability, remember that $P(B^c|A^c) = P(A^c \cap B^c)/P(A^c)$. So, P(Nonsmoker and Non Cancer)/P(Nonsmoker) = 0.72/0.75 = 0.96. **(Table 3.2e)**

Table 3.2e

| | Developing Cancer | | |
	Cancer	Non Cancer	Totals
Smoker	0.05	0.20	0.25
Nonsmoker	0.03	0.72	0.75
Totals	0.08	0.92	1

3.3   Random Sampling

In a random sample, each individual, object, or measurement has an equal probability of being selected from a population. Although a random number table does appear in the appendices of the text, PHStat provides the user with an alternative method of selecting samples randomly from a known population. Here are the steps for conducting random samples in PHStat.

Step 1: Open a new workbook and be sure the cursor is in the upper left-hand corner of the first cell.

Figure 3.2

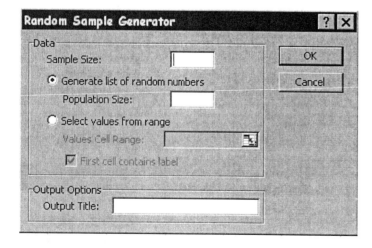

Step 2: Selecting PHStat-Data Preparation from the menu at the top of the screen

62

Step 3: Select Random Sample Generator.

Step 4: Click OK to finish.

Step 5: After opening the Random Sample Generator Menu, notice that PHStat provides the user with two methods for selecting that random sample. First, the user may generate a list of random numbers by selecting a sample size and population size that is appropriate for the data set.

Step 6: Second, the user may select a sample size from a set of data that already are contained with the Excel spreadsheet. The following example will illustrate how to conduct random sampling in PHStat.

Example 3.2: Suppose a researcher wishes to sample 15 items from a list of 350 items. Which objects should be sampled?

Solution

Please follow the steps provided to answer this question.

Step 1: Select PHStat-Data Preparation-Random Sample Generator

Step 2: Enter a **Sample Size of 15**.

Step 3: Enter a **Population Size of 350**.

Figure 3.3

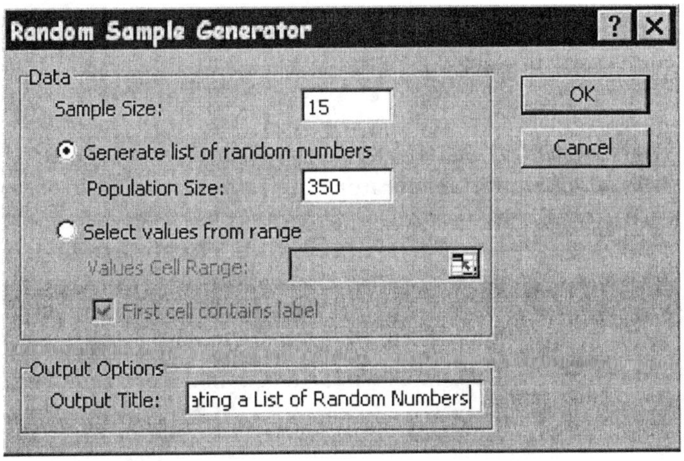

Step 4: Enter an **Output Title**, if desired. In this example, the output title is **"Generating a List of Random Numbers"**.

Step 5: Click OK to finish.

Notice that if the user were to sample the same n items from a population of size N, the output will differ from trial to trial. The output reflected in this example corresponds to the number of the observations within the population that have been selected randomly. Once these observations have been selected, the researcher will then use that sample to conduct an experiment.

63

Table 3.3

Generating a List of Random Numbers
65
192
62
199
76
251
18
335
350
184
238
11
94
254
296

Example 3.3 Please refer to Example 2.4, (Table 2.2), page 30 (EPA Mileage Rating on 100 Cars)

Use the data to randomly select 20 EPA Mileage Ratings for Automobiles from the 100 automobiles presented in the previous data set.

Solution

Step 1: Open a new worksheet and be sure the cursor is placed in the upper left-hand cell of the spreadsheet.

Step 2: Enter the data manually or open the prepared data set with EPA Mileage Ratings.

Step 3: Select PHStat - Data Preparation - Random Sample Generator.

Step 4: Enter the value **20** for the **sample size**.

Step 5: Select the **select values from range** option.

Step 6: Enter the **cell range** containing the data from which a sample will be taken. In this example, EPA mileage ratings are located in cells **B2:B101**.

Step 7: Title the output if desired.

Step 8: Click OK to finish.

Figure 3.4

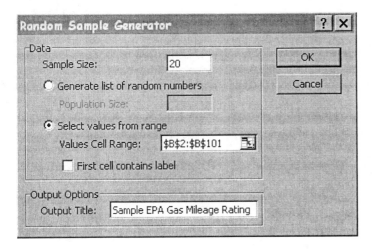

The output generated in the table represents the 20 EPA Gas Mileage Ratings that were selected randomly from the 100 listed in the data set. Note that if the user were to run the random sample generator again, different samples may be taken.

Table 3.5

Sample EPA Gas Mileage Rating
37.6
39.3
36.9
38.2
41.0
38.0
37.9
39.9
37.6
35.7
31.8
36.6
32.5
41.2
40.1
38.7
36.7
34.5
41.0
35.2

3.4　Sample Exercises

The following exercises were selected from the textbook to test the reader's understanding of PHStat and its usage within the chapter. Try to replicate the output in PHStat for the following problems.

65

Problem 3.158, page 181 (Study of aggressiveness and birth order)

Psychologists tend to believe that there is a relationship between aggressiveness and order of birth. To test this belief, a psychologist chose 500 elementary school students at random and administered each a test designed to measure the student's aggressiveness. Each student was classified according to one of four categories. The percentage of students falling in the categories is shown here.

    a. If one student is chosen at random from the 500, what is the probability that the student is firstborn?

    b. What is the probability that the student is aggressive?

    c. What is the probability that the student is aggressive, given the student was firstborn?

	Firstborn	Not Firstborn
Aggressive	15%	15%
Not Aggressive	25%	45%

Solution to Problem 3.158

Probabilities Calculations

Sample Space

		Event B		
		Firstborn	not Firstborn	Totals
Event A	Aggressive	0.15	0.15	0.3
	Not Aggress	0.25	0.45	0.7
	Totals	0.4	0.6	1

Simple Probabilities

P(Aggressive)	0.30
P(Not Aggressive)	0.70
P(Firstborn)	0.40
P(not Firstborn)	0.60

PHStat User Note:
Enter the values for both events in the cell range C5:D6.

Enter replacement labels for A1, A2, B1, and B2, in cell ranges B5:B6 and C4:D4, if desired.

Select this note and then select Edit | Cut to delete this note from the worksheet.

Joint Probabilities

P(Aggressive and Firstborn)	0.15
P(Aggressive and not Firstborn)	0.15
P(Not Aggressive and Firstborn)	0.25
P(Not Aggressive and not Firstborn)	0.45

Addition Rule

P(Aggressive or Firstborn)	0.55
P(Aggressive or not Firstborn)	0.75
P(Not Aggressive or Firstborn)	0.85
P(Not Aggressive or not Firstborn)	0.85

Problem 3.102, page 164 (Auditing an accounting system)

In auditing a firm's financial statements, an auditor is required to assess the operational effectiveness of the accounting system. In performing the assessment, the auditor frequently relies on a random sample of actual transactions (Stickney and Weil, *Financial Accounting: An Introduction to Concepts, Methods, and*

Uses, 1994). A particular firm has 5,382 customer accounts that are number 0001 to 5382. Draw a random sample of 10 accounts and explain in detail the procedure you used.

Solution to Problem 3.102

Randomly Selected Values		
576		
679		
2212		
5202		
4055		
5193		
4303		
1740		
1935		
1986		

67

4.1 Introduction

In chapter 4 of the textbook, the authors present the two major types of random variables discussed throughout the chapter: discrete and continuous variables. By definition, discrete variables are countable and can assume only certain distinct values whereas continuous variables can assume any value within a specified range. When analyzing discrete variables, specific discrete probability distributions (such as the binomial, Poisson, and hypergeometric distributions) may be utilized in order to calculate the probabilities associated with a specific variable. Notice that the appendices in the textbook present cumulative probability tables for only **specified** discrete variables for the binomial and Poisson distributions.

Excel 2003 and PHStat can be used to find cumulative and exact probabilities for both the binomial and the Poisson distribution. The major difference between PHStat and Excel 2000 is that PHStat provides an easy-to-use method for interpreting "language" asymmetries most often encountered by students first learning about probabilities. PHStat allows the user to interpret "at least", "at most", "equal to", "more than" and "less than" questions for both the binomial and Poisson distributions by generating all possible cumulative and exact probabilities in a single table contained within the generated output. The hypergeometric distribution will only be demonstrated in plain Excel.

The following examples will be used to illustrate techniques in PHStat and Excel.
  Example 4.9, page 206 (Heart Association)
  Example 4.13, page 219 (Rare Animal Species)
  Example 4.14, page 225 (New Teaching Assistants)
  Sample Exercise #1: Problem 4.64, page 217 (Chickens with fecal contamination)
  Sample Exercise #2: Problem 4.79, page 223 (Eye fixation problem)
  Sample Exercise #3: Problem 4.97, page 227 (On-site disposal of waste)

4.2   Calculating Binomial Probabilities Using PHStat.

To use the binomial probability distribution within PHStat, please refer to the following steps.

Step 1: Open a new workbook and be sure the cursor is located within the upper left-hand cell of the spreadsheet.

Figure 4.1

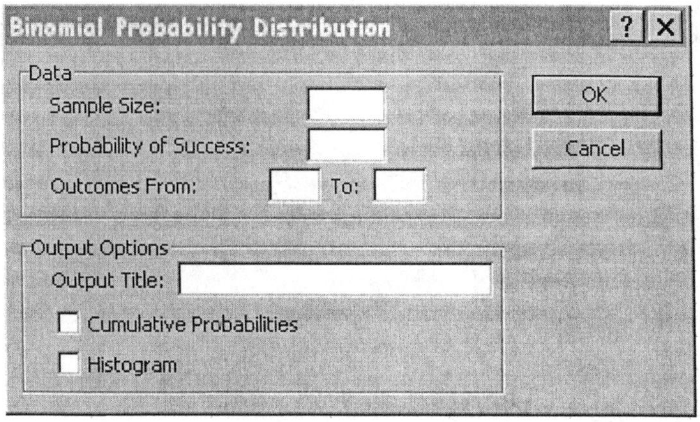

Step 2: Click on the **PHStat** menu near the top of the screen.

Step 3: Select the **Probability Distribution** option from the choices available.

Step 4: Open the **Binomial** option from the list.

Step 5: Notice that the user is asked to enter in the **sample size** (symbol n), the **probability of successes** (symbol p) and the **outcomes** that specify the values for which probabilities may be found.

Step 6: Within output options, the user may choose to create a descriptive **output title**. .

Step 7: In addition, PHStat allows the statistician to generate **cumulative probabilities** and a **histogram** if necessary.

Example 4.1   Please refer to Example 4.9, page 206 (Heart Association) of the text.

The Heart Association claims that only 10% of U.S. adults over 30 can pass the President's Physical Fitness Commission's minimum requirements. Suppose four adults are randomly selected, and each is given the fitness test.
   a.   Find the probability that none of the four adults pass the test.
   b.   Find the probability that three of the four adults pass the test.
   c.   Calculate the mean and standard deviation, respectively, of the number of the four adults who pass the test.

Solution

Step 1: Open a new workbook and be sure the cursor is located within the upper left-hand cell of the spreadsheet.

Step 2: Click on the **PHStat** menu near the top of the screen and **select Probability Distribution-Binomial** from the list provided

Figure 4.2

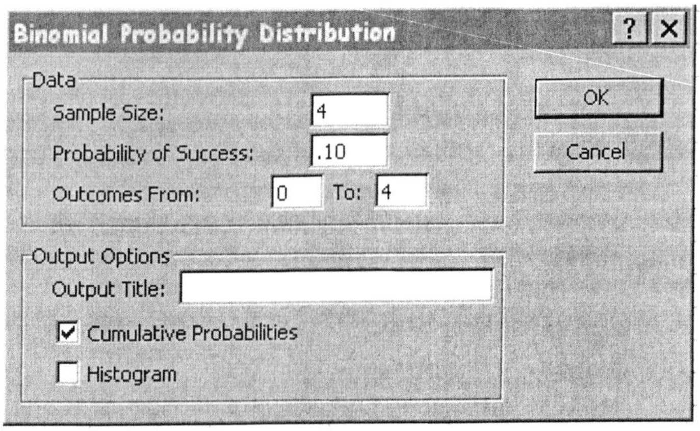

Step 3: In the example from the textbook, the problem specifies that the **sample size is 4** (n=4), the **probability of success is 0.10** (p=0.10).

Step 4: For illustrative purposes, enter the **outcomes from 0 to 4** in order to generate a complete probability table. **Check the Cumulative Probability** box and **click OK**. The output generated is shown below.

Table 4.1

Binomial Probabilities						
Sample size	4					
Probability of success	0.1					
Mean	0.4					
Variance	0.36					
Standard deviation	0.6					
Binomial Probabilities Table						
	X	P(X)	P(<=X)	P(<X)	P(>X)	P(>=X)
	0	0.6561	0.6561	0	0.3439	1
	1	0.2916	0.9477	0.6561	0.0523	0.3439
	2	0.0486	0.9963	0.9477	0.0037	0.0523
	3	0.0036	0.9999	0.9963	1E-04	0.0037
	4	0.0001	1	0.9999	0	1E-04

Step 5: After reviewing the data, the following answers are derived.
  a. The probability that none of the four adults pass the test is 0.6561 or 65.51%.
  b. The probability that three of the four adults pass the test is 0.0036 or 0.36%
  c. The mean number of individuals that passed the test is 0.4 and the standard deviation is 0.6.

4.3 Calculating Binomial Probabilities using Excel

It is also possible to calculate Binomial Probabilities without using the PHStat option in Excel. In order to do so, please follow the steps provided.

Step 1: Open a new worksheet and be sure the cursor is located within the upper left hand cell.

Step 2: Click on the **f$_x$** icon at the top of the screen.

Step 3: In **Function Category**, highlight **Statistical**.

Step 4: In **Function Name**, click on **Binomdist**.

Step 5: Insert the number of successes, number of trials, probability of success, and cumulative. Under the cumulative command, "TRUE" signifies the use of the cumulative distribution function. "FALSE" signifies the use of the probability mass function.

Step 6: Click OK to finish.

Using Example 4.8 from the text, we will recalculate the probabilities using the function (**f$_x$**) key located at the top of the toolbar in Excel and calculate the probabilities of derivations of this problem. Notice that the Binomial Distribution function in Excel will not calculate the mean and variance for the user. In addition, the Binomial Distribution function is defined as a "less-than or equal to" function. If the user requires calculating a probability of "more-than" a certain number, the equation of the Binomial Distribution was be redefined by the user. An example will follow.

Figure 4.3

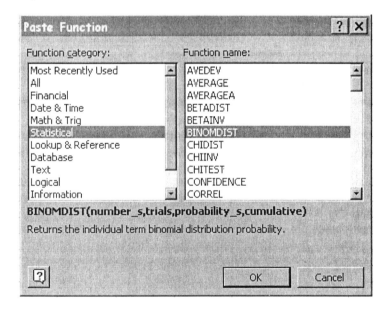

Example 4.2   Please refer to Example 4.9, page 206 (Heart Association) of the textbook.

The Heart Association claims that only 10% of U.S. adults over 30 can pass the President's Physical Fitness Commission's minimum requirements. Suppose four adults are randomly selected, and each is given the fitness test.
    a.   Find the probability that none of the four adults pass the test.
    b.   Find the probability that three of the four adults pass the test.
    c.   Find the probability that more than two adults pass the test.
    d.   Find the probability that less than two adults pass the test.

Solution

Please refer to the following steps in order to answer the problem. Find the probability that none of the four adults pass the test.

Step 1: Open a new workbook and **click** on the **f_x** near the top of the screen.

Step 2: After selecting the **Statistical** option from the choices available, scroll down the list and highlight **BINOMDIST** and **click OK.**

Step 3: Enter **Binomial distribution with Number_s = 0, Trials =4, Probability_s = 0.10, and Cumulative = False.** (Answer should be 65.61%). **(Figure 4.4a)**

Figure 4.4a

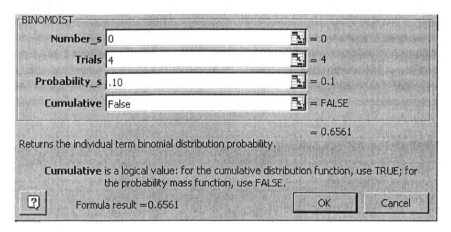

Step 4: Find the probability that three of the four adults pass the test. Enter **Number_s = 3, Trials = 4, Probability_s = 0.10, and Cumulative = False.** Answer should be 0.36%. **(Figure 4.4b)**

Figure 4.4b

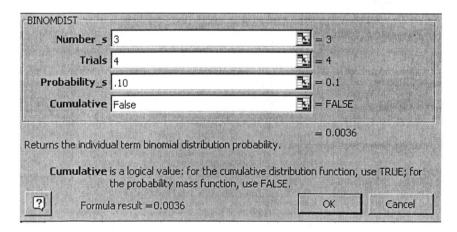

Step 5: Find the probability that three or less adults pass the test. Enter **Number _s = 3, Trials = 4, Probability _s = 0.10 and Cumulative = True.** Answer should be 99.99%. **(Figure 4.4c)**

72

Figure 4.4c

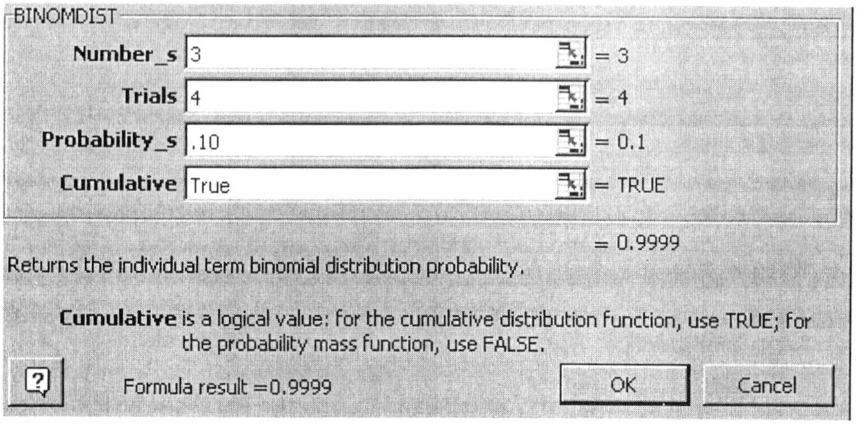

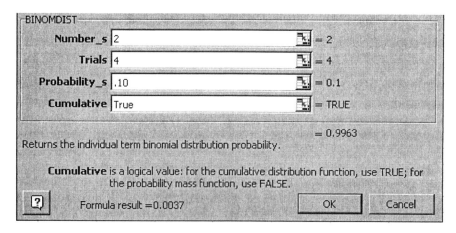

Step 6: Find the probability that more than 2 pass the test. Enter **Number_s = 2, Trials =4, Probability_s = 0.10, Cumulative = True** and recalculate formula on equation line to 1-Binomdist = 1-0.9963 = 0.0037. Answer is .37%. **(Figure 4.4d)**

Figure 4.4d

Step 7: Find the probability that less than 2 pass the test. Enter **Number_s = 1, Trials = 4, Probability_s = 0.10, and Cumulative = True.** Answer should be 94.77%. **(Figure 4.4e)**

Figure 4.4e

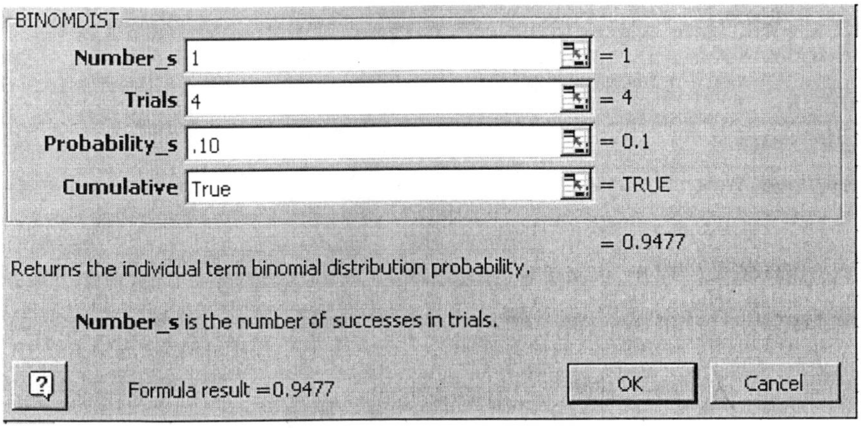

4.4 Calculating Poisson Probabilities using PHStat

According to the authors, Poisson distributions are best used to describe the number of "successes" or events that will occur within a specified time period. To use the Poisson probability distribution within PHStat, please refer to the following steps.

Step 1: Open a new workbook

Step 2: Click on the **PHStat** menu near the top of the screen.

Step 3: Select the **Probability Distribution** option from the choices available and then **click** open the **Poisson** option from the list.

Step 4: Insert the **average or expected number of successes**.

Step 5: Enter an output title if desired.

Step 6: Check **Cumulative Probabilities and Histogram (if desired).**

Step 7: Click OK to finish.

Figure 4.5

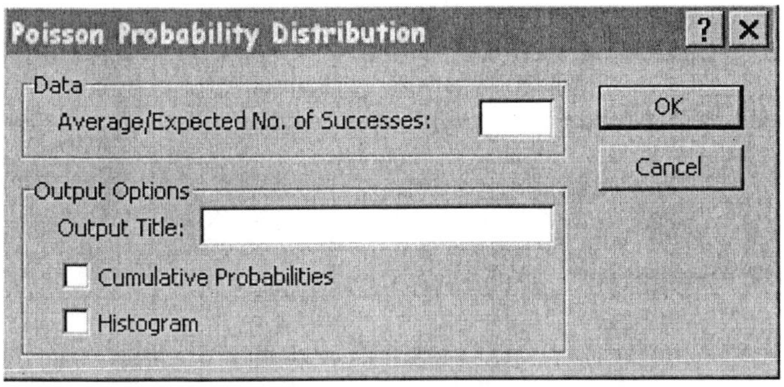

74

Please refer to the following example to illustrate the use of the Poisson Distribution in PHStat.

Example 4.3   Please refer to **Example 4.13, page 219 (Rare Animal Species)** of the textbook.
Ecologists often use the number of reported sightings of a rare species of animal to estimate the remaining population size. For example, suppose the number, x; of reported sightings per week of blue whales is recorded. Assume that x has (approximately) a Poisson probability distribution. Furthermore, assume that the average number of weekly sightings is 2.6.

    a.   Find the mean and standard deviation of x, the number of blue whale sightings per week. Interpret the results.
    b.   Find the probability that fewer than two sightings are made during a given week
    c.   Find the probability that more than five sightings are made during a given week Find the probability that exactly five sightings are made during a given week.

Solution

Step 1: Open a new workbook and be sure the cursor is located within the upper left-hand corner of the first cell.

Step 2: Click on the **PHStat** menu near the top of the screen.

Step 3: Select the **Probability Distribution** option from the choices available and then **click** open the **Poisson** option from the list.

Step 4: Insert the **average or expected number of successes as 2.6**.

Step 5: Enter an output title if desired.

Step 6: Check **Cumulative Probabilities**.

Step 7: Click OK to finish. Refer to Figure 4.6 to ensure the data were entered correctly.

Figure 4.6

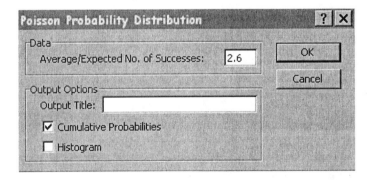

PHStat generates a Poisson distribution similar to the one displayed in the appendix at the back of the textbook. Notice that the third row generates the mean (average/expected number of successes) necessary to answer this question but not the variance or standard deviation as requested by this question.

Notice that the Poisson probabilities for P (x), P (<=x), P (<x), P (>x), and P (x>=) are all listed in the sixth row within the Excel worksheet. Please refer to the appropriate column when answering the preceding question.

Table 4.2

Poisson Probabilities for Customer Arrivals					
Average/Expected number of successes:	2.6				
Poisson Probabilities Table					
X	P(X)	P(<=X)	P(<X)	P(>X)	P(>=X)
0	0.074274	0.074274	0.000000	0.925726	1.000000
1	0.193111	0.267385	0.074274	0.732615	0.925726
2	0.251045	0.518430	0.267385	0.481570	0.732615
3	0.217572	0.736002	0.518430	0.263998	0.481570
4	0.141422	0.877423	0.736002	0.122577	0.263998
5	0.073539	0.950963	0.877423	0.049037	0.122577
6	0.031867	0.982830	0.950963	0.017170	0.049037
7	0.011836	0.994666	0.982830	0.005334	0.017170
8	0.003847	0.998513	0.994666	0.001487	0.005334
9	0.001111	0.999624	0.998513	0.000376	0.001487
10	0.000289	0.999913	0.999624	0.000087	0.000376
11	0.000068	0.999982	0.999913	0.000018	0.000087
12	0.000015	0.999996	0.999982	0.000004	0.000018
13	0.000003	0.999999	0.999996	0.000001	0.000004
14	0.000001	1.000000	0.999999	0.000000	0.000001
15	0.000000	1.000000	1.000000	0.000000	0.000000
16	0.000000	1.000000	1.000000	0.000000	0.000000
17	0.000000	1.000000	1.000000	0.000000	0.000000
18	0.000000	1.000000	1.000000	0.000000	0.000000
19	0.000000	1.000000	1.000000	0.000000	0.000000
20	0.000000	1.000000	1.000000	0.000000	0.000000

Step 8: The solutions to the questions are provided below
   a. Find the mean and standard deviation of x, the number of blue whale sightings per week. Interpret the results. **Square root of 2.6 or 1.1625**
   b. Find the probability that fewer than two sightings are made during a given week **26.7%**
   c. Find the probability that more than five sightings are made during a given week **4.9%**
   d. Find the probability that exactly five sightings are made during a given week. **7.4%**

4.5 Calculating Poisson Probabilities using Excel

Using the same example, recalculate the Poisson probabilities using Excel 97/2000. To use the binomial probability distribution, refer to the following steps.

Step 1: Open a new workbook and **click** on the f_x near the top of the screen.

Step 2: After selecting the **Statistical** option from the choices available, scroll down the list and highlight **POISSON**.

Step 3: Click OK to finish.

Step 4: Once the Poisson menu appears, the user is asked to define **X** (the number of events), the **mean**, and **cumulative** (true for the cumulative Poisson probability function and false for the Poisson probability mass function).

Step 5: Notice, like the binomial distribution, the Poisson distribution is designed as a "less than or equal to" distribution in Excel. In order to calculate probabilities for a value "greater than" or "greater than or equal to" the variable x, we must change the original binomial equation by hand.

Step 6: Once the data are entered, **click OK** to finish.

Figure 4.7

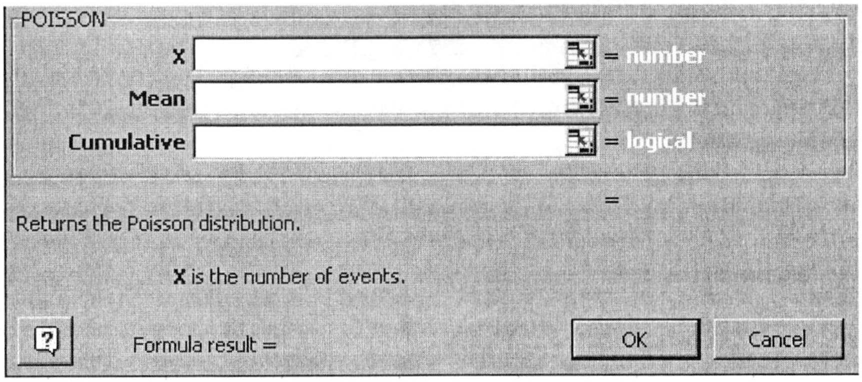

We will now rework the following probabilities for the previous example. As with the Binomial distribution in Excel, the Poisson distribution will not calculate the mean and standard deviation automatically; however, the user may do so by writing a formula for the given data.

Example 4.4   Please refer to **Example 4.13, page 219 (Rare Animal Species)** of the textbook.

Ecologists often use the number of reported sightings of a rare species of animal to estimate the remaining population size. For example, suppose the number, x; of reported sightings per week of blue whales is recorded. Assume that x has (approximately) a Poisson probability distribution. Furthermore, assume that the average number of weekly sightings is 2.6.
   a. Find the mean and standard deviation of x, the number of blue whale sightings per week. Interpret the results.
   b. Find the probability that fewer than two sightings are made during a given week
   c. Find the probability that more than five sightings are made during a given week
   d. Find the probability that exactly five sightings are made during a given week.

Solution

Please refer to the following steps to find the probability that fewer than two sightings are made during a given week

Step 1: Select f$_x$ - Statistical - Poisson

Step 2: Enter X=1, Mean =2.6, Cumulative TRUE.

Step 3: Click OK to finish. The probability is 26.7% **(Figure 4.8a)**

Figure 4.8a

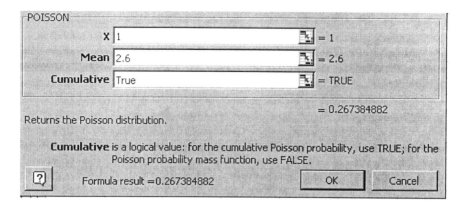

Please refer to the following steps to find the probability that more than five sightings are made during a given week

Step 1: Select f_x - Statistical - Poisson

Step 2: Enter X=1, Mean =2.6, Cumulative TRUE.

Step 3: Click OK to finish.

Step 4: Enter 1-Poisson = 1-0.9509 =.0049 or 4.9%

Figure 4.8b

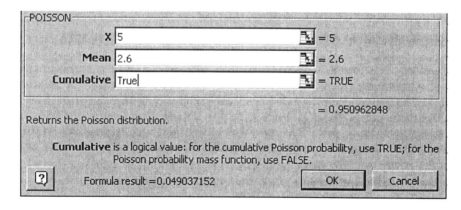

Please refer to the following steps to find the probability that exactly five sightings are made during a given week. (7.4%)

Step 1: Select f_x - Statistical - Poisson

Step 2: Enter X=1, Mean =2.6, Cumulative TRUE.

Step 3: Click OK to finish. The probability is 7.4% **(Figure 4.8c)**

78

Figure 4.8c

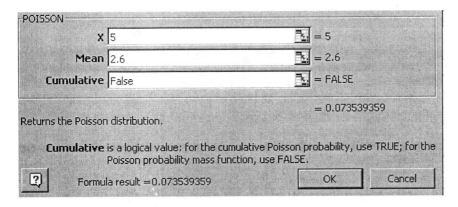

4.6   Calculating Hypergeometric Distributions using Excel

The hypergeometric distribution can be found using the fx function in Excel.

Step 1: Click on fx, HYPGEOMDIST(

Step 2: Number_s is the number of successes in the sample

Step 3: Number_sample is the size of the sample

Step 4: Population_s is the number of successes in the population

Step 5: Number_pop is the population size.

Figure 4.9

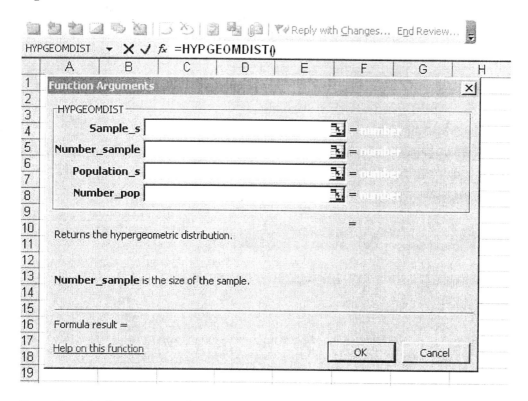

Example 4.14 Suppose a professor randomly selects three new teaching assistants from a total of 10 applicants, six males and four female students. Let x be the number of females who are hired. Find the probability that no females are hired.

Solution: Note that the question is phrased such that our focus is female students. Therefore, we are selecting 0 females from 4. Thus, the remaining 3 (from 6) are therefore, male.

1. Click on fx, HYPGEOMDIST(
2. Number_s is 0 (selecting 0 females from 4)
3. Number_sample is 4 (sample of interest is females)
4. Population_s is 3 (still need 3 teaching assistants.)
5. Number_pop is 10 (4 males plus 3 females)

Figure 4.10 illustrates that our answer is 0.1667 or 16.67%. This is the equivalent of 1/6, which is shown in the text.

Figure 4.10

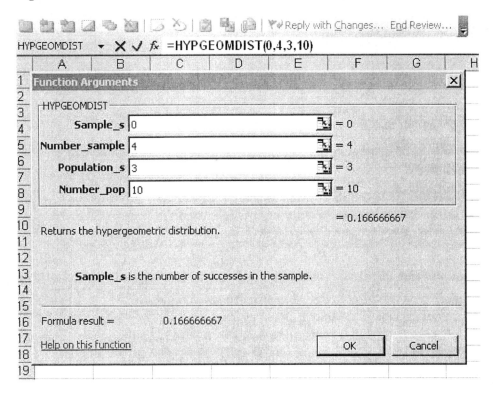

4.7 Sample Exercises

The following exercises were selected from the textbook to test the reader's understanding of PHStat and its usage within the chapter. Try to replicate the output in PHStat for the following problems.

Problem 4.64, page 217 (Chickens with fecal contamination)

The United States Department of Agricultural (USDA) reports that, under its standard inspection system, one in every 100 slaughtered chickens pass inspection with fecal contamination (*Tampa Tribune*, Mar. 31, 2000). Find the probability that, in a random sample of five slaughtered chickens, at least one passes inspection with fecal contamination.

Problem 4.79, page 223 (Eye fixation problem)
Cognitive scientists at the University of Massachusetts designed an experiment to measure x, the number of times a reader's eye fixated on a single word before moving past that word (*Memory and Cognition*, Sept. 1997). For this experiment, x was found to have a mean of 1. Suppose one of the readers in the experiment is randomly selected and assume that x has a Poisson distribution.
   a.   Find $P(x=0)$.
   b.   Find $P(x>1)$.
   c.   Find $P(x \le 2)$.

Test and CI for One Proportion: Time

```
Test of p = 0.5 vs p not = 0.5

Event = N

Variable   X   N  Sample p       95% CI        Z-Value  P-Value
Time      26  37  0.702703  (0.555428, 0.849978)   2.47    0.014
```

From this, we see that the proportion of suicides committed at night is between .56 and .85, with 95% confidence.

(c) Use **Stat>Basic Statistics>1-Sample t** (as in example 7.2 above)
(d) Use **Stat>Basic Statistics>1 Proportion** (as in part a. above)

A final chapter note: Individual Value Plots – new in Minitab Release 14

In the 1-sample *t* test of example 7.2, Minitab gives you 3 choices for graphing the data. If you click on **Graphs...: Individual value plot** you get the plot on the left below, which is a dotplot with ***random jitter*** applied to all the points (in the vertical direction).

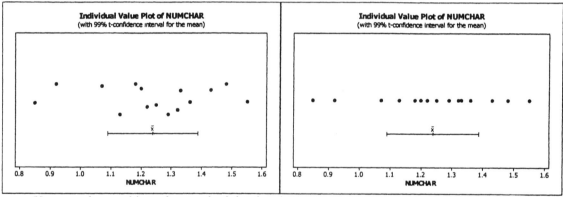

You can change this to the standard dotplot shown on the right by reducing the jitter to zero – just ***double-click on any of the dots in the plot*** to produce the pop-up dialog box shown below, and de-select **Add jitter**:

Why did Minitab introduce this plot, in release 14? The *jitter* will reveal *overlapping points*, which would not be visible at all in the dotplot on the right. You can also choose some number smaller than the default 0.025, like 0.01, for a lesser degree of jitter. You can try generating the plot several times, to see which illuminates the data most clearly.

Solution to Problem 4.79, page 223 (Eye fixation problem), part a.

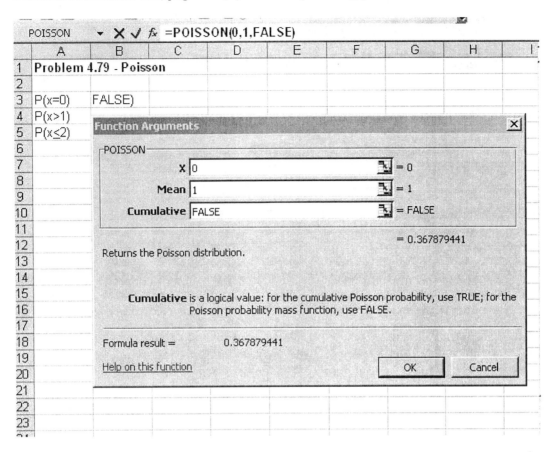

Solution to Problem 4.79, part b

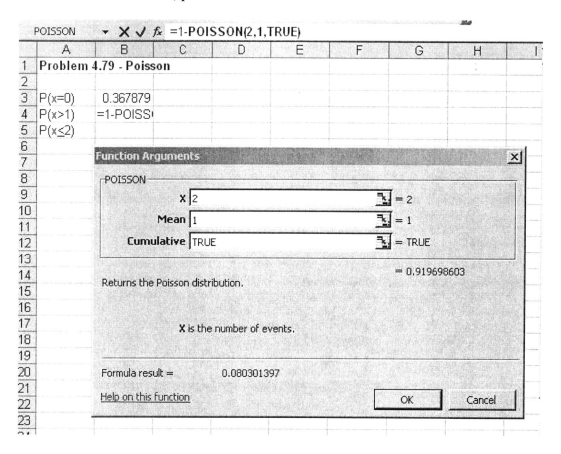

POISSON   ▾ ✗ ✓ *fx* =1-POISSON(2,1,TRUE)

	A	B	C	D	E	F	G	H	I
1	Problem 4.79 - Poisson								
2									
3	P(x=0)	0.367879							
4	P(x>1)	=1-POISS							
5	P(x≤2)								
6									
7									

Function Arguments ✗

POISSON

X |2| = 2

Mean |1| = 1

Cumulative |TRUE| = TRUE

= 0.919698603

Returns the Poisson distribution.

X is the number of events.

Formula result =   0.080301397

Help on this function   OK   Cancel

Solution to Problem 4.79, part c.

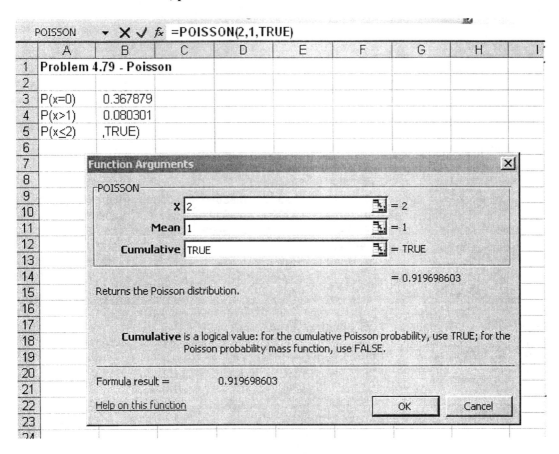

Final Answer to Problem 4.79

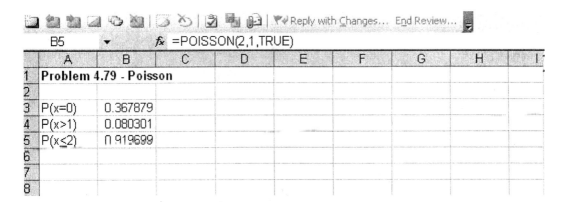

85

Solution to Problem 4.64

Binomial Probabilities

Sample size	5
Probability of success	0.01
Mean	0.05
Variance	0.0495
Standard deviation	0.222486

Binomial Probabilities Table

X	P(X)	P(<=X)	P(<X)	P(>X)	P(>=X)
0	0.95099	0.95099	0	0.04901	1
1	0.04803	0.99902	0.95099	0.00098	0.04901
2	0.00097	0.99999	0.99902	9.85E-06	0.00098
3	9.8E-06	1	0.99999	4.96E-08	9.85E-06
4	4.95E-08	1	1	1E-10	4.96E-08
5	1E-10	1	1	0	1E-10

Solution to Problem 4.97, page 227 (On-site disposal of waste)

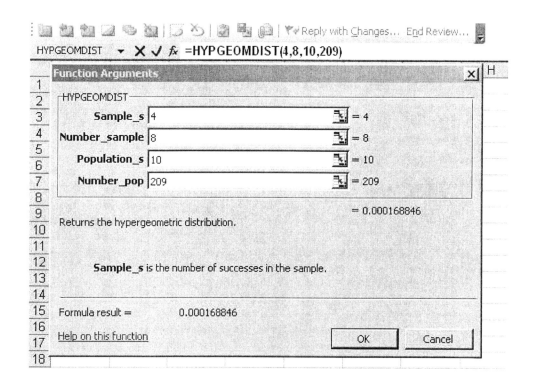

86

5.1   Introduction

In Chapter 4 of the text, discrete random variables were introduced and the distributions necessary to derive probabilities for these variables (Binomial and Poisson distributions) were discussed in detail. In Chapter 5, the second type of random variable, the continuous random variable, will be discussed. The most common distributions used for deriving the probabilities of continuous variables are the uniform, normal, and exponential distributions that may be utilized under certain, distinct conditions. Remember that a continuous variable is defined as one that can assume an infinite number of values contained in one or more intervals and a continuous distribution is graphically depicted as a smooth curve in which the area under the curve is equivalent to probabilities in the distribution.

Although the text does present statistical tables for the normal and exponential distribution that may be utilized to solve for probabilities, PHStat offers an alternative method for calculating those probabilities.

The following examples will be used to illustrate techniques in PHStat and Excel.

> Example 5.7, page 251 (Normal Distribution – Automobile Manufacturer)
> Example 5.11, page 262 (EPA Gas Mileage Ratings for 100 Cars)
> Example 5.13, page 277 (Emergency Arrivals at Hospitals)
> Sample Exercise #1: Problem 5.115, page 283 (Lengths of sardines)
> Sample Exercise #2: Problem 5.60, page 267 (Dart throwing errors)

5.2   Calculating Normal Probabilities using PHStat

To use the normal probability distribution within PHStat, please follow the steps provided.

Step 1: Open a new workbook and **click** on the **PHStat** menu near the top of the screen.

Step 2: Select the **Probability Distribution** option from the choices available and then **click** open the **Normal** option (not Normal Probability Plot) from the list.

Step 3: Then click **OK**. If done correctly, the **Normal Probability Distribution** menu should appear.

Figure 5.1

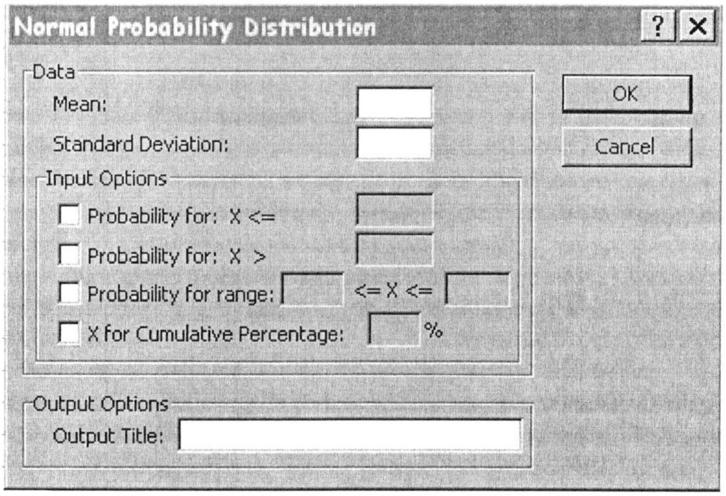

Step 4: Notice that user must enter the population **mean**, μ, and the population **standard deviation**, σ, from a specified normal distribution.

Step 5: Within **Input Options**, the user may specify three types of probability options and may also use this option to find a specified value of x.

Step 6: Specifying an **output title** is optional; however, if the user is working with several different types of data sets, it is always recommended to create titles to organize work.

Step 7: Click OK to finish.

The following example from the textbook will illustrate the use of the normal distribution.

Example 5.1   Please refer to Example 5.7, 251 (Normal Distribution – Automobile Manufacturer) from the textbook

Suppose an automobile manufacturer introduces a new model that has an advertised mean in-city mileage of 27 miles per gallon. Although such advertisements seldom report any measure of variability, suppose you write the manufacturer for the details of the tests and you find that the standard deviation is 3 miles per gallon. This information leads you to formulate a probability model for the random variable, x, the in-city mileage for this car model. You believe that the probability distribution of x can be approximated by a normal distribution with a mean of 27 and a standard deviation of 3. If you were to buy this model of automobile, what is the probability that you would purchase one that averages less than 20 miles per gallon for in-city driving? In other words, find P (x<20).

Solution

Please refer to the following steps to answer this question.

Step 1: Using **PHStat**, select the **Normal Probability Distribution** from the list of options under **Probability Distribution**.

88

Step 2: The problem specified that the **population mean, μ,** is **27** and the **population standard deviation, σ, is 3**.

Figure 5.2

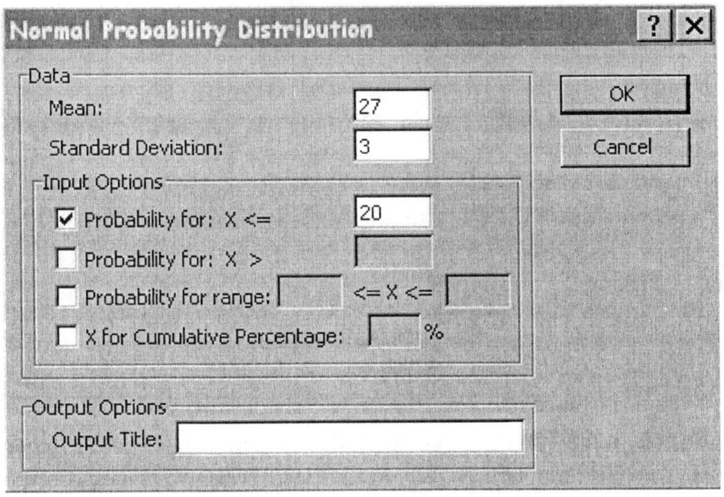

Step 3: After entering this data into the Normal Probability Distribution menu, locate **Input Options** and select **Probability for: X<=** and input the value "20" in the appropriate box. Remember that the problem specifies that the user find the probability for purchasing **less than** 20 miles per gallon; however, since this is a continuous function, this is equivalent to calculating the probability of 20 or less miles per gallon!

Step 4: Enter in an **Output Title** if desired and

Step 5: Click OK to finish.

Step 6: PHStat generates the output found in Table 5.1. The conclusion is that the probability that an automobile would average less than 20 miles per gallon is 0.0098 or 0.98%. Please verify, using the appendix in the textbook, that 0.0098 is the correct answer.

Table 5.1

Normal Probabilities	
Mean	27
Standard Deviation	3
Probability for X<=	20
Z Value	-2.333333333
P(X<=20)	0.009815307

Example 5.2 Please refer to Example 5.10 (page 255) from the textbook.

Suppose the scores, x, on a college entrance examination are normally distributed with a mean of 550 and a standard deviation of 100. A certain prestigious university will consider for admission only those

89

applicants whose scores exceed the 90th percentile of the distribution. Find the minimum score, an applicant must achieve in order to receive consideration for admission to the university.

Solution

Step 1: Using **PHStat**, select the **Normal Probability Distribution** from the list of options under **Probability Distribution**.

Step 2: The problem specified that the **population mean, μ, is 550** and **the population standard deviation, σ, is 100**.

Step 3: After entering this data into the Normal Probability Distribution menu, locate **Input Options** and select **X for Cumulative Percentage** and input the value "**90**" in the appropriate box. Notice that in this example, the specified x value that indicates the boundary between the 90th percentile and the rest of the distribution is the variable of interest.

Step 4: Enter in an **Output Title** if desired and click **OK.**

Step 5: The output generated appears in Table 5.2. Notice that a minimum score an applicant must achieve in order to receive consideration for admission into a prestigious university is 678. Using the statistical tables accompanying the textbook, the formulas may be utilized in order to derive this answer.

Figure 5.3

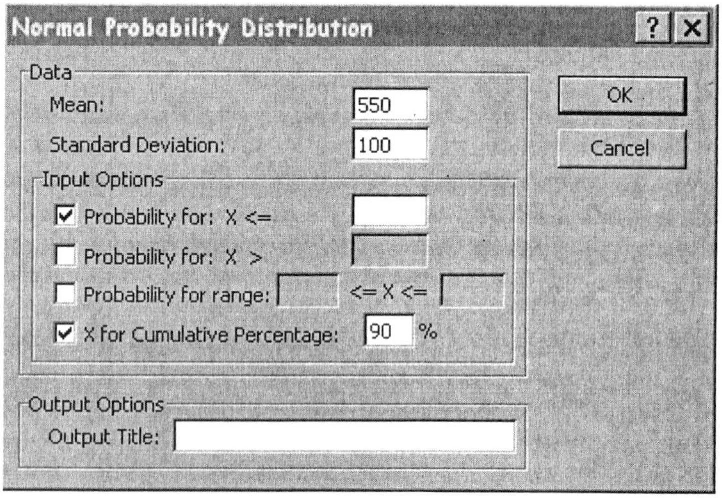

90

And we get:

Cumulative Distribution Function

```
Chi-Square with 3 DF

   x  P( X <= x )
2.09     0.446062
```

Suppose we set $\alpha=.05$. The alternative is two sided ($\sigma^2 \neq 1$), so the rejection region has $.05/2 = .025$ in each tail. Does the chi-square statistic here (2.09) catch either the upper or lower .025 tail of the chi-square distribution? Since the probability above is not less than .025 or greater than .975, the answer is no. We may not reject the null hypothesis at the 5% level.

We may also choose to report the P-value here, which equals 2 x .4461 or .891, far exceeding any possible designated α.

If you want to find the exact critical values for a 5% level test, select **Inverse Cumulative Probability** from **Calc>Probability Distributions: Chi-Square**, and input the constants *.025* and *.975*. There is no need to do this here, since the p-value is more informative.

A final chapter note re Minitab's confidence intervals:

One-sided confidence intervals: Note that the one-sample *t* output in examples 8.5 & 8.6 contains something named the **95% Upper Bound** = *19.2814*. Also, if you look at the histogram and boxplot displayed there, you will see an interval drawn from that upper bound of 19.2814 to negative infinity. The interval (- ∞, 19.2814) is called a *one-sided confidence interval*. If you select a one-sided alternative, in this, and other dialog boxes, Minitab will assume that you also desire a one-sided confidence interval. Such intervals are interpreted similarly to two-sided ones, but are not relevant to this course. You can ignore it. *If you want a two-sided CI, you should go back and redo the procedure, after selecting the two-sided alternative.*

Figure 5.4

	A	B	C
1	EPA Mileage Rating in 100 cars		
2	36.3		
3	32.7		
4	40.5		
5	36.2		
6	38.5		
7	36.3		
8	41.0		
9	37.0		
10	37.1		
11	39.9		

Solution

Please refer to the following steps in order to answer this question

Step 1: Open a new worksheet and input the data manually or access the worksheet that contains the data for EPA gas mileage.

Step 2: Click PHStat and select **Normal Probability Plot** from the list of options under **Probability Distribution**.

Step 3: Specify the **Variable Cell Range** by highlighting and clicking on the appropriate cells where the data is contained.

Step 4: If a title is contained in the first row, check the **First cell contains label** box.

Step 5: Again, an **Output Title** may be created if the user desires.

Figure 5.5:

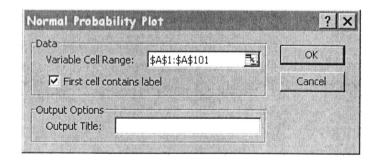

Step 6: When finished, click **OK**.

The following figure (5.6) presents the normality plot for the EPA mileage ratings on 100 cars. Notice that PHStat arranged the data from smallest to largest and then plotted each data point against its expected z score. Since the data appear to fall in a straight line, it does appear that this data set approximates the normal distribution.

Figure 5.6

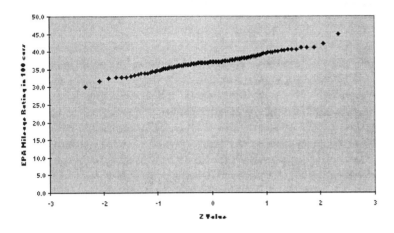

EPA Mileage

5.4 Calculating Exponential Probabilities using PHStat

Often referred to as the "waiting time distribution", the exponential probability distribution is best used to describe the amount of time or distance between occurrences of **random** events. Please refer to the following steps in order to calculate exponential probabilities in PHStat.

Step 1: Open a new workbook and **click** on **PHStat** at the top of the screen.

Step 2: Select Exponential from the **Probability Distribution** options listed in the menu.

Step 3: Then click **OK**.

Step 4: Notice that the user is required to enter the **Mean per unit** of λ ($1/\sigma$) and the **X Value** of interest. Note that the probabilities generated using this function are "less than or equal to" probabilities. To find a "greater than" or "greater than or equal to" probability, the user must adjust the formula.

Step 5: Again, the user may create a title for the output being generated. An example will illustrate this concept. **Click OK** to finish.

Figure 5.7

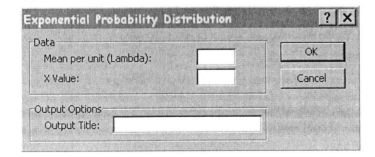

93

Example 5.4   Please refer to Example 5.13, page 277 (Emergency Arrivals at Hospitals) from the text.

Suppose the length of time (in hours) between emergency arrivals at a certain hospital is modeled as an exponential distribution with θ =2. What is the probability that more than 5 hours pass without an emergency room arrival?

Solution

Please refer to the following steps in order to answer this question.

Step 1: Open a new worksheet and place the cursor in the upper left-hand cell.

Step 2: PHStat - Probability Distribution- Exponential

Step 3: In this example, the **Mean per unit of lambda** (λ) is 0.5 (1/θ= 1/2) and the **X-value** is 5.

Step 4: Enter an **Output Title** if desired.

Step 5: Click OK. The output generated is shown below.

Figure 5.8

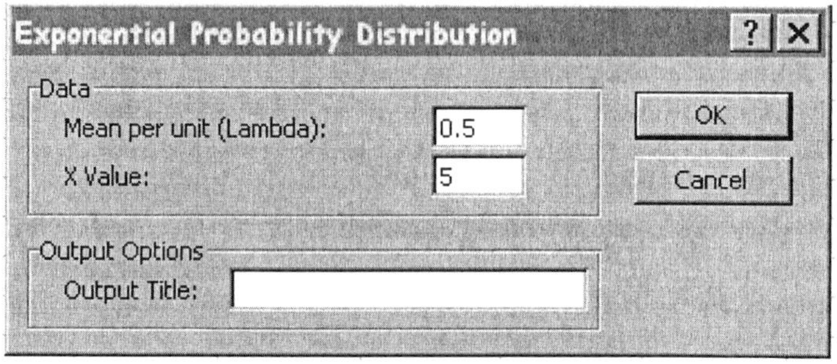

Step 6: PHStat calculated the probability for P (x ≤ 5) = 0.917915. In order to find the probability of P (x>5); subtract this value from one. Thus, P (x>5) = P (x ≥ 6) = 1 - P (x ≤ 5) = 1-0.917915 = 0.082085 or 8.2%. Using the statistical tables at the back of the textbook, this answer may be verified.

Table 5.3

Exponential Probabilities	
Mean	0.5
X Value	5
P(<=X)	0.917915

5.5   Sample Exercises

The following exercises were selected from the textbook to test the reader's understanding of PHStat and its usage within the chapter. Try to replicate the output in PHStat for the following problems.

Problem 5.115, page 283 (Lengths of sardines)

Fisheries Science (Feb. 1995) published a study of the length distributions of sardines inhabiting Japanese waters. At two years of age, fish have a length distribution that is approximately normal with a mean of 20.20 centimeters (cm) and a standard deviation of 0.65 cm.

    a. Find the probability that a two-year-old sardine inhabiting Japanese waters is between 20 and 21 cm long.

    b. Find the probability that a two-year-old sardine in Japanese waters is less than 19.84 cm long.

    c. Find the probability that a two-year-old sardine in Japanese waters is greater than 22.01 cm long.

Solution to Problem 5.115 part a:

Normal Probabilities	
Mean	24.1
Standard Deviation	6.3
Probability for range	20 <= X <= 21
Z Value for 20	-0.650793651
Z Value for 21	-0.492063492
P(X<=20)	0.257589782
P(X<=21)	0.311337228
P(20<=X<=21)	0.053747446

Solution to Problem 5.115 part b:

Normal Probabilities	
Mean	24.1
Standard Deviation	6.3
Probability for X<=	19.84
Z Value	-0.676190476
P(X<=19.84)	0.249459788

Solution to Problem 5.115 part c:

Normal Probabilities	
Mean	24.1
Standard Deviation	6.3
Probability for X>	22.01
Z Value	-0.331746032
P(X>22.01)	0.629959416

Problem 5.60, page 267 (Dart throwing errors)

How accurate are you at the game of darts? Researchers at Iowa State University attempted to develop a probability model for dart throws (*Chance*, Summer 1997). For each of 590 throws made at certain targets on a dartboard, the distance from the dart to the target point was measured (to the nearest millimeter). The error distribution for the dart throws is described by the frequency table shown in the text. Construct a normal probability plot for the data and use it to assess whether the error distribution is approximately normal.

Solution to Problem 5.60

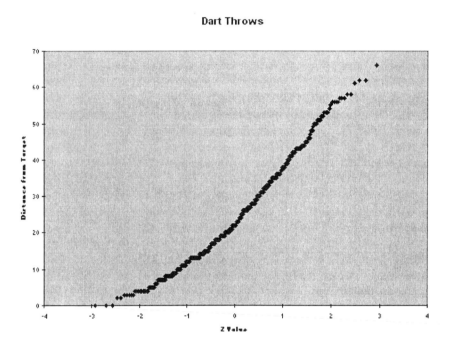

96

6.1 Introduction

Chapter 6 in the textbook begins to introduce the basic properties of a sampling distribution which is used to estimate a population parameter such as the population mean (μ) or the population proportion (p). In this section, PHStat will be used to derive various probabilities using the sampling distribution of the sample mean ($\bar{x}$). In the previous chapter, the Normal Probability Distribution was discussed and probabilities were calculated using specific examples from the textbook. Be sure and reread that section before continuing with the following examples.

The following examples will be used to illustrate techniques in PHStat and Excel.

> Example 6.8, page 307, (Automobile Batteries)
> Sample Exercise #1: Problem 6.36, page 311 (Research on eating disorders)

6.2 Calculating Probabilities Using the Sampling Distribution ($\bar{x}$) using PHStat

Recall from the text that the sampling distribution of the sample means is a probability distribution of possible sample means of a given sample size and the Central Limit Theorem is an application to the sampling distribution of the sample means. According to this theorem, the sampling distribution of the sample means is approximately a normal distribution if samples of a particular size are selected from any population. Although the approximation improves as the sample size increases, this theorem allows the researcher to make statements about the population using a sample, without any information about the shape of the original distribution from which the sample is taken.

Using the Central Limit Theorem, probabilities for sample distributions may be estimated using the **Normal Probability Distribution** function within PHStat.

Step 1: Open a new workbook in Excel.

Step 2: Click on the **PHStat** menu and select the **Probability Distribution** option.

Step 3: Click on the **Normal Probability Distribution** option and a menu should appear that looks similar to Figure 6.1.

Figure 6.1a

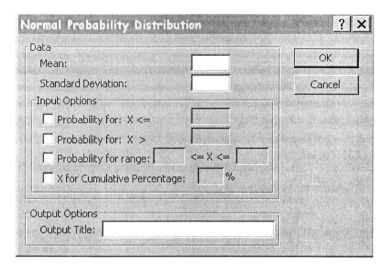

Step 4: Within the **Data** menu, the researcher must enter the population **mean**, μ, and the **standard deviation of the sampling distribution**, σ, for the normal probability distribution. Note that when sampling from a distribution, the appropriate standard deviation is $\sigma/\sqrt{n}$.

Step 5: In **Input Options**, specify the type of probability needed or the specified X value. Although the user may opt to specify an output title, if desired, it is usually recommended, especially when working with several data sets. An example from the textbook will illustrate its usage. **Click OK** to finish.

The Excel function (NORMDIST) will also provide solutions for the normal distribution; however, the user must clearly define what area is being derived. Figure 6.1b shows the complement to PHStat in terms of the Normal Distribution.
1. X is the value for the distribution
2. Mean is the arithmetic mean of the distribution
3. Standard deviation is the standard deviation of the distribution.
4. Cumulative is defined as "TRUE" for the cumulative distribution and "FALSE" for the probability mass function.

Figure 6.1c provides us with the function "NORMINV" which will find "x"
1. Probability refers to the probability corresponding to the normal distribution.
2. Mean and the standard deviation must also be included. Remember to change the standard deviation so that you divided it by the square root of the sample size, n.

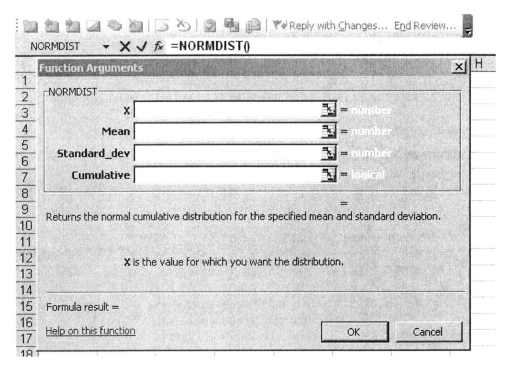

Figure 6.1c

Example 6.1   Please refer to Example 6.8, page 307, (Automobile Batteries) from the textbook.

A manufacturer of automobile batteries claims that the distribution of the lengths of life of its best battery has a mean of 54 months and a standard deviation of 6 months. Suppose a consumer group decides to check the claim by purchasing a sample of 50 of these batteries and subjecting them to tests that determine battery life. Assuming that the manufacturer's claim is true, what is the probability the consumer group's sample has a mean life of 52 or fewer months?

Solution

Step 1: Before utilizing the Normal Probability Distribution option in PHStat, notice that we must derive the standard deviation of the sampling distribution ($\sigma/\sqrt{n}$).

Step 2: In this example, the standard deviation of the sampling distribution is ($6/\sqrt{50}$) 0.85 months. Verify this answer with the textbook.

Step 3: Next, open the **Normal Probability Distribution** menu and input a **mean** of 54 and a **standard deviation** of 0.85.

Figure 6.2

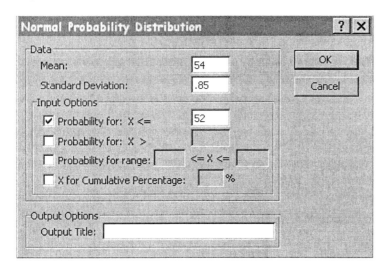

Step 4: In **Input Options**, choose **Probability for: X<=** and input the number 52.

Step 5: Specify an **Output Title** if desired

Step 6: Click OK to finish.

Step 7: Table 6.1 presents the values entered into **Data** and **Input Options** within the **Normal Probability Distribution** menu. Notice that the mean, standard deviation of the sampling distribution and the probability that was defined are all repeated in the output derived. the z value of -2.35 is defined as well as the probability of interest. Thus, the **probability the consumer group will observe a sample mean of 52 or less is only 0.0093 or 0.93% if the manufacturer's claims is true.**

Table 6.1

Normal Probabilities	
Mean	54
Standard Deviation	0.85
Probability for X<=	52
Z Value	-2.352941176
P(X<=52)	0.009312773

6. 3   Solving the Normal Distribution using Excel

Example 6.2 will be demonstrated through the use of normal Excel as opposed to PHStat.

Step 1: Find the standard deviation of the sampling distribution (0.85 months, see above).

Step 2: Go to fx =NORMDIST
Step 3: Input X as 52

Step 4: Input the Mean as 54

Step 5: Input the standard deviation as 0.848528 (the more precise the decimal, the more precise the answer).

Step 6: Insert true for cumulative. This will find the "less than" distribution desired in this problem.

Figure 6.3

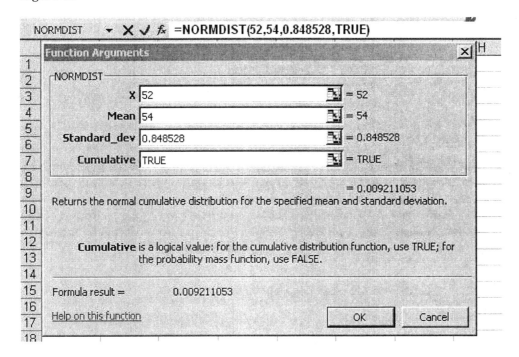

6. 4   Sample Exercises

The following exercises were selected from the textbook to test the reader's understanding of PHStat and its usage within the chapter. Try to replicate the output in PHStat for the following problems.

Problem 6.36, page 311 (Research on eating disorders)

Refer to *The American Statistician* (May, 2001) study of female students who suffer from bulimia. Recall that each student completed a questionnaire from which a "fear of negative evaluation" (FNE) score was produced. (The higher the score, the greater the fear of negative evaluation.) Suppose the FNE scores of bulimic students have a distribution with mean of 18 and a standard deviation of 5. Now consider a random sample of 45 female students with bulimia.

    a.   What is the probability that the sample mean FNE score is greater than 17.5?
    b.   What is the probability that the sample mean FNE score is between 18 and 18.5?
    c.   What is the probability that the sample mean FNE score is less than 18.5?

Solution to Problem 6.36

Solution for Exercise 6.36 part a	
Mean	18
Standard Deviation	0.7454
Probability for X>	17.5
Z Value	-0.670780789
P(X>17.5)	0.748819977

Solution for Exercise 6.36 part b		
Mean	18	
Standard Deviation	0.7454	
Probability for range	18	<= X <= 18.5
Z Value for 18	0	
Z Value for 18.5	0.670780789	
P(X<=18)	0.5	
P(X<=18.5)	0.748819977	
P(18<=X<=18.5)	0.248819977	

Solution for Exercise 6.36 part c	
Mean	18
Standard Deviation	0.7454
Probability for X<=	18.5
Z Value	0.670780789
P(X<=18.5)	0.748819977

7.1   Introduction

In Chapter 7 of the textbook, population means and parameters are estimated using statistics generated from a single sample. Using the Central Limit Theorem, the sampling distribution of the sample mean can be approximated as a normal distribution for large samples and an interval can be created to incorporate the population mean, μ, given a certain level of error. In this chapter, examples will be used to estimate
1. Population means
2. Proportions
3. The sample size necessary to estimate population means and proportions

PHStat can be used to calculate confidence intervals for population means, μ, and population parameters, p. When calculating intervals estimating population means, remember to determine whether:
1. The population standard deviation, σ, is known
2. The population standard deviation, σ, is unknown
When the population standard deviation is unknown, the confidence interval for the population mean can be estimated using statistics generated from the sample data.

In order to estimate a population proportion, p, enter both the number of successes and the sample size into the PHStat menu. At this time, PHStat does not provide an option for entering a specific data set to use to estimate proportions.

The following examples will be used to demonstrate techniques in PHStat and Excel.

> Example 7.1: Table 7.1, page 323, (Length of Stay (in Days) for 100 Patients
> Example 7.3, page 343, (Public Opinion Polls)
> Example 7.6, page 352, (NFL Footballs)
> Example 7.7, page 354, (Telephone Manufacturer – Customer Complaints)
> Sample Exercise #1: Problem 7.18, page 331 (Sentence complexity study)
> Sample Exercise #2: Problem 7.41, page 342 (Study on waking sleepers early)
> Sample Exercise #3: Problem 7.58, page 350 (FTC checkout scanner study)
> Sample Exercise #4: Problem 7.75, page 357 (Bacteria in bottled water)

7.2   Estimation of the Population Mean: Sigma Unknown

Using PHStat, **open** a new workbook and place the cursor in the upper left cell of the workbook.

Step 1: Click PHStat menu at the top of the screen

Figure 7.1

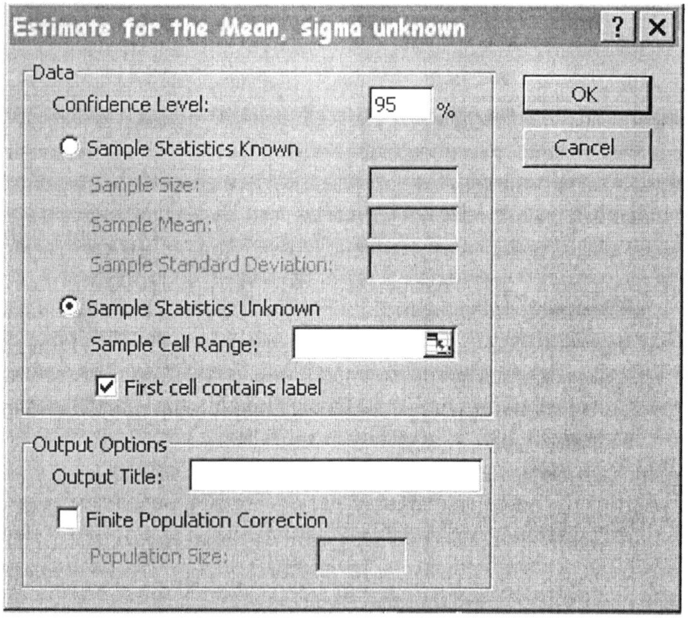

Estimate for the Mean, sigma unknown

Data
Confidence Level: 95 %
○ Sample Statistics Known
  Sample Size:
  Sample Mean:
  Sample Standard Deviation:
◉ Sample Statistics Unknown
  Sample Cell Range:
  ☑ First cell contains label

Output Options
Output Title:
☐ Finite Population Correction
  Population Size:

OK   Cancel

Step 2: Select Confidence Interval (Figure 7.1)

Step 3: Select Estimate for the Mean, sigma unknown

Step 4: Enter the Confidence Level

Step 5: Click Sample Statistics Known or Sample Statistics unknown

Step 6: Type in an Output Title if desired

Step 7: Click OK

Example 7.1   Please refer to Table 7.1 (page 323) in the textbook.

Table 7.1 of the text presents data on lengths of stay (in days) for 100 patients. Using this data set, calculate a 95% confidence interval for the mean length of patient stay in a hospital.

Solution

Table 7.1 presents the data as it appears in the textbook for the 100 patients. In order to create the confidence interval, notice that the population standard deviation is unknown and the sample statistics are unknown. This implies that, given raw data, sample statistics must be generated.

Step 1: Click on the **PHStat** menu at the top of the screen

Step 2: Select **Confidence Intervals**

Step 3: Select **Estimate for the Mean, sigma unknown**

104

Step 4: Under **Data**, the confidence interval calculated should be 95% by default. If not, type in 95.

Step 5: Select **Sample Statistics Unknown**.

Step 6: In **Sample Cell Range**, highlight the range of the data. In this example, it is A1:A101.

Step 7: In cell A1, a title is present, so the **first cell contains label** box is checked.

Step 8: Title the output if desired. In this case, the output label is "Length of Stay in Days."

Step 9: Click **OK** to finish.

Figure 7.2

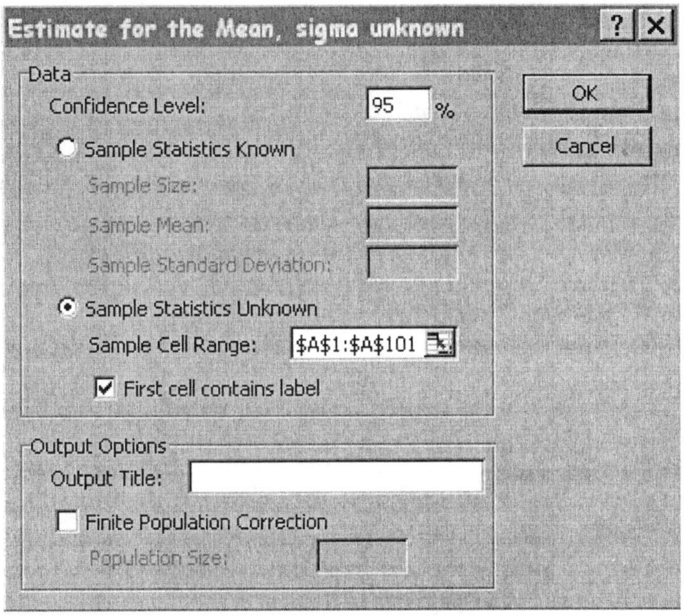

Table 7.1

Confidence Interval Estimate for the Mean	
Sample Standard Deviation	3.677545844
Sample Mean	4.53
Sample Size	100
Confidence Level	95%
Standard Error of the Mean	0.367754584
Degrees of Freedom	99
t Value	1.984217306
Interval Half Width	0.729705011
Interval Lower Limit	3.80
Interval Upper Limit	5.26

Step 10: The preceding figures present the output needed to answer the original question. Using this output, the mean length of stay for ALL hospital patients will be between 3.79 days and 5.27 days about 95% of the time. This implies that about 5% of the time, the mean length of stay for ALL hospital patients does not fall within that interval.

7.3   Estimation of the Population Proportion

Before estimating the population proportion using PHStat, both the sample size and the number of "successes" in the sample need to be known. The following steps present the procedure necessary to generate confidence intervals for population proportions:

Step 1: Open a new workbook and place the cursor in the upper left cell of the worksheet

Step 2: Click on the **PHStat** menu.

Step 3: Select Confidence Interval from the choices available. **(Figure 7.3)**

Step 4: Select Estimate for the Proportion from the options available.

Step 5: The user is then required to enter the **sample size**, **number of successes**, and the **confidence level**. Notice that the confidence level is usually set at 95% by default.

Step 6: An **output title** may be selected

Step 7: Click OK to finish

The Estimate for the Proportion menu should appear as follows:

Figure 7.3

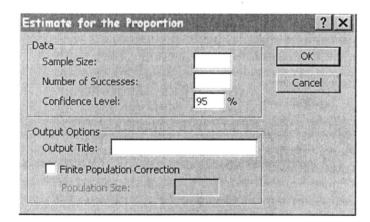

Example 7.2: Use Example 7.3 (page 343) from the text to derive an interval for the proportion.

Public opinion polls are conducted regularly to estimate the fraction of U.S. citizens who trust the president. Suppose 1,000 people are randomly chosen and 637 answer that they trust the president. Using a 95% confidence level, how would you estimate the true fraction of ALL U.S. citizens who trust the president?

Solution

The following are the steps necessary to solve this question.

Step 1: Click **PHStat-Confidence Intervals-Estimate for the Proportion**.

Stpe 2: First, identify that the sample size (n) is 1,000. The number of successes (x) is 637 and the confidence level is 95%. Enter this information in the spaces provided. **(Figure 7.4)**

Step 3: Enter in an output title. The one selected for this example is "Public Opinion Poll".

Step 4: Click **OK** to finish.

Figure 7.4

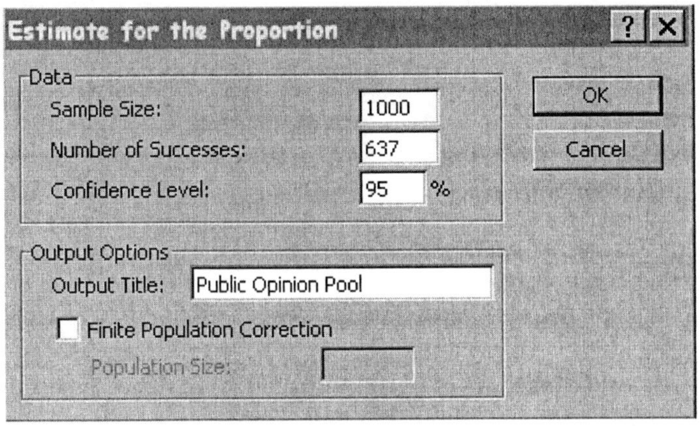

Step 5: Refer to Table 7.2 below. Therefore, using the intervals derived in PHStat, we are 95% confident that the true proportion of ALL U.S. citizens who trust the president are between 60.7% and 66.7%.

Table 7.2

Public Opinion Poll	
Sample Size	1000
Number of Successes	637
Confidence Level	95%
Sample Proportion	0.637
Z Value	-1.95996108
Standard Error of the Proportion	0.015206282
Interval Half Width	0.02980372
Interval Lower Limit	0.60719628
Interval Upper Limit	0.66680372

7.4   Determining the Sample Size

In Section 7.4 of the textbook, the authors discuss the importance of determining the sample size necessary for making an inference about a population parameter if a specific level of confidence is required! Suppose that a researcher requires the confidence interval generated to be within +/- x units so

107

that the interval is "neat and clean". In order to do so, it is necessary to find the sample size needed to estimate the upper and lower bounds of the interval.

The general steps necessary to find the appropriate sample size using PHStat are shown below.

Step 1: Click PHStat at the top of the screen

Step 2: Select Sample Size Option

Step 3: Notice that there is a choice in selecting **Determination for the Mean** or **Determination for the Proportion**. Both menus are shown below in Figure 7.5a and Figure 7.5b.

Figure 7.5a

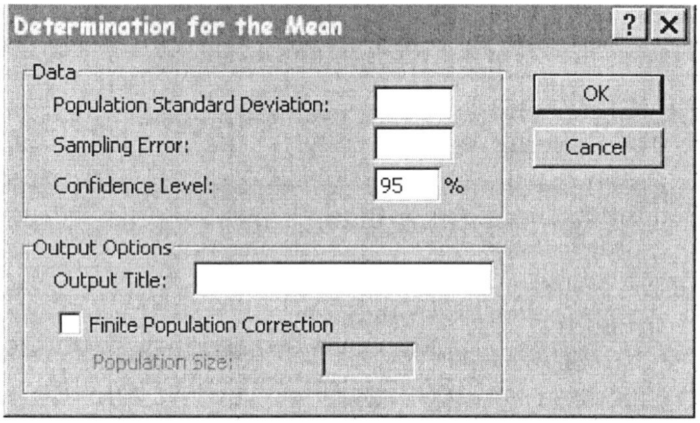

Figure 7.5b

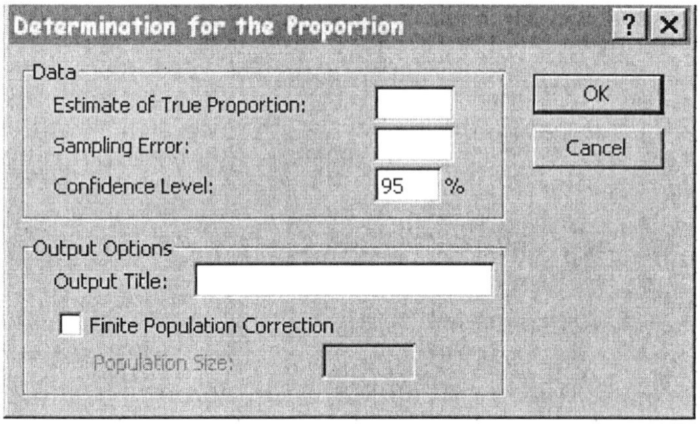

7.4.1 Determining the Sample Size for the Mean

Before determining the sample size for the mean, the user needs to know the value of the
1. **Population standard deviation** (or some estimate of it)
2. **Sampling Error**
3. **Confidence Level** (by default, usually set at 95%)

108

Example 7.3   Example 7.6 (page 352) from the text is used to calculate the sample size for the mean

Suppose the manufacturer of official NFL footballs uses a machine to inflate the new balls to a pressure of 13.5 pounds. When the machine is properly calibrated, the mean inflation pressure is 13.5 pounds but uncontrollable factors cause the pressures of individual footballs to vary randomly from about 13.3 to 13.7 pounds. For quality control purposes, the manufacturer wishes to estimate the mean inflation pressure to within 0.025 pound of its true value with a 99% confidence interval. What sample size should be specified for this experiment?

Solution

In order to find the appropriate sample size, the following steps are necessary.

Step 1: PHStat-Sample Size-Determination for the Mean

Step 2: A menu box will appear asking for the input of the population standard deviation, sampling error, and confidence level.

Step 3: To find the population standard deviation, note that the range required is 13.7-13.3 =.4 so the population standard deviation can be estimated as σ = Range/4 = .4/4 = .1 **Enter population standard deviation as .1 (Figure 7.6)**

Figure 7.6

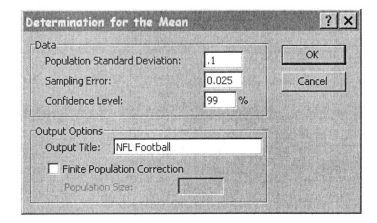

Step 4: Enter **sampling error as 0.025 pounds** as specified by the problem.

Step 5: Enter **confidence level** as **99%**.

Step 6: Enter an Output Title if desired. In this example, the title is "NFL Football".

Step 7: Click OK to finish. In answer to the question, the sample size needed to estimate the mean inflation pressure to within 0.025 pounds of its true value with 99% confidence is 107 footballs!

Table 7.3

NFL Football	
Population Standard Deviation	**0.1**
Sampling Error	**0.025**
Confidence Level	**99%**
Z Value	-2.57583451
Calculated Sample Size	106.1587751
Sample Size Needed	**107**

7.4.2 Determining the Sample Size for Proportions

Before determining the sample size for a proportion, the user needs to know the value of
1. **Estimate of the true proportion**
2. **Sampling Error**
3. **Confidence Level**

Example 7.4   Example 7.7 (page 354) from the text will be used to find the sample size for proportions.

Suppose a large telephone manufacturer that entered the post-regulation market quickly has an initial problem with excessive customer complaints and consequent returns of the phones for repair or replacement. The manufacturer wants to estimate the magnitude of the problem in order to design a quality control program. How many telephones should be sampled and checked in order to estimate the fraction defective, p, to be within 0.01 with 90% confidence?

Solution

In order to find the appropriate sample size, the following steps are necessary.

Step 1: Select PHStat-Sample Size-Determination for the Proportion.

Step 2: Enter the **Estimate of True Proportion as .1** as specified by the problem According to the problem, the value of .1 which corresponds to 10% defective will be a conservative estimate.

110

Figure 7.7

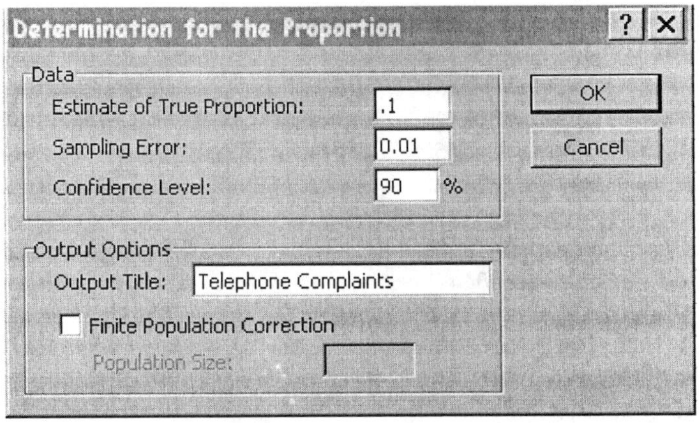

Step 3: Enter the **Sampling Error as .01** as specified by the problem. (**Figure 7.7**)

Step 4: Enter the **Confidence Level** as 90.

Step 5: Specify an **Output Title** as desired. In this case, the title is "Telephone Complaints".

Step 6: Click OK to finish.

PHStat calculates the correct sample size needed (n) to be 2,435. Notice, however, that the book solution is 2,436. This discrepancy is attributed to the fact that PHStat calculates a more accurate Z value (1.644853) than the textbook table (Z = 1.645). Therefore, 2,435 telephones should be samples and checked in order to estimate the proportion of defectives to be within 0.01 with 90% confidence.

Table 7.4

Telephone Complaints	
Estimate of True Proportion	**0.1**
Sampling Error	**0.01**
Confidence Level	**90%**
Z Value	-1.644853
Calculated Sample Size	2434.987254
Sample Size Needed	**2435**

7.5   Sample Exercises

The following exercises were selected from the textbook to test the reader's understanding of PHStat and its usage within the chapter. Try to replicate the output in PHStat for the following problems.

Problem 7.18, page 331 (Sentence complexity study)

Refer to the *Applied Psycholinguistics* (June 1998) study of language skills of low-income children. Each in a sample of 65 low-income children were administered the Communicative Development Inventory (CDI) exam. The sentence complexity scores had a mean of 7.62 and a standard deviation of 8.91. Construct a 90% confidence interval for the mean sentence complexity score of all low-income children.

Solution for Problem 7.18

Solution for Exercise 7.18	
Population Standard Deviation	8.91
Sample Mean	8.91
Sample Size	65
Confidence Level	90%
Standard Error of the Mean	1.105149485
Z Value	-1.644853
Interval Half Width	1.817808447
Interval Lower Limit	**7.092191553**
Interval Upper Limit	**10.72780845**

Problem 7.41, page 342 (Study on waking sleepers early)

Scientists have discovered increased levels of the hormone adrenocorticotropin in people just before they awake from sleeping (*Nature*, Jan. 7, 1999). In the study, 15 subjects were monitored during their sleep after being told they would be woken at a particular time. One hour prior to the designated wake-up time, the adrenocorticotropin level (pg./ml) was measured in each, with a mean of 37.3 and a standard deviation of 13.9. Estimate the true mean adrenocorticotropin level of sleepers one-hour prior to waking using a 95% confidence level.

Solution for Problem 7.41

Solution for Exercise 7.41	
Sample Standard Deviation	13.9
Sample Mean	37.3
Sample Size	15
Confidence Level	95%
Standard Error of the Mean	3.588964567
Degrees of Freedom	14
t Value	2.144788596
Interval Half Width	7.697570274
Interval Lower Limit	**29.60**
Interval Upper Limit	**45.00**

Problem 7.58, page 350 (FTC checkout scanner study)

Refer to the Federal Trade Commission (FTC) 1998 "Price Check" study of electronic checkout scanners. The FTC inspected 1,669 scanners at retail stores and supermarkets by scanning a sample of items at each store and determining if the scanned price was accurate. The FTC gives a store a "passing" grade if 98% or more of the items are price accurately. OF the 1,669 stores in the study, 1,185 passed inspection. Find a 90% confidence interval for the true proportion of retail stores and supermarkets with electronic scanners that pass the FTC price-check test.

Solution

Solution for Exercise 7.58	
Sample Size	1669
Number of Successes	1185
Confidence Level	90%
Sample Proportion	0.710005992
Z Value	-1.644853
Standard Error of the Proportion	0.011107015
Interval Half Width	0.018269407
Interval Lower Limit	0.691736585
Interval Upper Limit	0.728275398

Problem 7.75, page 357 (Bacteria in bottled water)

Is the bottled water you drink safe? According to an article in *U.S. News & World Report* (April 12, 1999), the Natural Resources Defense Council warns that the bottled water you are drinking may contain more bacteria and other potentially carcinogenic chemicals than allowed by state and federal regulators. Of the more than 1,000 bottles studied, nearly one-third exceed government levels. Suppose that the Natural Resources Defense Council wants an updated estimate of the population proportion of bottled water that violates at least one government standard. Determine the sample size (number of bottles) needed to estimate this proportion to be within 0.01 with 99% confidence.

Solution

Solution for Exercise 7.75	
Estimate of True Proportion	0.3333
Sampling Error	0.01
Confidence Level	99%
Z Value	-2.575834515
Calculated Sample Size	14743.53704
Sample Size Needed	14744

8.1   Introduction

In Chapter 8 of the textbook, the authors define hypothesis testing as the utilization of sample information in order to test what the true value of the population mean, μ, or population proportion, p may be. From a single sample, a researcher may infer characteristics about the population within a specific level of error or level of significance. Remember that the procedures for conducting hypothesis tests on large or small samples are the same; however, the probability distributions necessary for determining the critical values that separate the rejection or not rejection regions are different.

PHStat can be used to conduct hypothesis tests for population means, μ, and population parameters, p. When calculating intervals estimating population means, remember to determine whether:
1. The population standard deviation, σ, is known
2. The population standard deviation, σ, is unknown

When the population standard deviation is unknown, the test statistic can be estimated using statistics generated from the sample data.

In order to estimate a population proportion, p, enter both the number of successes and the sample size into the PHStat menu. At this time, PHStat does not provide an option for entering a specific data set to use to estimate proportions.

The following examples will be used to illustrate techniques in PHStat and Excel.

Example 8.1 & 8.2, page 378 & 379, (Neurological response-test time)
Example 8.7, page 399 (Alkaline batteries and defective shipments)
Sample Exercise #1: Problem 8.35, page 383 (Social interaction of mental patients)
Sample Exercise #2: Problem 8.89, page 404 (Testing the placebo effect)

8.2   Tests of Hypothesis of a Population Mean: Sigma Unknown

In the majority of cases, the population standard deviation, σ, is not known and it will be necessary to substitute the sample standard deviation, s, as its estimate. Using PHStat, the steps necessary to conduct hypothesis tests are as follows:

Step 1: Open a new workbook and be sure the cursor is in the upper left cell of the worksheet.

Step 2: Click on the **PHStat** menu at the top of the screen.

Step 3: Select One-Sample Tests from the choices provided.

Step 4: Select t Test for the Mean, sigma unknown from the options listed. **(Figure 8.1)**

Step 5: Under **Data**, the user needs to enter the **Null Hypothesis** and **Level of Significance**.

Step 6: Under **Data,** the user may select **Sample Statistics Known** if the sample mean and standard deviation have been already provided. Otherwise select **Sample Statistics Unknown** if the raw data is provided only. Input the **Sample Cell Range** in the space provided.

Step 7: Under **Test Options**, indicate whether the test is two-tailed, upper-tailed, or lower-tailed. This indicates the direction of the rejection region

Step 8: An **Output Title** may be used to define the data.

Step 9: Click OK when finished.

The following figure illustrates the basic options provided for hypothesis testing of the mean when sigma is unknown.

Figure 8.1

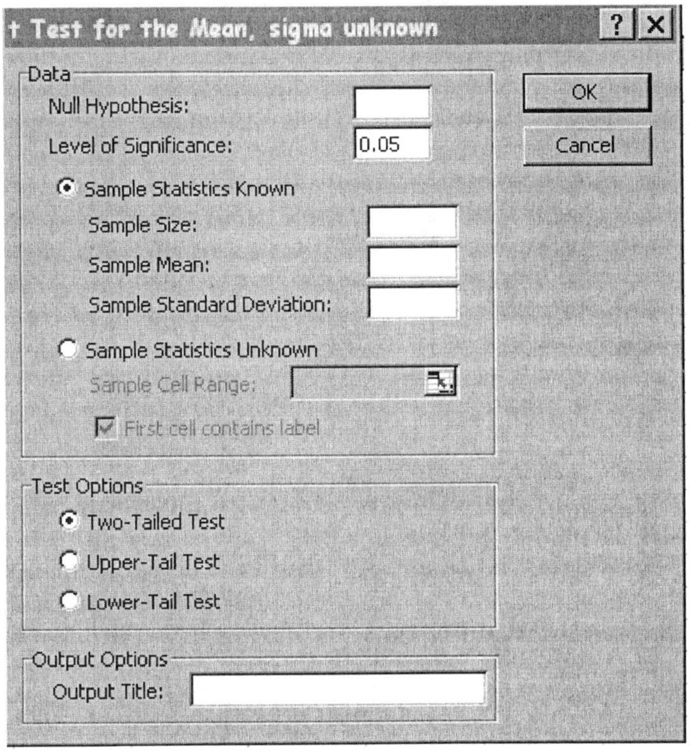

Example 8.1   Example 8.1 & 8.2 (page 379) will be used to illustrate hypothesis testing of the mean.

The effects of drugs and alcohol on the nervous system have been the subjects of considerable research. Suppose a research neurologist is testing the effect of a drug on response time by injecting 100 rats with a unit does of the drug, subjecting each to a neurological stimulus, and recording its response time. The neurologist knows that the mean response time for rats not injected with the drug (the "control" mean) is 1.2 seconds. She wishes to test whether the mean response time for drug-injected rats differs from 1.2 seconds. Set up the test of hypothesis for this experiment, using $\alpha=0.01$. Assume that the sample mean response time for 100 rats not injected with the drug is 1.05 seconds and the sample standard deviation is 0.5 seconds.

Solution:

In this example, notice that the sample mean and sample standard deviation have been provided directly; however, the sample size is large. Therefore, in order to replicate the answers within the text, it is acceptable to use the **One-Sample Tests-Z Test for the Mean, sigma known**. In order to show the disparity in answers, the preceding example will be completed using both the one sample z test and t test for the mean. Notice how the formula presented in Chapter 8 of the text for large sampled hypothesis testing. Here are the steps necessary for conducting this test.

Step 1: Select PHStat- One-Sample Tests-t Test for the Mean, sigma unknown (Figure 8.2)

Step 2: Enter the **Null Hypothesis as 1.2**.

Figure 8.2

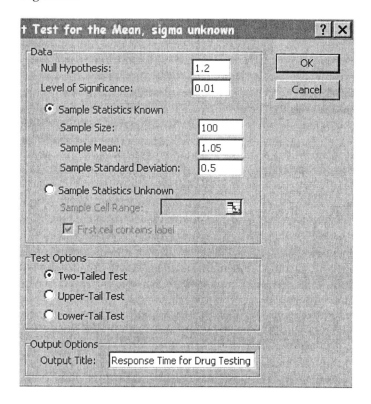

Step 3: Enter the **Level of Significance as 0.01**.

Step 4: Select Sample Statistics Known

Step 5: Enter the **Sample Size as 100**, **Sample Mean as 1.05**, and the **Sample Standard Deviation as 0.5**.

Step 6: Under **Test Options**, select **Two-Tailed Test**.

Step 7: In this example, the **Output Title** entered is **Response Time for Drug Testing**.

Step 8: Click OK to finish.

Table 8.1

Response Time for Drug Testing	
Null Hypothesis $\mu=$	1.2
Level of Significance	0.01
Sample Size	100
Sample Mean	1.05
Sample Standard Deviation	0.5
Standard Error of the Mean	0.05
Degrees of Freedom	99
t Test Statistic	-3
Two-Tailed Test	
Lower Critical Value	-2.62640242
Upper Critical Value	2.62640242
p-Value	0.003415508
Reject the null hypothesis	
Calculations Area	
For one-tailed tests:	
TDIST value	0.001707754
1-TDIST value	0.998292246

Here, the preceding example will be solved using the One-Sample Test-Z Test for the Mean, sigma known. Compare the differences in the answers to those in the textbook. Here are the steps necessary for conducting this test.

Example 8.1a   Example 8.1 & 8.2 (page 379) will be rerun using a different method.

Step 1: PHStat- One-Sample Tests-Z Test for the Mean, sigma known

Step 2: Enter the **Null Hypothesis as 1.2. (Figure 8.3)**

Step 3: Enter the **Level of Significance as 0.01**.

Step 4: Population Standard Deviation as 0.5.

Step 5: Select Sample Statistics Known

Step 6: Enter the **Sample Size as 100**, **Sample Mean as 1.05.**

Step 7: Under **Test Options**, select **Two-Tailed Test**.

Step 8: In this example, the **Output Title** entered is **Response Time for Drug Testing**.

Step 9: Click OK to finish. The null hypothesis is rejected so it appears that the mean response time differs from 1.2.

Figure 8.3

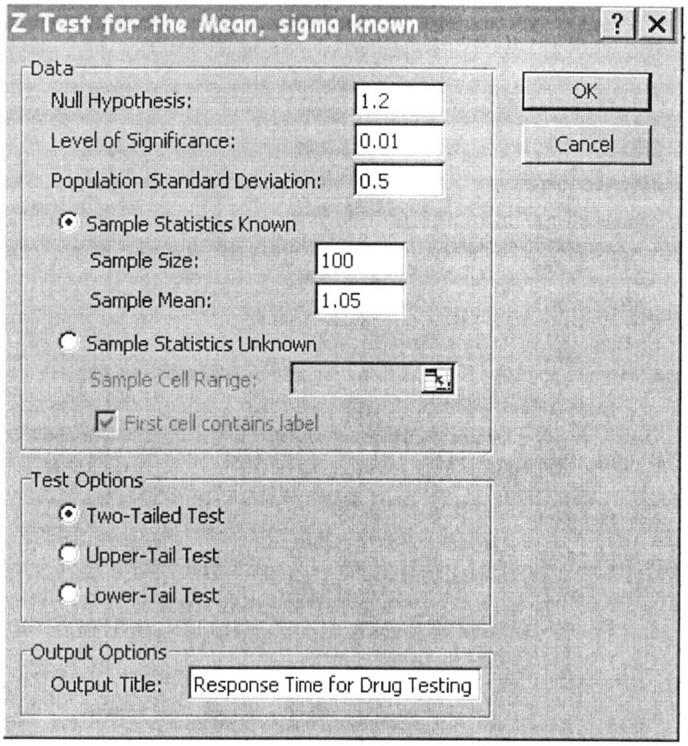

Table 8.1a

Response Time for Drug Testing		
Null Hypothesis $\mu=$		1.2
Level of Significance		0.01
Population Standard Deviation		0.5
Sample Size		100
Sample Mean		1.05
Standard Error of the Mean		0.05
Z Test Statistic		-3
Two-Tailed Test		
Lower Critical Value		-2.575834515
Upper Critical Value		2.575834515
p-Value		0.002699934
Reject the null hypothesis		

8.3   Tests of Hypothesis of a Population Proportion

In order to test a hypothesis involving a population proportion, both the sample size and the number of successes in the sample must be provided explicitly in order to use PHStat. Keying these values into the

118

appropriate places in PHStat will generate the appropriate test for a single proportion. The steps necessary to conduct the test are as follows.

Step 1: Open a new workbook and place the cursor in the upper left-hand cell of the worksheet.

Step 2: Click on the **PHStat** menu near the top of the screen.

Step 3: Select One-Sample Tests from the options available under PHStat.

Figure 8.4

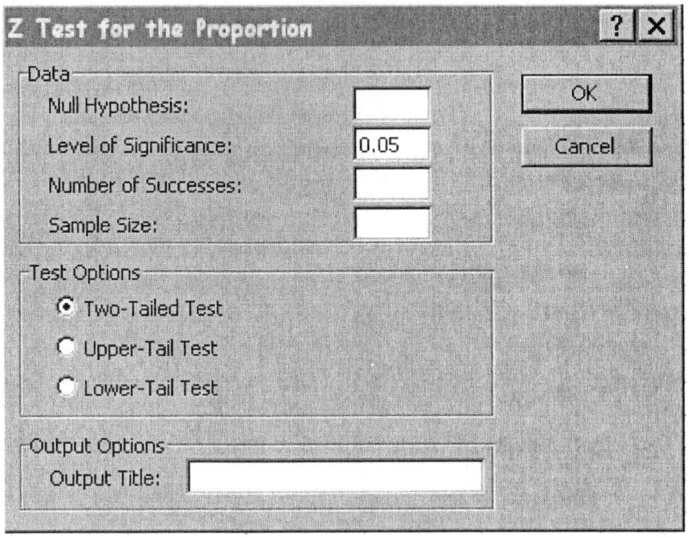

Step 4: Select Z Test for the Proportion option.

Step 5: Enter the Null Hypothesis, Level of Significance, Number of Successes, and Sample Size in the spaces provided.

Step 6: Select Two-Tailed Test, Upper-Tail Test, or Lower-Tail Test to indicate the direction of the rejection region.

Step 7: Enter an **Output Title** if desired to describe the data.

Step 8: Click OK to finish.

By selecting **PHStat-One-Sample Tests-Z Test for the Proportion**, the user should open a menu that appears to be similar to the following figure. Notice that by default, the level of significance is 0.05 and two-tailed test is selected. The next example will illustrate how to estimate a population proportion using PHStat.

Example 8.2   Example 8.7 (page 399) from the textbook will be used.

The reputations (and hence sales) of many businesses can be severely damaged by shipments of manufactured items that contain a large percentage of defectives. For example, a manufacturer of alkaline batteries may want to be reasonably certain that fewer than 5% of its batteries are defective. Suppose 300 batteries are randomly selected from a very large shipment; each is tested and 10 defective batteries are found. Does this provide sufficient evidence for the manufacturer to conclude that the fraction defective in the entire shipment is less than 0.05? Use $\alpha = 0.01$.

Solution:

According to the problem, notice that the sample size (n) is 300, and the number of successes (x) is 10. The null hypothesis is that the fraction (proportion) of defective batteries is **MORE THAN or EQUAL TO** 0.05. In order to reject the null, conduct a lower-tailed test at $\alpha = 0.01$. Please refer to the following steps in order to solve this problem.

Step 1: Select PHStat-One-Sample Test-Z Test for the Proportion.

Figure 8.5

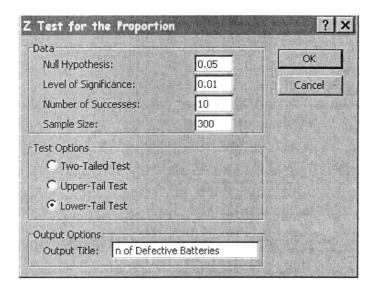

Step 2: Enter a **Null Hypothesis of 0.05**.

Step 3: Enter the level of significance as 0.01.

Step 4: Enter the Number of Successes as 10.

Step 5: Enter the Sample Size as 300.

Step 6: Select Lower-Tailed Test.

Step 7: Input an **Output Title** if desired. In this case, the title **"Proportion of Defective Batteries"** was used.

Step 8: Click OK to finish. The output generated by PHStat is shown in the following table. Notice that the Z test statistic of -2.326 does not fall in the rejection range for the null hypothesis. Therefore, the null hypothesis cannot be rejected. At a level of significance of 0.01, it is unlikely that the population proportion of defective batteries is less than 0.05.

Table 8.2

Proportion of Defective Bacteries	
Null Hypothesis *p*=	0.05
Level of Significance	0.01
Number of Successes	10
Sample Size	300
Sample Proportion	0.033333333
Standard Error	0.012583057
Z Test Statistic	-1.324532357
Lower-Tail Test	
Lower Critical Value	-2.326341928
p-Value	0.092663215
Do not reject the null hypothesis	

8.4   Sample Exercises

The following exercises were selected from the textbook to test the reader's understanding of PHStat and its usage within the chapter. Try to replicate the output in PHStat for the following problems.

Problem 8.35, page 383 (Social interaction of mental patients)

The *Community Mental Health Journal* (Aug. 2000) presented the results of a survey of over 6,000 clients of the Department of Mental Health and Addiction Services (DMHAS) in Connecticut. One of the many variables measured for each mental health patient was frequency of social interaction (on a 5-point scale, where 1 = very infrequently, 3 = occasionally, and 5 = very frequently). The 6,681 clients who were evaluated had a mean social interaction score of 2.95 with a standard deviation of 1.10. Conduct a hypothesis test ($\alpha = 0.01$) to determine if the true mean social interaction score of all Connecticut mental health patients differs from 3.

Solution for Problem 8.35:

Table 8.3

Solution for Exercise 8.35	
Null Hypothesis μ=	3
Level of Significance	0.01
Population Standard Deviation	1.1
Sample Size	6
Sample Mean	2.95
Standard Error of the Mean	0.44907312
Z Test Statistic	-0.111340443
Two-Tailed Test	
Lower Critical Value	-2.575834515
Upper Critical Value	2.575834515
p-Value	0.911346284
Do not reject the null hypothesis	

Problem 8.89, page 404 (Testing the placebo effect)

The *placebo effect* describes the phenomenon of improvement in the condition of a patient taking a placebo - a pill that looks and tastes real but contains no medically active chemicals. Physicians at a clinic in La Jolla, California, gave what they thought were drugs to 7,000 asthma, ulcer, and herpes patients. Although the doctors later learned that the drugs were really placebos, 70% of the patients reported an improved condition (*Forbes*, May 22, 1995). Use this information (at $\alpha = 0.05$) to test the placebo effect at the clinic. Assume that if the placebo is ineffective, the probability of a patient's condition improving is 0.5.

Solution:

Solution for Problem 8.89

Solution for Exercise 8.89	
Null Hypothesis p=	0.5
Level of Significance	0.05
Number of Successes	4900
Sample Size	7000
Sample Proportion	0.7
Standard Error	0.005976143
Z Test Statistic	33.46640106
Upper-Tail Test	
Upper Critical Value	1.644853
p-Value	0
Reject the null hypothesis	

9.1   Introduction

In Chapters 7 and 8 of the text, inferences about the population from a single sample were conducted by developing confidence intervals and conducting hypothesis tests were conducted Using these techniques, inferences about the population mean, μ, and the population parameter, p, were derived. In this chapter, inferences about a population will again be derived using multiple samples. Three types of parameters will be estimated: the population mean, μ, the population parameter, p, and the population variance, σ^2.

PHStat may be used only for a subset of the techniques defined within the textbook. In particular, PHStat will perform the following.
1. Tests of hypothesis for comparing population means.
2. Tests of hypothesis for comparing population proportions.
3. Tests of hypothesis for comparing population variances.

At this time, PHStat does not contain the necessary functions to conduct the following operations. Please consult other statistical software packages for assistance with the following:
1. Paired difference analysis.
2. Calculation of confidence intervals for populations means, population proportions, and population variances for multiple samples.
3. Sample size determination techniques.

Before conducting inferences on two samples, PHStat requires that descriptive statistics for the sample data already be provided. This way, the required information may be inputted directly into the main menus necessary for conducting analysis.

The following examples will be used to demonstrate techniques in PHStat and Excel.

> Example 9.2, Low-fat versus Regular Diet, page 434
> Example 9.6, Antismoking Campaigns, page 471
> Example 9.10, Metabolic Rates of White Mice, page 485
> Sample Exercise #1: Problem 9.65, "Tip of the Tongue" study, page 478
> Sample Exercise #2: Problem 9.98, Gender differences in math test scores, page 496

9.2   Tests for Differences in Two Means: Independent Sampling

In this section, PHStat will be used to compare two population means using two random and independent samples. To use the hypothesis test tool within PHStat, refer to the following steps.

Step 1: Open a new workbook and be sure the cursor is in the upper left cell of the worksheet.

Step 2: Click on the **PHStat** menu at the top of the screen

Step 3: Select the **Two-Sample Tests** option.

Step 4: Select the **t Test for Differences in Two Means**.

Step 5: Enter the **Hypothesized Difference, Level of Significance, Summary Statistics** for both samples, and the **Test Option**, which indicates the direction of the test.

Step 6: Input an **Output Title**, which describes the data, if desired.

Step 7: Click OK.

The following figure **(Figure 9.1)** displays the menu box for the **t Test for Differences in Two Means.** Remember that the t test is used for conducting tests for small samples, generally less than 30, whereas the z test is used for conducting tests for large samples. However, as the sample size increases beyond 30, the t distribution converges to the z distribution. Generally, the level of significance is 0.05 and the test options are set for a two-tailed test by default. The next example will illustrate how to test for differences in two means.

Figure 9.1

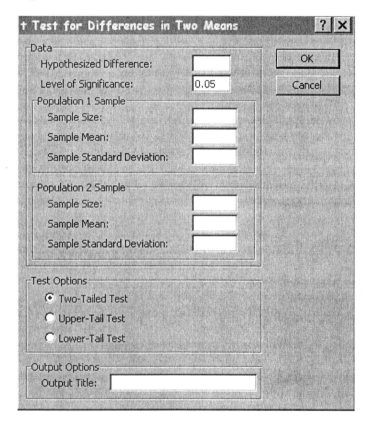

Example 9.2   Please refer to Example 9.2 (page 434) of the textbook.

A dietitian has developed a diet that is low in fats, carbohydrates, and cholesterol. Although the diet was initially intended to be used by people with heart disease, the dietitian wishes to examine the effect this diet has on the weights of obese people. Two random samples of 100 obese people each are selected and one group of 100 is placed on the low-fat diet. The other 100 are placed on a diet that contains approximately the same quantity of food but is not as low in fats, carbohydrates, and cholesterol. For each person, the amount of weight lost (or gained) in a 3-week period is recorded. Using a 0.05 level of significance, is there a difference in the mean weight loss between these two groups?

Solution:

Table 9.1 contains the summary statistics necessary to complete the problem. Notice again that although the samples are large, the t test may be used to answer this problem.

Table 9.1

	Low-Fat Diet	Regular Diet
Sample Size	100	100
Sample Mean	9.31	7.40
Sample Standard Deviation	4.67	4.04

The steps for solving the problem are as follows.

Step 1: Click on the **PHStat** menu at the top of the screen.

Step 2: Select the **Two-Sample Test** option.

Figure 9.2

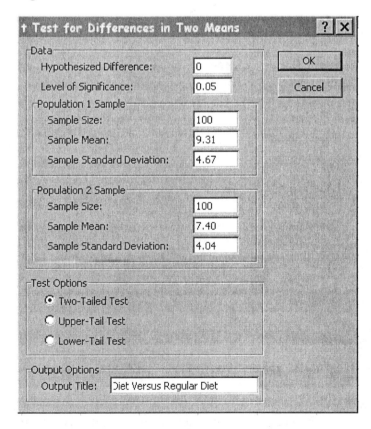

Step 3: Select t Test for Differences in Two Means.

Step 4: Specify a Hypothesized Value of 0 (two means are compared to see if they differ).

Step 5: Specify the Level of Significance as 0.05.

Step 6: Enter the Sample Sizes, Means, and Standard Deviations as provided in the table.

Step 7: Select the **Two-Tailed Test** option from the Tests Options.

Step 8: Specify an **Output Title**, if desired. In this example, the output title is **Low-Fat Diet versus Regular Diet**.

Step 9: Click OK to finish.

The output generated by this example is provided in Table 9.2. Notice that the test statistic generated is 3.10 and the p-value is approximately 0.002. Therefore, the null hypothesis is rejected so it appears that there is a difference in the mean weight losses for Low-Fat Dieters and Regular Dieters at the 0.05 level of significance.

Table 9.2

Low-Fat Diet Versus Regular Diet	
Hypothesized Difference	**0**
Level of Significance	**0.05**
Population 1 Sample	
Sample Mean	**9.31**
Sample Size	**100**
Sample Standard Deviation	**4.67**
Population 2 Sample	
Sample Mean	**7.4**
Sample Size	**100**
Sample Standard Deviation	**4.04**
Population 1 Sample Degrees of Freedom	99
Population 2 Sample Degrees of Freedom	99
Total Degrees of Freedom	198
Pooled Variance	19.06525
Difference in Sample Means	1.91
t-Test Statistic	3.093122479
Two-Tailed Test	
Lower Critical Value	-1.972016435
Upper Critical Value	1.972016435
p-Value	0.002266695
Reject the null hypothesis	

9.3   Tests for Differences in Two Proportions: Independent Sampling

In this section, PHStat will be used to compare two population proportions using two random and independent samples. Remember that the two samples represent binomial experiments. To use the hypothesis test tool within PHStat, refer to the following steps.

126

Step 1: Open a new workbook and be sure the cursor is in the upper left cell of the worksheet.

Step 2: Click on the **PHStat** menu at the top of the screen

Step 3: Select the **Two-Sample Tests** option.

Step 4: Select the **z Test for Differences in Two Proportions**.

Figure 9.3

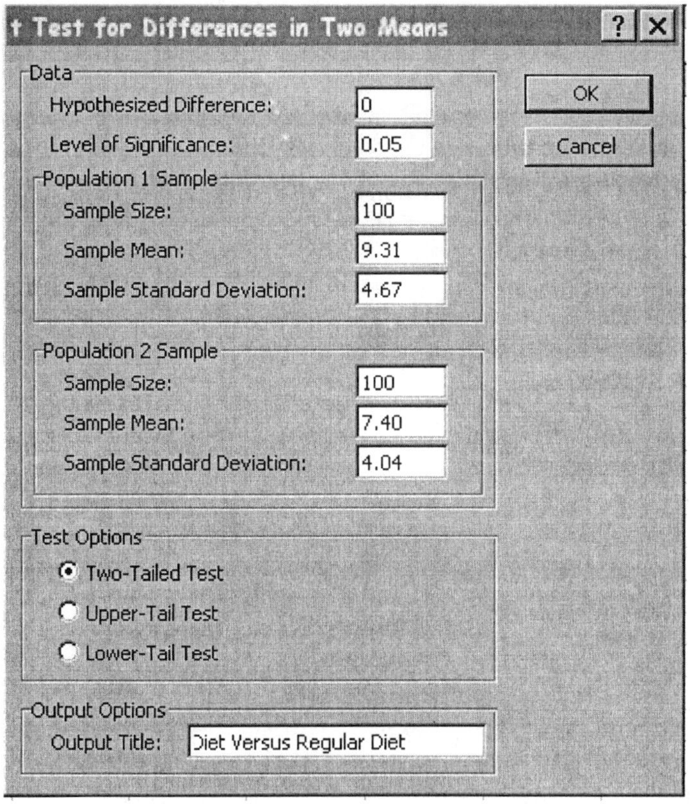

Step 5: Enter the **Hypothesized Difference**, **Level of Significance**, **Sample Size**, and **Number of Successes** from both samples collected, and the **Test Option**, which indicates the direction of the test.

Step 6: Input an **Output Title**, which describes the data, if desired.

Step 7: Click OK to finish.

Figure 9.3 presents the **Z Test for Differences in Two Proportions** menu. Notice that the level of significance is 0.05 and two-tailed test within Test Options is automatically selected. The next example will provide and example of how to test for differences in two proportions using independent sampling.

Example 9.2   Example 9.6 (page 471) will be used from the textbook.

In the past decade, intensive antismoking campaigns have been sponsored by both federal and private agencies. Suppose the American Cancer Society randomly sampled 1,500 adults in 1995 and then sampled 1,750 adults in 2005 to determine whether there was evidence that the percentage of smokers had decreased. The results of the two sample surveys are shown in the table below, where x_1 and x_2 represent the numbers of smokers in the 1995 and 2005 samples, respectively. Do these data indicate that the fraction of smokers decreased over this ten-year period? Use $\alpha=0.05$.

Table 9.3

Year	1995	2005
Sample Size	1,500	1,750
x_i	555	578
Sample Proportion	$555/1,500 = 0.37$	$578/1,750 = 0.33$

Solution:

In this example, the test is one-tailed. In order to determine whether the fraction of smokers decreased over the ten-year period, notice that $p_1 = 0.37$ is the sample proportion of those that decreased smoking in 1995. Also notice that $p_2 = 0.33$ is the sample proportion of those that decreased smoking in 2005. So the alternative hypothesis is thus H_a: $p_1 > p_2$.

The steps for solving the problem are as follows.

Step 1: Click on the **PHStat** menu at the top of the screen.

Step 2: Select the **Two-Sample Test** option.

Step 3: Select Z Test for Differences in Two Proportions.

Step 4: Specify a Hypothesized Value of 0.

Step 5: Specify the Level of Significance as 0.05.

Step 6: Enter the Number of Successes and sample size for each sample from the table above. Remember the first sample represents the data from 1995.

Step 7: Select the **Upper Tailed Test** option from the Tests Options.

Figure 9.4

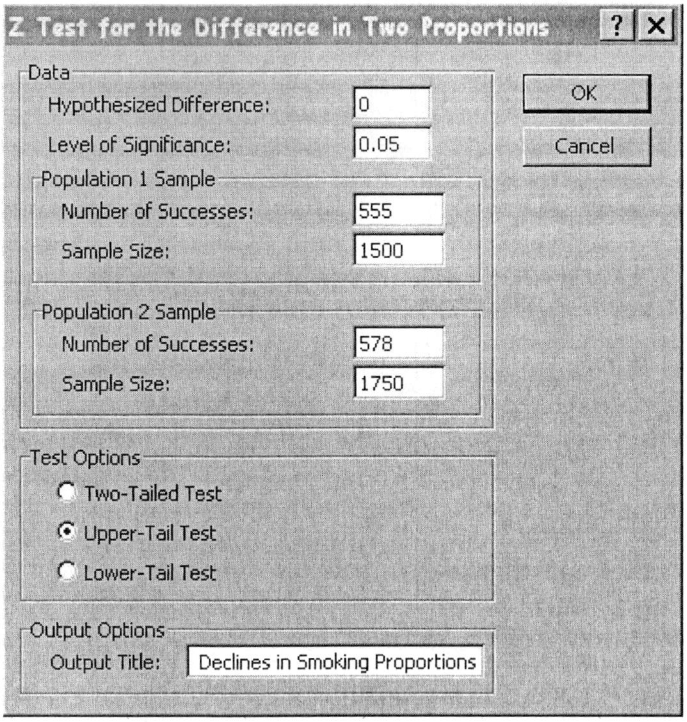

Z Test for the Difference in Two Proportions [?] [X]

Data
- Hypothesized Difference: `0`
- Level of Significance: `0.05`

Population 1 Sample
- Number of Successes: `555`
- Sample Size: `1500`

Population 2 Sample
- Number of Successes: `578`
- Sample Size: `1750`

[OK] [Cancel]

Test Options
- ◯ Two-Tailed Test
- ◉ Upper-Tail Test
- ◯ Lower-Tail Test

Output Options
- Output Title: `Declines in Smoking Proportions`

Step 8: Specify an **Output Title**, if desired. In this example, the output title is **1992 versus 2002 Declines in Smoking Proportions**.

Step 9: Before clicking OK, review Figure 9.4 and be sure that the data are inputted correctly. **Click OK** to finish.

The output generated by PHStat is generated in Table 9.4. Notice that the z statistic generated from the data is 2.37 in comparison to the critical value of 1.645. This indicates that the null hypothesis is rejected. Thus, it does appear that the proportion of adults who smoke did decrease over the 1995-2005 period, at the 0.05 level of significance.

Table 9.4

1995 versus 2005 Declines in Smoking Proportions	
Hypothesized Difference	0
Level of Significance	0.05
Group 1	
Number of Successes	555
Sample Size	1500
Group 2	
Number of Successes	578
Sample Size	1750
Group 1 Proportion	0.37
Group 2 Proportion	0.330285714
Difference in Two Proportions	0.039714286
Average Proportion	0.348615385
Z Test Statistic	2.368523545
Upper-Tail Test	
Upper Critical Value	1.644853
p-Value	0.008929608
Reject the null hypothesis	

9.4 Tests for Differences in Two Variances: Independent Sampling

In this section, PHStat will be used to compare population variances using two random and independent samples. Remember that in order to conduct this test, the two sampled populations must be normally distributed and the samples must be randomly and independently selected from their respective populations. To use the hypothesis test tool within PHStat, refer to the following steps.

Step 1: Open a new workbook and be sure the cursor is in the upper left cell of the worksheet.

Step 2: Click on the **PHStat** menu at the top of the screen

Step 3: Select the **Two-Sample Tests** option.

Step 4: Select the **F Test for Differences in Two Variances**.

Step 5: Enter the **Level of Significance**, and the **Sample Size and Sample Standard Deviation** for both samples.

Step 6: Enter the **Test Option**, which indicates the direction of the test.

Step 7: Input an **Output Title**, which describes the data, if desired.

Step 8: Click OK to finish.

The menu for the F Test for Differences in Two Variances should resemble Figure 9.5. Note that the level of significance is 0.05 and the two-tailed test option is selected by default. The following example will be used to illustrate the F test.

130

Figure 9.5 –

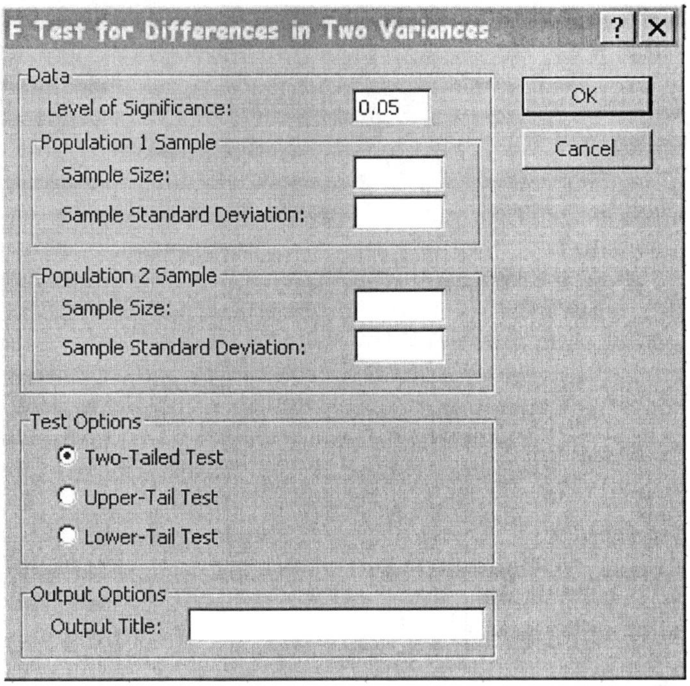

Example 9.3   Example 9.10 (page 485) will be used from the textbook.

An experimenter wants to compare the metabolic rates of white mice subjected to different drugs. The weights of the mice may affect their metabolic rates, and thus the experimenter wishes to obtain mice that are relatively homogeneous with respect to weight. Five hundred mice will be needed to complete the study. Currently, 13 mice from supplier 1 and another 18 mice from supplier 2 are available for comparison. The experimenter weighs these mice and obtains the following summary information.

Table 9.5

	Supplier 1	Supplier 2
Sample Size	13	18
Sample Standard Deviation	0.2021	0.0982

Do these data provide sufficient evidence to indicate a difference in the variability of weights of mice obtained from the two suppliers? (Use $\alpha=0.10$)

Solution:

If there is no difference in the variabilities, then $\sigma_1^2 = \sigma_2^2$ or the ratio of $\sigma_1^2/\sigma_2^2=1$. This is the null hypothesis. Also note that when using the F Test for Differences in Two Variances, the first sample refers to the one with the largest standard deviation.

The steps for solving the problem are as follows.

Step 1: Click on the **PHStat** menu at the top of the screen.

Step 2: Select the **Two-Sample Test** option.

Step 3: Select F **Test for Differences in Two Variances.**

Figure 9.6

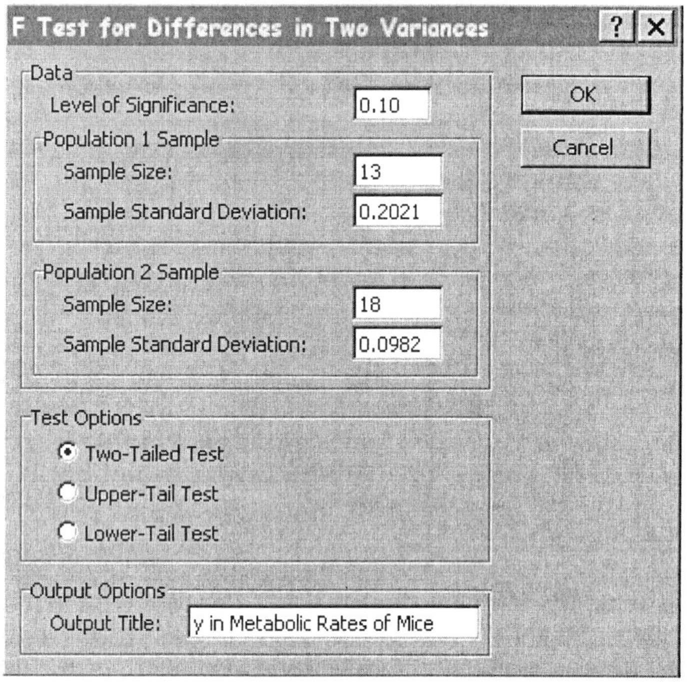

Step 4: Enter a **Level of Significance as 0.10.**

Step 5: Specify for Sample 1 a Sample Size of 18 and a sample standard deviation of 0.0982.

Step 6: Specify for Sample 2 a Sample Size of 13 and a sample standard deviation of 0.2021

Step 7: Select the **Two-Tailed Test** option from the Tests Options.

Step 8: Specify an **Output Title**, if desired. In this example, the output title is **Variability in Metabolic Rates of Mice**.

Step 9: Figure 9.6 presents the data entered into the F Test for Differences in Two Variances. Please check and be sure all data are entered correctly. **Click OK** to finish.

From Table 9.6, the F-Test statistic generated from the data is 4.24, which falls into the rejection region. Therefore, with a 0.10 level of significance, it appears that the population variances differ and the null hypothesis is rejected. Thus, the weights of mice obtained from supplier 1 tend to be more homogeneous than the weights of mice obtained from supplier 2.

Table 9.6

Variability in Metabolic Rates of Mice				
Level of Significance	0.1			
Population 1 Sample				
Sample Size	13			
Sample Standard Deviation	0.2021			
Population 2 Sample				
Sample Size	18			
Sample Standard Deviation	0.0982			
F-Test Statistic	4.235548426			
Population 1 Sample Degrees of Freedom	12			
Population 2 Sample Degrees of Freedom	17			
			Calculations Area	
Two-Tailed Test			FDIST value	0.003567946
Lower Critical Value	0.387171184		1-FDIST value	0.996432054
Upper Critical Value	2.38065212			
p-Value	0.007135892			
Reject the null hypothesis				

9.5   Sample Exercises

The following exercises were selected from the textbook to test the reader's understanding of PHStat and its usage within the chapter. Try to replicate the output in PHStat for the following problems.

Problem 9.65, "Tip of the tongue" Study (page 478)

Trying to think of a word you know, but can't instantly retrieve is called the "tip of the tongue" phenomenon. *Psychology and Aging* (Sep. 2001) published a study of this phenomenon in senior citizens. The researchers compared 40 people between 60 and 72 years of age with 40 between 73 and 83 years of age. When primed with initial syllable of a missing word (e.g., seeing the word "include" to help recall the word "incisor"), the younger seniors had a higher recall rate. Suppose 31 of the 40 seniors in the younger group could recall the word when primed with the initial syllable, while only 22 of the 40 seniors could recall the word. The recall rates of the two groups were compared. Using the 0.05 level of significance, does one group of elderly have a significantly higher recall rate than the other? (Is there a difference in the recall rate between groups?)

Table 9.9

Solution to Exercise 9.65	
Hypothesized Difference	0
Level of Significance	0.05
Group 1	
Number of Successes	31
Sample Size	40
Group 2	
Number of Successes	22
Sample Size	40
Group 1 Proportion	0.775
Group 2 Proportion	0.55
Difference in Two Proportions	0.225
Average Proportion	0.6625
Z Test Statistic	2.127980706
Two-Tailed Test	
Lower Critical Value	-1.959961082
Upper Critical Value	1.959961082
p-Value	0.033338564
Reject the null hypothesis	

Problem 9.98, Gender differences in math test scores (page 496)

Refer to the *American Educational Journal* (Fall, 1998) study to compare the mathematics achievement test scores of male and female students. The researchers hypothesized that the distribution of test scores for males is more variable than the corresponding distribution for females. Use the summary information reproduced in the table to test this claim at $\alpha = 0.01$.

	Males	Females
Sample Size	1,764	1,739
Mean	48.9	48.4
Standard Deviation	12.96	11.85

134

Solution for Problem 9.98

Solution to Exercise 9.98			
Level of Significance	**0.01**		
Population 1 Sample			
Sample Size	**1764**		
Sample Standard Deviation	**12.96**		
Population 2 Sample			
Sample Size	**1739**		
Sample Standard Deviation	**11.85**		
F-Test Statistic	1.196116007		
Population 1 Sample Degrees of Freedom	1763		
Population 2 Sample Degrees of Freedom	1738		
		Calculations Area	
Upper-Tail Test		FDIST value	9.1202E-05
Upper Critical Value	**1.117699711**	1-FDIST value	0.999908798
p-**Value**	**9.1202E-05**		
Reject the null hypothesis			

10.1   Introduction

In Chapter 10 of the textbook, the authors discuss how to design and analyze experiments in order to compare two (or more) population means. The goal of ANOVA (analysis of variance) is to identify factors that contribute information to the response variable of interest. Remember that factor levels are the values of the factor utilized in the experiment. Treatments of an experiment are the factor-level combinations utilized. The text presents several methods of comparing the multiple means of the experiment.

Although analysis of variance software does not currently exist in PHStat, Excel will be used in order to solve problems within this chapter. In particular, Excel may be used to assist in the analysis of completely randomized and factorial designs. Notice that when regression analysis discussed, basic ANOVA calculations will be performed within the regression analysis provided both by Excel and PHStat.

Remember from Section 10.2 of the textbook that a completely randomized design is one for which independent random samples of experimental units are selected for each treatment. In ANOVA, the null hypothesis tests whether the treatment means are all equal. The alternative hypothesis tests the hypothesis that at least two of the treatment means differ.

The following examples will be used to illustrate techniques in PHStat and Excel.

> Example 10.2, USGA and Golf, page 517
> Example 10.10, USGA Factorial ANOVA, page 565
> Sample Exercise #1: Problem 10.32, Hair Color and Pain, page 535
> Sample Exercise #2: Problem 10.87. Impact of Vitamin B Supplement, page 576

10.2   Completely Randomized Design

When constructing a randomized design, compare the mean responses of the various treatments in an experimental design. Since there is only one factor in the design, the various levels of the factor are the treatments of the design. Usually, the null hypothesis is a test to determine if the treatment means are all equal whereas the alternative hypothesis is a test to determine whether the treatment means (at least one) differ.

When solving for completely randomized design problems, the basic function in Excel 97/2000 can be found by using **Tools-Data Analysis-ANOVA: Single Factor**. The steps in Excel are as follows. An example from the textbook will follow.

Step 1: Open a new workbook and be sure the cursor is placed in the upper left-hand cell of the current worksheet.

Step 2: Open the appropriate file where the data is located or input the data manually.

Step 3: Select Tools from the menu and **Data Analysis** from the list of options available. If Data Analysis is not currently listed under Tools, please add it in (select **Add-Ins**) manually.

Step 4: From Data Analysis, select ANOVA: Single Factor.

136

Step 5: Enter the Input Range (indicating if the data are grouped by rows or columns) **and Alpha** (level of significance) required by the problem. Notice that the default setting is 0.05.

Step 6: If the first cell of data contains a descriptive nonnumeric title, please check the **Label in the First Row** box.

Step 7: Select where the data should be imported: either in a **new workbook**, **new worksheet**, or the **output range** within the current worksheet. If selecting an output range within the current worksheet, please be sure that the range does not contain numeric data.

Step 8: Click OK to finish.

Figure 10.1

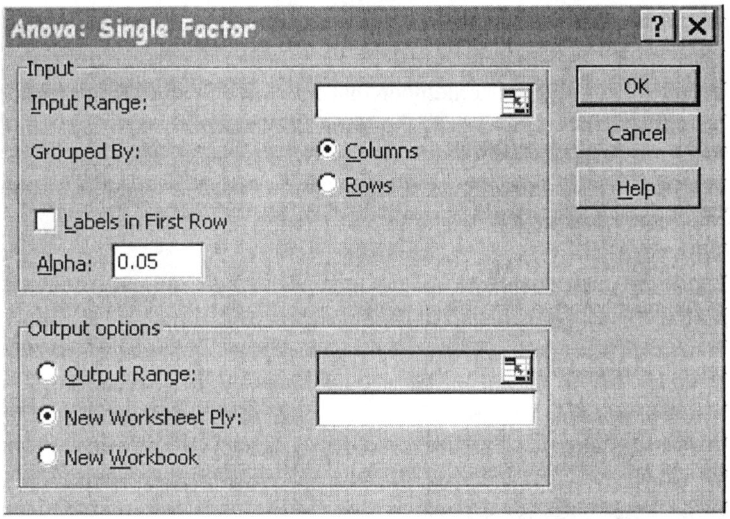

Example 10.1   Please refer to Example 10.2 (page 517) of the textbook.

Suppose the USGA wants to compare the mean distances associated with four different brands of golf balls when struck with a driver. A completely randomized design is employed, with Iron Byron, the USGA's robotic golfer, using a driver to hit a random sample of 10 balls of each brand in a random sequence. The distance is recorded for each hit, and the results are shown in the table below, organized by brand. Set up a test to compare the mean distances for the four brands. Use $\alpha = 0.10$

Table 10.1

Golfer	Brand A	Brand B	Brand C	Brand D
1	251.2	263.2	269.7	251.6
2	245.1	262.9	263.2	248.6
3	248.0	265.0	277.5	249.4
4	251.1	254.5	267.4	242.0
5	260.5	264.3	270.5	246.5
6	250.0	257.0	265.5	251.3
7	253.9	262.8	270.7	261.8
8	244.6	264.4	272.9	249.0
9	254.6	260.6	275.6	247.1
10	248.8	255.9	266.5	245.9

Solution:

Table 10.1 presents the data for each brand of golf balls. Remember that the test statistic compares the variation among the four treatments (brand of golf balls) means to the sampling variability within each of the treatments.

In order to solve this problem, please follow the steps below.

Step 1: Open a new workbook and be sure the cursor is placed in the upper left-hand cell of the current worksheet.

Step 2: Open the appropriate file where the data is located or input the data manually.

Figure 10.2

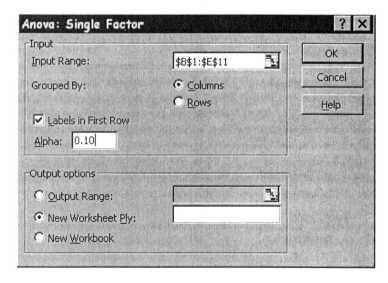

Step 3: Select Tools from the menu and **Data Analysis** from the list of options available. If Data Analysis is not currently listed under Tools, please add it in (select **Add-Ins**) manually.

Step 4: From Data Analysis, select ANOVA: Single Factor.

Step 5: Enter the Input Range as $B$1: $E$11.

Step 6: Indicate that the Data is **Grouped by Columns**.

Step 7: Select Labels in First Row since the first row contains the brand labels of golf balls.

Step 8: Enter Alpha as 0.10.

Step 9: Select where the output should be placed. In this example, output is placed in a **New Worksheet Ply**.

Step 10: Click OK to finish.

In order to derive an answer to this question, please refer to the output generated by Excel (**Table 10.2**). First, notice that the ANOVA output has tow main components. The first component displays the statistical summary of the various treatments in the analysis. In this case, Excel provides the mean distances traveled for each brand of golf ball.

In the second section of output, ANOVA statistics are generated. Notice that the critical value for F at an alpha of 0.10 is 2.25 for 3 and 36 degrees of freedom. Reject the null hypothesis (there is no difference in the mean distances of the four brands of golf balls) if F > 2.25. Since the test statistic generated is 43.99, the null hypothesis is rejected. Therefore, there is a difference in the mean distances of the four brands of golf balls.

Table 10.2

Anova: Single Factor						
SUMMARY						
Groups	*Count*	*Sum*	*Average*	*Variance*		
Brand A	10	2507.8	250.78	22.42177778		
Brand B	10	2610.6	261.06	14.94711111		
Brand C	10	2699.5	269.95	20.25833333		
Brand D	10	2493.2	249.32	27.07288889		
ANOVA						
Source of Variation	*SS*	*df*	*MS*	*F*	*P value*	*F crit*
Between Groups	2794.38875	3	931.4629	43.98874592	3.97E-12	2.242608
Within Groups	762.301	36	21.17503			
Total	3556.68975	39				

10.3   The Factorial Design

In Factorial Design, an additional factor is added to the original design experiment. Using Excel, the output generated for Factorial Design problems may be replicated using **Tools-Data Analysis-ANOVA: Two-Factor With Replication**. Remember, if Data Analysis does not appear under Tools, please select **Add-Ins** and select the statistical tool packages available. The following exercise will be used to illustrate factorial design problems.

139

The steps in Excel are as follows. An example from the textbook will follow.

Step 1: Open a new workbook and be sure the cursor is placed in the upper left-hand cell of the current worksheet.

Step 2: Open the appropriate file where the data is located or input the data manually.

Step 3: Select Tools from the menu and **Data Analysis** from the list of options available. If Data Analysis is not currently listed under Tools, please add it in (select **Add-Ins**) manually.

Step 4: From Data Analysis, select ANOVA: Two-Factor With Replication.

Step 5: Enter the Input Range.

Step 6: Enter the Rows per Sample and Alpha (level of significance).

Step 7: Select where the data should be imported: either in a **new workbook, new worksheet**, or the **output range** within the current worksheet. If selecting an output range within the current worksheet, please be sure that the range does not contain numeric data.

Step 8: Click OK to finish.

Figure 10.3

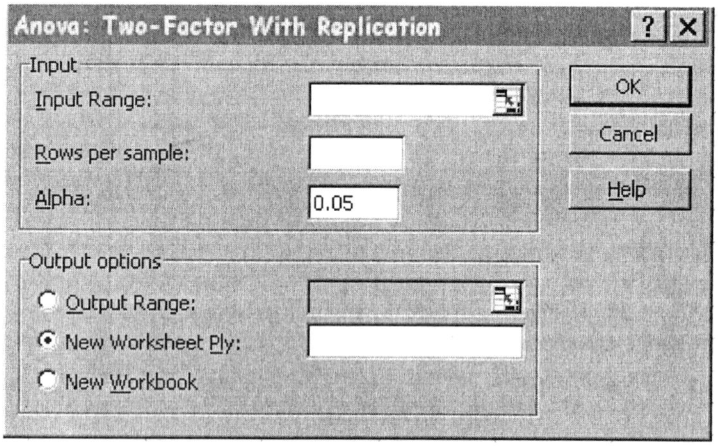

Example 10.2 Please refer to Example 10.10 (page 565) of the textbook

Suppose the USGA tests four different brands (A, B, C, D) of golf balls and two different clubs (driver, five-iron) in a completely randomized design. Each of the eight Brand-Club combinations (treatments) is randomly and independently assigned to four experimental units, each experimental unit consisting of a specific position in the sequence of hits by Iron Byron. The distance response is recorded for each of the 32 hits, and the results are shown in the table below.
  a. Use Excel to partition the Total Sum of Squares into the components necessary to analyze this 4x2 factorial experiment.
  b. Follow the steps for analyzing a two-factor factorial experiment and interpret the results of your analysis. Use $\alpha = 0.10$ for the tests conducted.

140

Table 10.3

		Brand			
		A	B	C	D
	Driver	226.4	238.3	240.5	219.8
	Driver	232.6	231.7	246.9	228.7
	Driver	234.0	227.7	240.3	232.9
Club	Driver	220.7	237.2	244.7	237.6
	Five-Iron	163.8	184.4	179.0	157.8
	Five-Iron	179.4	180.6	168.0	161.8
	Five-Iron	168.6	179.5	165.2	162.1
	Five-Iron	173.4	186.2	156.5	160.3

Solution:

Please follow the steps below to generate the output in Excel.

Step 1: Open a new workbook and be sure the cursor is placed in the upper left-hand cell of the current worksheet.

Step 2: Open the appropriate file where the data is located or input the data manually.

Step 3: Select Tools from the menu and **Data Analysis** from the list of options available. If Data Analysis is not currently listed under Tools, please add it in (select **Add-Ins**) manually.

Step 4: From Data Analysis, select ANOVA: Two- Factor with Replication.

Step 5: Enter the Input Range which in this case is **$B$2:$F$10**. Please be sure and highlight columns referring to the four brands of golf balls AND the description of whether it was hit by a driver or a five-iron.

Step 6: Rows per sample is 4.

Step 7: Enter an Alpha of 0.10.

Step 8: Select where the data should be imported: either in a **new workbook**, **new worksheet**, or the **output range** within the current worksheet. If selecting an output range within the current worksheet, please be sure that the range does not contain numeric data. In this case, **New Worksheet Ply** is selected.

Step 9: Click OK to finish.

Figure 10.4

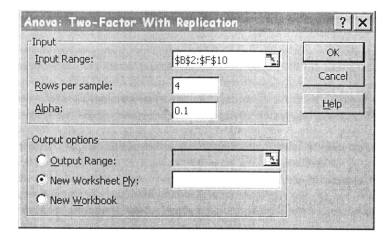

The ANOVA output generated for factorial design is separated into two sections. First, a statistical summary is provided for the various treatments. For each of the eight Brand-Club treatments, Excel provides the average distance and variance of each drive. Please refer to the textbook to interpret this data.

The second section contains the analysis of variance output necessary to answer the question. Once partitioning is complete and it is determined that at least two of the Brand-Club combinations differ in mean distance (see text for details), determine how the factors affect the mean response rate. In this example,

1. H_0: The factors Brand and Club do not interact to affect the mean response rate
2. H_1: The factors Brand and Club do interact to affect the mean response rate

Notice that the test statistic, F = 7.4524 exceed the F critical value of 2.32739 and the p-value, 0.001079, does not exceed $\alpha = 0.10$. Thus, the null is rejected so that the Brand and Club do interact to affect mean distance.

Table 10.4

Anova: Two-Factor With Replication					
SUMMARY	A	B	C	D	Total
Driver					
Count	4	4	4	4	16
Sum	913.7	934.9	972.4	919	3740
Average	228.425	233.725	243.1	229.75	233.75
Variance	37.429167	24.469167	10.533333	57.216667	61.0706667
Five-Iron					
Count	4	4	4	4	16
Sum	685.2	730.7	668.7	642	2726.6
Average	171.3	182.675	167.175	160.5	170.4125
Variance	44.52	9.9291667	86.1225	3.86	98.1918333
Total					
Count	8	8	8	8	
Sum	1598.9	1665.6	1641.1	1561	
Average	199.8625	208.2	205.1375	195.125	
Variance	967.48268	759.34286	1688.4541	1396.3364	

ANOVA						
Source of Variation	*SS*	*df*	*MS*	*F*	*P-value*	*F crit*
Sample	32093.111	1	32093.111	936.75164	9.629E-21	2.92711633
Columns	800.73625	3	266.91208	7.7907788	0.00084013	2.32738984
Interaction	765.96125	3	255.32042	7.4524348	0.00107892	2.32738984
Within	822.24	24	34.26			
Total	34482.049	31				

10.4   Sample Exercises

The following exercises were selected from the textbook to test the reader's understanding of Excel and its usage within the chapter. Try to replicate the output for the following problems.

Problem 10.32, Hair Color and Pain (page 535)

Studies conducted at the University of Melbourne (Australia) indicate that there may be a difference between the pain thresholds of blondes and brunettes. Men and women of various ages were divided into four categories according to hair color: light blond, dark blond, light brunette, and dark brunette. The purpose of the experiment was to determine whether hair color is related to the amount of pain evoked by common types of mishaps and assorted types of trauma. Each person in the experiment was given a pain threshold score based on his or her performance in a pain sensitivity test (the higher the score, the higher the person's pain tolerance). Conduct a test to determine whether the mean pain thresholds differ among people possessing the four types of hair color. Use $\alpha = 0.05$.

143

Light Blonde	Dark Blonde	Light Brunette	Dark Brunette
62	63	42	32
60	57	50	39
71	52	41	51
55	41	37	30
48	43		35

Output Table for Problem 10.32

Anova: Single Factor

SUMMARY

Groups	Count	Sum	Average	Variance
Light Blonde	5	296	59.2	72.7
Dark Blonde	5	256	51.2	86.2
Light Brunette	4	170	42.5	29.66667
Dark Brunette	5	187	37.4	69.3

ANOVA

Source of Variation	SS	df	MS	F	P-value	F crit
Between Groups	1360.726	3	453.5754	6.791407	0.004114	3.287383
Within Groups	1001.8	15	66.78667			
Total	2362.526	18				

Problem 10.87, Impact of Vitamin B Supplement (page 576)

In the *Journal of Nutrition* (July 1995), University of Georgia researchers examined the impact of a vitamin-B supplement (nicotinamide) on the kidney. The experimental subjects were 28 Zucker rats-a species that tends to develop kidney problems. Half of the rats were classified as obese and half as lean. Within each group, half were randomly assigned to receive a vitamin-B-supplemented diet and half were not. Thus, a 2x2 factorial experiment was conducted with seven rats assigned to each of the four combinations of size (lean or obese) and diet (supplemental or not). One of the response variables measured was weight (in grams) of the kidney at the end of a 20-week feeding period. The data (simulated from the summary information provided in the journal article) are shown in the table. Conduct an analysis of variance on the data. Summarize the results in an ANOVA table. Conduct the appropriate ANOVA F-tests at $\alpha = 01$.

		Diet	
		Regular	Vitamin B
	Lean	1.62	1.51
	Lean	1.80	1.65
	Lean	1.71	1.45
	Lean	1.81	1.44
	Lean	1.47	1.63
	Lean	1.37	1.35
	Lean	1.71	1.66
Rat Size	Obese	2.35	2.93
	Obese	2.97	2.72
	Obese	2.54	2.99
	Obese	2.93	2.19
	Obese	2.84	2.63
	Obese	2.05	2.61
	Obese	2.82	2.64

Solution for Problem 10.65

Anova: Two-Factor With Replication

SUMMARY	Regular	Vitamin B	Total
Lean			
Count	2	2	4
Sum	3.42	3.16	6.58
Average	1.71	1.58	1.645
Variance	0.0162	0.0098	0.0143
Lean			
Count	2	2	4
Sum	3.52	2.89	6.41
Average	1.76	1.445	1.6025
Variance	0.005	5E-05	0.034758
Lean			
Count	2	2	4
Sum	2.84	2.98	5.82
Average	1.42	1.49	1.455
Variance	0.005	0.0392	0.016367
Lean			
Count	2	2	4
Sum	4.06	4.59	8.65
Average	2.03	2.295	2.1625
Variance	0.2048	0.80645	0.360492
Obese			
Count	2	2	4
Sum	5.51	5.71	11.22
Average	2.755	2.855	2.805
Variance	0.09245	0.03645	0.0463
Obese			
Count	2	2	4
Sum	5.77	4.82	10.59
Average	2.885	2.41	2.6475
Variance	0.00405	0.0968	0.108825
Obese			
Count	2	2	4
Sum	4.87	5.25	10.12
Average	2.435	2.625	2.53
Variance	0.29645	0.00045	0.111
Total			
Count	14	14	
Sum	29.99	29.4	
Average	2.142143	2.1	
Variance	0.337234	0.391677	

ANOVA

Source of Variation	SS	df	MS	F	P-value	F crit
Sample	7.412143	6	1.235357	10.72126	0.000149	4.455842
Columns	0.012432	1	0.012432	0.107894	0.747416	8.861662

146

Interaction	0.450543	6	0.07509	0.651686	0.688826	4.455842
Within	1.61315	14	0.115225			
Total	9.488268	27				

11.1   Introduction

In Chapters 11-13, McClave and Sincich begin to introduce students to the basics of forecasting, starting with the simple linear regression model presented in Chapter 11. For the first time, students are given methods to allow for comparisons between variables in order to predict an outcome. In the simplest case, two variables are studied in which the independent variable, x, is used to define or predict the dependent variable, y.

PHStat can be used to derive coefficients of correlation, coefficients of determination, and the output necessary to analyze the validity of the regression and the regression coefficients. In order to use regression analysis in PHStat, please refer to the following steps.

Step 1: Open a new workbook and place the cursor in the upper left-hand cell of the spreadsheet.

Step 2: Click on PHStat at the top of the menu bar.

Step 3: Under PHStat, **select Regression**.

Figure 11.1

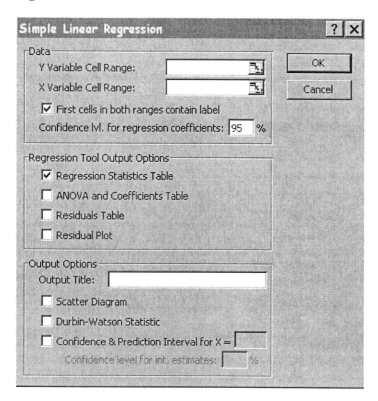

Step 4: Select Simple Linear Regression from the choices listed.

Step 5: Highlight the Y and X variable ranges for the dataset.

148

Step 6: Input the **Confidence Level for Regression Coefficients**. Notice that, by default, the 95% confidence level is selected.

Step 7: The user may choose from a variety of regression tool output options including **Regression Statistics Table, ANOVA and Coefficients Table, Residuals Table**, and **Residuals Plot**.

Step 8: Output Options include **Scatter Diagram, Durbin-Watson Statistic**, and **Confidence and Prediction Level for X**.

Step 9: Enter an **Output Title** if desired.

Step 10: Click OK to finish.

The following examples will be used to illustrate techniques in PHStat and Excel.

Example 11.1, Reaction Drug Data, page 598
Example 11.2, Reaction Drug Data, page 613
Example 11.9, Fire Insurance Company, page 644 (Table 11.7 data)
Sample Exercise #1: Problem 11.118, Study of Fertility Rates, page 655
Sample Exercise #2: Problem 11.19, College Protests of Labor, page 607

11.2   The Coefficient of Correlation

In Section 11.6 of the text, the coefficient of correlation, r, is used as a proxy for covariance in order to describe the relationship between two variables. Also known as the Pearson product moment coefficient of correlation, it measures the strength of the relationship between two variables as well as the direction of the relationship (is there a direct relationship between x and y or an inverse relationship between x and y). In order to calculate the coefficient of correlation; please refer to the following steps.

Step 1: Open a new workbook and place the cursor in the upper left-hand cell.

Step 2: Click PHStat - Regression - Simple Linear Regression.

Step 3: Highlight the Y and X variable ranges for the dataset. This can be done by typing the location of the data manually or by clicking and dragging over the appropriate data cells. If the first row of data contains a nonnumeric, descriptive title, click **First cells in both ranges contain labels.**

Step 4: Check the Regression Statistics Table.

Step 5: Enter an Output Title if desired.

Step 6: Click OK to finish.

Example 11.1   Please refer to Example 11.1 (page 598)

An important consideration when taking a drug is how it may affect one's perception or general awareness. Suppose you want to model the length of time it takes to respond to a stimulus (a measure of awareness) as a function of the percentage of a certain drug in the bloodstream. The first question to be answered is this: "Do you think an exact relationship exists between these two variables?" That is, do you think it is possible to state the exact length of time it takes an individual (subject) to respond if the amount of the drug in the bloodstream is known?

a. Use the method of least squares to estimate the values of the coefficients.
b. Predict the reaction time when x=2%.
c. Find SSE for the analysis
d. Give practical interpretations for the coefficients.

Table 11.1

Amount of Drug x (%)	Reaction Time y (seconds)
1	1
2	1
3	2
4	2
5	4

Solution:

Please refer to the following steps in order to answer this question.

Step 1: Open a new workbook and place the cursor in the upper left-hand cell.

Step 2: Open the data file or input the data manually.

Step 3: Click PHStat - Regression - Simple Linear Regression.

Figure 11.2

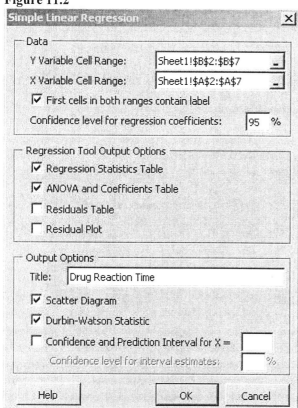

Step 4: Highlight the Y variable range for the dataset. In this case the y variable is the Crime Rate which is C1:C11. This can be done by typing the location C1:C11 in the Y variable range or by clicking and dragging over the range of data.

Step 5: Highlight the X variable range for the dataset. In this case the x variable is the Number of Casino Employees which is B1:B11. This can be done by typing the location B1:B11 in the X variable range or by clicking and dragging over the appropriate data cells.

Step 6: Click the First Cell in both ranges contain label because B1 and C1 contain a descriptive, nonnumeric title.

Step 7: Check the Regression Statistics Table.

Step 8: Enter an Output Title if desired.

Step 9: Click OK to finish.

Once the steps are completed, the output generated should appear as in Table 11.2

Table 11.2

Drug Reaction Time					
Regression Statistics					
Multiple R	0.9037				
R Square	0.8167				
Adjusted R Square	0.7556				
Standard Error	0.6055				
Observations	5				
ANOVA					
	df	*SS*	*MS*	*F*	*Significance F*
Regression	1	4.9	4.9	13.36	0.035
Residual	3	1.1	0.367		
Total	4	6			
	Coefficients	*Standard Error*	*t Stat*	*P-value*	*Lower 95%*	*Upper 95%*
Intercept	-0.1	0.6351	-0.1575	0.8849	-2.1211	1.9211
Amount of Drug x (%)	0.7	0.1915	3.6556	0.0354	0.0906	1.3094

Notice that the coefficient of correlation is also referred to as Multiple R on the regression output. In this example, r = 0.98703.

11.3   The Coefficient of Determination

The coefficient of determination is the square of the coefficient of correlation, r. When calculated, it refers to the percentage of sample variation in the dependent variable (y) that can be explained by the independent variable (x) in a linear regression model. In order to calculate the coefficient of determination; please refer to the following steps.

151

Step 1: Open a new workbook and place the cursor in the upper left-hand cell.

Step 2: Click PHStat - Regression - Simple Linear Regression.

Step 3: Highlight the Y and X variable ranges for the dataset. This can be done by typing the location of the data manually or by clicking and dragging over the appropriate data cells. If the first row of data contains a nonnumeric, descriptive title, click **First cells in both ranges contain labels.**

Step 4: Check the Regression Statistics Table.

Step 5: Enter an Output Title if desired.

Step 6: Click OK to finish.

An example from the text will now be used to illustrate how to calculate coefficients of correlation.

Example 11.2   Please refer to Exercise 11.2 from the text.

Calculate the coefficient of determination for the drug reaction problem. The data appears in the following problem.

Table 11.3

Amount of Drug x (%)	Reaction Time y (seconds)
1	1
2	1
3	2
4	2
5	4

Solution:

Please refer to the following steps in order to answer this question.

Step 1: Open a new workbook and place the cursor in the upper left-hand cell.

Step 2: Open the dataset or input the data manually. The data should be contained within A1:B6.

Step 3: Click PHStat - Regression - Simple Linear Regression.

Step 4: Highlight the Y variable range for the dataset. This can be done by typing the location of the data manually (in this case B1:B6) or by clicking and dragging over the appropriate data cells.

Step 5: Highlight the X variable range for this dataset. This can be done by typing the location of the data manually (in this case A1:A6) or by clicking and dragging over the appropriate data cells.

Step 6: If the first row of data contains a nonnumeric, descriptive title, click **First cells in both ranges contain labels.** In this case, labels appear in A1 and B1.

Step 7: Check the Regression Statistics Table.

Step 8: Enter an Output Title if desired. In this case, the title selected is Solution for Example 11.2 of the text.

Figure 11.3

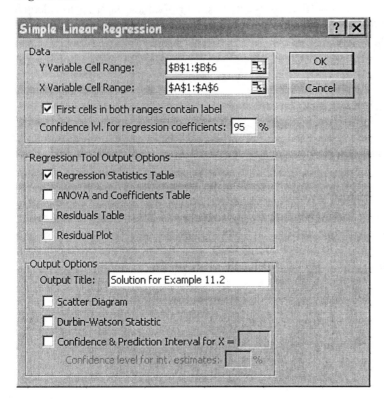

Step 9: Click OK to finish.

Notice that the output generates a coefficient of determination, R^2, of 0.8167. Thus 81.67% of the changes in the reaction time (y) can be explained by the amount of drug (x) being administered.

Table 11.4

Solution	
Regression Statistics	
Multiple R	0.903696114
R Square	**0.816666667**
Adjusted R Square	0.755555556
Standard Error	0.605530071
Observations	5

11.4   Simple Linear Regression: A Complete Model

In this section, a complete analysis of regression will be provided. This includes not only the derivation of the coefficients of correlation and determination, but also combines ANOVA analysis developed in Chapter 10 of the textbook. The goal is first to analyze all elements necessary to determine if a straight-line regression is the best model to interpret data. Second, once determining that the model is appropriate, the second goal is to interpret the regression coefficients, t-statistics, and p-values. In order to calculate regression output; please refer to the following steps.

Step 1: Open a new workbook and place the cursor in the upper left-hand cell.

Step 2: Open the dataset or input the data manually.

Step 3: Click PHStat - Regression - Simple Linear Regression.

Step 4: Highlight the Y variable range for the dataset. This can be done by typing the location of the data manually or by clicking and dragging over the appropriate data cells.

Step 5: Highlight the X variable range for this dataset. This can be done by typing the location of the data manually or by clicking and dragging over the appropriate data cells.

Step 6: If the first row of data contains a nonnumeric, descriptive title, click **First cells in both ranges contain labels.** In this case, labels appear in A1 and B1.

Step 7: Check the Regression Statistics Table and ANOVA and Coefficients Table.

Step 8: Check Scatter Diagram if asked to graph the relationship between the independent and dependent variables.

Step 9: Enter an Output Title if desired. In this case, the title selected is Solution for Example 11.2 of the text.

Step 10: Click OK to finish.

An example will be used in order to illustrate how PHStat generates regression statistics.

Example 11.3   No example is provided in Section 11.9. Use data in Table 11.7 (page 644) for analysis

Suppose a fire insurance company wants to relate the amount of fire damage in major residential fires to the distance between the burning house and the nearest fire station. The study is to be conducted in a large suburb of a major city; a sample of 15 recent fires in this suburb is selected. The amount of damage, y, and the distance between the fire and the nearest fire station, x, are recorded for each fire. Generate regression statistics.

154

Table 11.5

Distance from Fire Station x (miles)	Fire Damage y (thousands of dollars)
3.4	26.2
1.8	17.8
4.6	31.3
2.3	23.1
3.1	27.5
5.5	36.0
0.7	14.1
3.0	22.3
2.6	19.6
4.3	31.3
2.1	24.0
1.1	17.3
6.1	43.2
4.8	36.4
3.8	26.1

Solution:

Please refer to the following steps in order to answer this question.

Step 1: Open a new workbook and place the cursor in the upper left-hand cell.

Step 2: Open the dataset or input the data manually. Data should be contained within A1:B16.

Step 3: Click PHStat - Regression - Simple Linear Regression.

Step 4: Highlight the Y variable range for the dataset. This can be done by typing the location of the data manually (in this case B1:B16) or by clicking and dragging over the appropriate data cells.

Step 5: Highlight the X variable range for this dataset. This can be done by typing the location of the data manually (in this case A1:A16) or by clicking and dragging over the appropriate data cells.

Step 6: If the first row of data contains a nonnumeric, descriptive title, click **First cells in both ranges contain labels.** In this case, labels appear in A1 and B1.

Step 7: Check the Regression Statistics Table and ANOVA and Coefficients Table.

Step 8: Click Scatter Diagram if asked to graph the relationship between the independent and dependent variables.

Step 9: Enter an Output Title if desired. In this case, the title selected is Solution for Table 11.7 from the text.

Step 10 Click OK to finish.

155

Figure 11.4

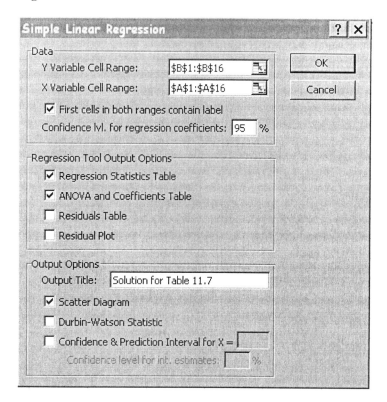

Table 11.6

Output for Table 11.7 from the Text						
Regression Statistics						
Multiple R	0.960977715					
R Square	0.923478169					
Adjusted R Square	0.917591874					
Standard Error	2.316346184					
Observations	15					
ANOVA						
	df	*SS*	*MS*	*F*	*Significance F*	
Regression	1	841.766358	841.766358	156.8861596	1.2478E-08	
Residual	13	69.75097535	5.365459643			
Total	14	911.5173333				
	Coefficients	*Standard Error*	*t Stat*	*P-value*	*Lower 95%*	*Upper 95%*
Intercept	10.27792855	1.420277811	7.236562082	6.58556E-06	7.209605476	13.34625162
Distance from Fire Stati·	4.919330727	0.392747749	12.52542054	1.2478E-08	4.070850963	5.767810491

Please refer to the output generated in the table and compare it to Figure 11.25 from the text. Notice how the STATISTIX printout differs in arrangement from PHStat. Now, refer to the steps presented in Section 11.9 of the text.

1. In Step 2, find the slope and intercept on the printout. In this case the slope of the line is 4.91 (the coefficient on the distance from fire station variable, x) and the intercept is 10.28.
2. In Step 3, the estimate of the standard deviation is 2.31635 (standard error in PHStat).

156

3. In Step 4, R^2 = 0.923478169 or 92.35% of the changes in the dollar value of fire damage, y, can be explained by the distance from the fire station, x.
4. In Step 4, the coefficient of correlation, r, or multiple R in PHStat is 0.960977715 or 96.10%.

11.5   Sample Exercises

The following exercises were selected from the textbook to test the reader's understanding of PHStat and its usage within the chapter. Try to replicate the output in PHStat for the following problems.

Problem 11.118, Study of Fertility Rates, page 655.

The fertility rate of a country is defined as the number of children a woman citizen bears, on average, in her lifetime. *Scientific American* (Dec. 1993) reported on the declining fertility rate in developing countries. The researchers found that family planning can have a great effect on fertility rate. The accompanying table gives the fertility rate, y, and contraceptive prevalence, x, (measured as the percentage of married women who use contraception for each of 27 developing countries. According to the researchers, "the data reveal that differences in contraceptive prevalence explain about 90% of the variation in fertility rates." Do you concur?

Country	Contraceptive Prevalence x	Fertility Rate y
Mauritius	76	2.2
Thailand	69	2.3
Colombia	66	2.9
Costa Rica	71	3.5
Sri Lanka	63	2.7
Turkey	62	3.4
Peru	60	3.5
Mexico	55	4.0
Jamaica	55	2.9
Indonesia	50	3.1
Tunisia	51	4.3
El Salvador	48	4.5
Morocco	42	4.0
Zimbabwe	46	5.4
Egypt	40	4.5
Bangladesh	40	5.5
Botswana	35	4.8
Jordan	35	5.5
Kenya	28	6.5
Guatemala	24	5.5
Cameroon	16	5.8
Ghana	14	6.0
Pakistan	13	5.0
Senegal	13	6.5
Sudan	10	4.8
Yemen	9	7.0
Nigeria	7	5.7

Solution:

Solution for Exercise 11.118 of the text	
Regression Statistics	
Multiple R	0.865019842
R Square	0.748259327
Adjusted R Square	0.738189701
Standard Error	0.695710716
Observations	27

Problem 11.19, College protests of labor exploitations, page 607

Refer to the *Journal of World-Systems Research* (Winter 2004) study of student "sit-ins" for a sweat free campus" at universities. Recall that the SITIN file contains data on the duration (in days) of each sit-in as well as the number of student arrests. The data for 5 sit-ins where there was at least one arrest are shown below. Let y=number of arrests and x=duration.

a. Give the equation of a straight-line model relating y to x.
b. Fit the model to the data for the 5 sit-ins using the methods of least squares. Give the equation of the least squares prediction equation.

Sit-In	University	Duration	Number of Arrests
12	Wisconsin	4	54
14	SUNY Albany	1	11
15	Oregon	3	14
17	Iowa	4	16
18	Kentucky	1	12

Solution:

SUMMARY OUTPUT	
Regression Statistics	
Multiple R	0.6009
R Square	0.3611
Adjusted R Square	0.1481
Standard Error	16.9132
Observations	5

ANOVA

	df	SS	MS	F	Significance F
Regression	1	485.0261	485.0261	1.6956	0.2838
Residual	3	858.1739	286.0580		
Total	4	1343.2000			

	Coefficients	Standard Error	t Stat	P-value	Lower 95%	Upper 95%
Intercept	2.52	16.35	0.15	0.89	-49.52	54.56
Duration	7.26	5.58	1.30	0.28	-10.48	25.01

158

12.1 Introduction

In Chapter 11, a straight-lined regression model was developed in order to define the relationship between a dependent variable, y, and a single independent variable, x. In Chapter 12, the basic regression model is extended in order to analyze the relationship between the dependent variable, y, and multiple independent variables. In this section, PHStat and Excel will be used to illustrate how to build regression using quantitative and qualitative independent variables and to test the utility of the model.

In order to use PHStat to develop multiple regression, please refer to the following steps. Please refer to Figure 12.1, which illustrates the available options in the multiple regression.

Step 1: Open a new workbook and be sure the cursor is located in the first cell of the spreadsheet.

Step 2: Click PHStat at the top of the menu bar.

Step 3: Select Regression from the pull-down menu under PHStat.

Step 4: Select Multiple Regression from the choices available.

Step 5: Under data, select the appropriate **y variable and x variable cell ranges**. This may be done by inputting the data manually or highlighting and dragging the data into the appropriate cells.

Step 6: Click first cells in both ranges contain label if the first cell has a nonnumeric descriptive title for all variables specified in the regression.

Step 7: The Confidence Level for regression coefficients is set at 95% by default. Change as desired.

Step 8: Under Regression Tools Output Options, select how the data is presented. Select **Regression Statistics Table, ANOVA and Coefficients Table, Residuals Table and/or Residual Plot**. If all boxes are checked, a complete analysis will be derived for the multiple regression.

Step 9: Under Output Options, an **output title** may be entered to describe the data set being analyzed.

Step 10: Under Output Options, do NOT select Durbin-Watson statistics, Coefficients of Partial Determination, VIF, and **Confidence & Prediction Interval Estimates** for this analysis.

Step 11: Click OK to finish.

Figure 12.1

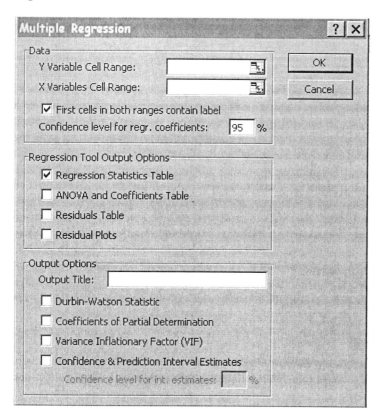

Excel may also be used to develop multiple regression. Please refer to the following steps and Figure 12.1, which illustrates the available options in the multiple regression.

Step 1: Open a new workbook and be sure the cursor is located in the first cell of the spreadsheet.

Step 2: Select Tools from the top of the tool bar.

Step 3: Select Data Analysis-Regression from the options listed.

Step 4: Select the appropriate y variable and x variable ranges. Notice that all independent variables may be selected to run a multiple regression if they reside in contingent columns (next to each other)

Step 5: Click Labels if the first cell contains a descriptive heading.

Step 6: Click Constant is zero if the regression does not contain a constant term.

Step 7: Select the confidence level of the multiple regression (default is 95%).

Step 8: Select Output Range (a new worksheet, a new workbook, or within the existing worksheet). If generating data within the current worksheet, remember to select a range that is empty so that pre-existing data will not be erased.

160

Step 9: Select residuals, standardized residuals, residual plots, line fit plots, and normal probability plots as desired.
Step 10: Click OK to finish.

Figure 12.1a refers to the multiple regression menu in Excel. Please compare the differences between PHStat and Excel.

Figure 12.1a

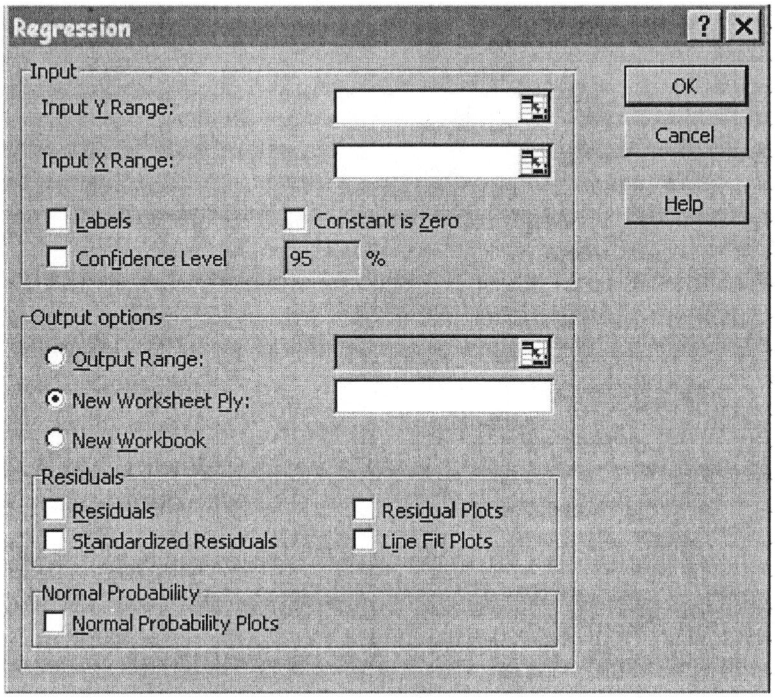

The following examples will be used to illustrate techniques in PHStat and Excel.

Example 12.1, Property Appraiser, page 665
Example 12.2, Antique Grandfather Clocks, page 669
Example 12.3, Antique Grandfather Clocks, part 2, page 671
Discussion from Section 12.4, Property Appraiser Adaptation, page 684
Example 12.5, Antique Grandfather Clocks, page 685
Example 12.7, Energy Conservation, page 698
Sample Exercise #1: Problem 12.19, Incomes of Mexican Street Vendors, page 680
Sample Exercise #2: Problem 12.30, Incomes of Mexican Street Vendors, page 688
Sample Exercise #4: Problem 12.144, Carp Experiment, page 773

12.2 Multiple Regression Models

Please refer to the six-step process in the textbook to review how to analyze a multiple regression model. In this section, a first-order model will be developed for independent variables that are quantitative in nature.

161

Example 12.1 Please refer to Example 12.1 (page 665) in the textbook.

Suppose a property appraiser wants to model the relationship between the sale price of a residential property in a mid-size city and the following three independent variables: (1) appraised land value of the property, (2) appraised value of improvements (i.e., home value) on the property, and (3) area of living space on the property (i.e., home size). Consider the first-order model

$$y = \beta_0 + \beta_1 x_1 + \beta_2 x_2 + \beta_3 x_3 + \epsilon$$

where
y = Sale price (dollars)
x_1 = Appraised land value (dollars)
x_2 = Appraised improvements (dollars)
x_3 = Area (square feet)

To fit the model, the appraiser selected a random sample of n=20 properties from the thousands of properties that were sold in a particular year.
    a.   Use scattergrams to plot the sample data. Interpret the plots.
    b.   Use the method of least squares to estimate the unknown parameters in the model.
    c.   Find the value of SSE that is minimized by the least squares method.

Solution:

Table 12.1 presents the data within Excel. Notice that row A1:A5 contains descriptive titles for the data as presented in the text.

Table 12.1

	A	B	C	D	E
1	Property # (Obs.)	Sale Price y	Land Value x1	Improvements x2	Area x3
2	1	68900	5960	44967	1873
3	2	48500	9000	27860	928
4	3	55500	9500	31439	1126
5	4	62000	10000	39592	1265
6	5	116500	18000	72827	2214

Please refer to the following steps in order to answer the preceding questions.

Step 1: Open a new workbook and be sure the cursor is located in the first cell of the spreadsheet.

Step 2: Enter the data manually or open the data from the database in the text.

Step 3: Click PHStat at the top of the menu bar.

Step 4: Select Regression from the pull-down menu under PHStat.

Step 5: Select Multiple Regression from the choices available.

Step 6: The y variable cell ranges are B1:B21 and x variable cell ranges are C1:E21. This may be done by inputting the data manually or highlighting and dragging the data into the appropriate cells.

Step 7: Click first cells in both ranges contain label. The first cells in A1:E1 contain descriptive titles of the data.

Step 8: The Confidence Level for regression coefficients is set at 95% by default. Since this problem does not specify a level of confidence, please keep the default setting.

Step 9: Under Regression Tools Output Options, select **Regression Statistics Table, ANOVA and Coefficients Table, Residuals Table and Residual Plot**. If all boxes are checked, a complete analysis will be derived for the multiple regression.

Step 10: Under Output Options, an **output title** may be entered to describe the data set being analyzed. In this case, the output title is **Solution for Example 12.1 from the text.**

Step 11: Click OK to finish.

Figure 12.2 displays the parameter specifications for this problem. Please refer to what options are selected before continuing with the problem. Notice that when the output is displayed, the full regression output is presented in sections within the spreadsheets. The scatterplots are located on the side of the spreadsheet.

Figure 12.2

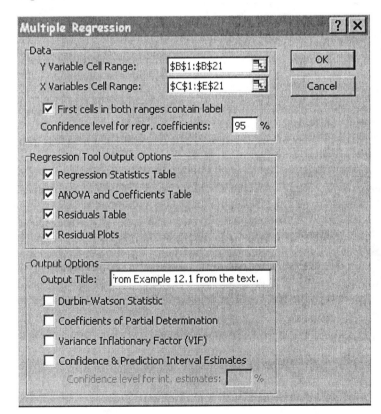

Solution for part a:

The scatterplots for examining the relationships between y and each independent variable (appraised land value, appraised improvements, and area) are presented in Figure 12.3. Please compare these to the ones derived in the text-using Minitab.

Figure 12.3

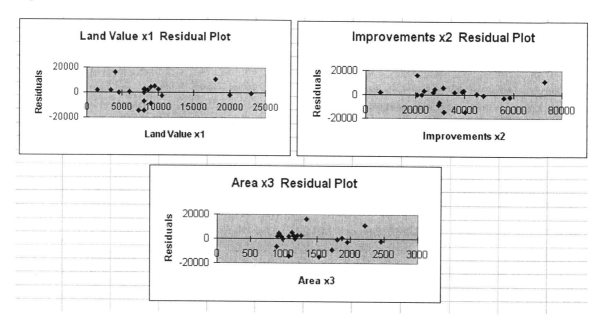

Solution for part b:

The model hypothesized is fit to the data presented in Table 12.1. The location of the estimates is presented in Figure 12.4. The estimate of the intercept (β_0) is 1,470.275919. The estimate of β_1 is 0.814490116. The estimate of β_2 is 0.820444695. And the estimate of β_3 is 13.52864993. The equation is therefore

$$y = 1,470.275919 + 0.814490116x_1 + 0.820444695x_2 + 13.52864993x_3$$

Figure 12.4

Solution for Example 12.1 from text						
Regression Statistics						
Multiple R	0.94732611					
R Square	0.897426758					
Adjusted R Square	0.878194275					
Standard Error	7919.482541					
Observations	20					

ANOVA						
	df	*SS*	*MS*	*F*	*Significance F*	
Regression	3	8779676741	2.93E+09	46.66203	3.9E-08	
Residual	16	1003491259	62718204			
Total	19	9783168000				

	Coefficients	*Standard Error*	*t Stat*	*P-value*	*Lower 95%*	*Upper 95%*
Intercept	1470.275919	5746.32458	0.255864	0.801316	-10711.39	13651.94
Land Value x1	0.814490116	0.51221871	1.590122	0.13137	-0.271365	1.900345
Improvements x2	0.820444695	0.21118494	3.884958	0.001315	0.3727527	1.268137

Solution for part c:

The location of the minimum value of SSE is located under the ANOVA statistics within the regression in Figure 12.5. The minimum value of the SSE is 1,003,491,259.

Figure 12.5

ANOVA						
	df	*SS*	*MS*	*F*	*Significance F*	
Regression	3	8779676741	2.93E+09	46.66203	3.9E-08	
Residual	16	1003491259	62718204			
Total	19	9783168000				

	Coefficients	*Standard Error*	*t Stat*	*P-value*	*Lower 95%*	*Upper 95%*
Intercept	1470.275919	5746.32458	0.255864	0.801316	-10711.39	13651.94
Land Value x1	0.814490116	0.51221871	1.590122	0.13137	-0.271365	1.900345
Improvements x2	0.820444695	0.21118494	3.884958	0.001315	0.3727527	1.268137
Area x3	13.52864993	6.58568006	2.054253	0.056664	-0.432365	27.48966

12.3   Inferences About the β Parameters

According to the text, β parameters may be estimated using confidence intervals or a test of hypothesis. The multiple regression package within PHStat will be used to highlight how to estimate β parameters.

Example 12.2 Please refer to Example 12.3 (page 669) in the textbook

A collector of antique grandfather clocks knows that the price received for the clocks increases linearly with the age of the clock. Moreover, the collector hypothesizes that the auction price of the clocks will increase linearly as the number of bidders increases. Thus, the following first-order model is hypothesized:

$$y = \beta_0 + \beta_1 x_1 + \beta_2 x_2 + \epsilon$$

where
y = Auction price
x_1 = Age of clock (years)
x_2 = Number of bidders

A sample of 32 auction prices of grandfather clocks, and the number of bidders, is given in Table 12.2.
 a. Test the hypothesis that the mean auction price of a clock increases as the number of bidders increases when age is held constant, that is, $\beta_2 > 0$. Use $\alpha = 0.05$.
 b. Form a 90% confidence interval for β_1 and interpret the results.

Solution:

Table 12.2 presents the data necessary for completing this problem. Notice that the data must be contained in three columns (unlike the text) in order for the regression to be run.

Table 12.2

	A	B	C
1	Age x1	# of bidders x2	Auction price,y
2	127	13	1235
3	115	12	1080
4	127	7	845
5	150	9	1522
6	156	6	1047

Please refer to the following steps in order to answer the preceding questions.

Step 1: Open a new workbook and be sure the cursor is located in the first cell of the spreadsheet.

Step 2: Enter the data manually or open the data from the database in the text.

Step 3: Click PHStat at the top of the menu bar.

Step 4: Select Regression from the pull-down menu under PHStat.

Step 5: Select Multiple Regression from the choices available.

Step 6: The y variable cell ranges are C1:C33 and x variable cell ranges are A1:B33. This may be done by inputting the data manually or highlighting and dragging the data into the appropriate cells.

Step 7: Click first cells in both ranges contain label. The first cells in A1:C1 contain descriptive titles of the data.

Step 8: The Confidence Level for regression coefficients is set at 95% by default. Since this problem specifies a level of confidence of 90%, please change that setting to 90%.

Step 9: Under Regression Tools Output Options, select **Regression Statistics Table, ANOVA and Coefficients Table only.** A scatter diagram is not required so the **Residuals Table and Residual Plot boxes** need NOT be checked.

Step 10: Under Output Options, an **output title** may be entered to describe the data set being analyzed. In this case, the output title is **Solution for Example 12.3 from the text.**

Step 11: Click OK to finish.

Figure 12.6 presents the data displayed in the multiple regression model for this problem. Be sure that the data entered corresponds to this figure before continuing with the problem.

Figure 12.6

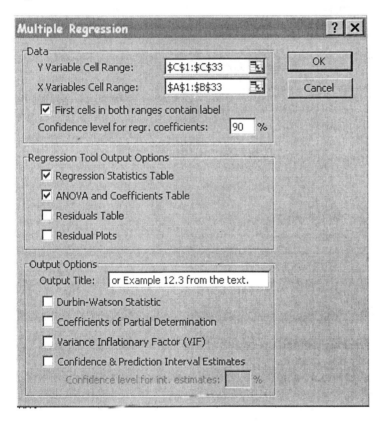

Figure 12.7 presents the output generated for this problem. Please refer to the highlighted sections in order to answer the questions provided.

Figure 12.7

Solution for Example 12.3 from the text.

Regression Statistics	
Multiple R	0.9446396
R Square	0.8923439
Adjusted R Square	0.8849194
Standard Error	133.48467
Observations	32

ANOVA

	df	SS	MS	F	Significance F
Regression	2	4283062.96	2141531	120.188	9.21636E-15
Residual	29	516726.54	17818.2		
Total	31	4799789.5			

	Coefficients	Standard Error	t Stat	P-value	Lower 95%	Upper 95%	Lower 90.0%	Upper 90.0%
Intercept	-1338.951	173.809471	-7.7036	1.7E-08	-1694.43182	-983.47087	-1634.276	-1043.627
Age x1	12.740574	0.90474031	14.082	1.7E-14	10.89017139	14.5909768	11.203305	14.277843
# of bidders x2	85.952984	8.72852329	9.84737	9.3E-11	68.10114007	103.804829	71.122114	100.78385

Solution for part a:

The test statistic is a t statistic for the estimate of the parameter β_2. In this case, the t statistic is 9.847368395. This test statistic value falls within the rejection region computed by the text (critical value of 1.699) therefore reject the null and conclude that the mean auction price of a clock increases with the number of bidders. Refer to figure 12.8 to locate the correct t statistic using PHStat.

Figure 12.8

	Coefficients	Standard Error	t Stat	P-value
Intercept	-1338.951	173.809471	-7.7036	1.7E-08
Age x1	12.740574	0.90474031	14.082	1.7E-14
# of bidders x2	85.952984	8.72852329	9.84737	9.3E-11

Solution for part b:

The 90% confidence interval for β_1 can be obtained from the PHStat output. In this case, a researcher is 90% confident that β_1 falls between 11.21 and 14.27. Figure 12.9 indicates where to locate the confidence interval generated in PHStat. Notice that the 95% interval is always generated; in order to generate another interval, specify the percentage interval in Data Options.

Figure 12.9

	Coefficients	Standard Error	t Stat	P-value	Lower 95%	Upper 95%	Lower 90.0%	Upper 90.0%
Intercept	-1338.9513	173.8094707	-7.703558	1.706E-08	-1694.4318	-983.4709	-1634.28	-1043.63
Age x1	12.7405741	0.904740307	14.08202	1.693E-14	10.8901714	14.590977	11.20331	14.27784
# of bidders x2	85.9529844	8.728523289	9.847368	9.345E-11	68.1011401	103.80483	71.12211	100.7839

12.4   Checking the Overall Utility of a Model

According to the text, the output generated from the multiple regression models in PHStat can be used to determine whether the overall model is helpful in predicting y. Use the multiple regression model in PHStat to find the adjusted coefficient of determination and to isolate the global F-test within the ANOVA output in order to determine the usefulness of the model in predicting y.

Example 12.3   Please refer to Example 12.4 (page 676) from the textbook.

Refer to Example 12.3 in the textbook in which antique collector modeled the auction price y of grandfather clocks as a function of the age of the clock, x_1, and the number of bidders, x_2. The hypothesized first-order model is

$$y = \beta_0 + \beta_1 x_1 + \beta_2 x_2 + \epsilon$$

    a.   Find and interpret the adjusted coefficient of determination for this example.
    b.   Conduct the global F-test of model usefulness at the $\alpha = 0.05$ level of significance.

Solution:

Please refer to the following steps in order to generate output in PHStat

Step 1: Open a new worksheet and be sure the cursor is located within the first cell of the spreadsheet.

Step 2: Input the data or open the data from the prepackaged data set located with the text.

Step 3: Select PHStat-Regression-Multiple Regression

Step 4: Highlight y range as C1:C33

Step 5: Highlight x range as A1:B33

Step 6: Click first cells in both ranges contain labels.

Step 7: Confidence level for regression coefficients is 95%

Step 8: Select Regression Statistics Table and ANOVA and Coefficients Table.

Step 9: Enter an Output Title if desired.

Step 10: Click OK to finish.

Please refer to Figure 12.10, which illustrates how the Multiple Regression menu should appear prior to executing the program. The output generated will now be used to answer the question.

Figure 12.10

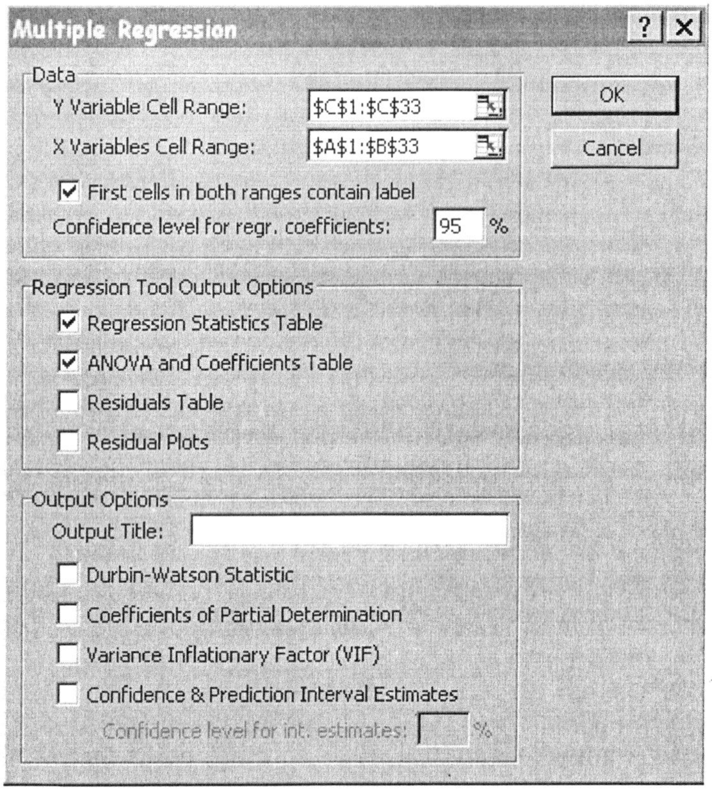

Solution for part a:

Figure 12.11 presents all output generated using the Multiple Regression option in PHStat. Notice that the adjusted R^2 is 0.884919359, which is located under Regression Statistics. Therefore, the model has explained about 88.49% of the total sample variation in auction prices after adjusting for sample size and the number of independent variables in the model.

Solution for part b:

Figure 12.11 presents all output generated using the Multiple Regression option in PHStat. Please refer to the ANOVA output within PHStat. The significance F is synonymous to the p-value for the F statistic generated in Minitab. In this example, significance F is 9.21636E-15 or 0.00000000000000921636 or essentially zero. Compare this to $\alpha = 0.05$ and conclude that at least one of the model coefficients is nonzero. So the model appears to be useful in predicting auction prices.

Figure 12.11

Solution for Example 12.3

Regression Statistics	
Multiple R	0.9446396
R Square	0.8923439
Adjusted R Square	0.8849194
Standard Error	133.48467
Observations	32

ANOVA

	df	SS	MS	F	Significance F
Regression	2	4283062.96	2141531	120.188	9.2164E-15
Residual	29	516726.5399	17818.2		
Total	31	4799789.5			

	Coefficients	Standard Error	t Stat	P-value	Lower 95%	Upper 95%
Intercept	-1338.9513	173.8094707	-7.7036	1.7E-08	-1694.43182	-983.470865
Age x1	12.740574	0.904740307	14.082	1.7E-14	10.8901714	14.5909768
# of bidders x2	85.952984	8.728523289	9.84737	9.3E-11	68.1011401	103.8048287

12.5   Using the Model for Estimation and Prediction

In this section, PHStat and Excel will be used to estimate the mean value of the dependent variable, $E(y)$, once a particular value of the independent variable, x, is substituted into the prediction equation. In order to use PHStat to estimate and predict the mean value of y, refer to the following steps.

Step 1: Select PHStat-Regression-Multiple Regression

Step 2: Select Y variable cell range and X variable cell range

Step 3: Select Confidence and Prediction Level Estimates. This generates the prediction for the mean value of y.

Step 4: Select the Confidence and Prediction Interval Estimates specified by the problem.

Step 5: Click OK to finish.

The following example is adapted from Section 12.4 of the textbook.

Example 12.4   Please refer to section 12.4 (page 684) of the textbook.

Recall that in Example 12.2, data were provided on the mean sale price of a residential property given land value (x_1), improvements (x_2), and home size (x_3).
   a. Find the estimated mean values and corresponding 95% confidence interval for selected independent variables. Assume that the mean sale price for a given property will be estimated if $x_1 = \$15,000$, $x_2 = \$50,000$, and $x_3 = 1,800$ square feet.
   b. Find the corresponding 95% confidence interval for the estimated mean sale price
   c. Find the corresponding 95% prediction interval for the predicted sale price.

Solution:

Please refer to the following steps in order to generate output for Example 12.4

Step 1: Open a new workbook and be sure the cursor is located in the upper left-hand cell of the spreadsheet.

Step 2: Input the data for the example or input the data manually.

Step 3: Select PHStat-Regression-Multiple Regression

Step 4: Y variable cell range is B1:B21

Step 5: X variable cell range is C1:E21

Step 6: Click first cell in both ranges contain labels

Step 7: Click Regression Statistics

Step 8: Click Confidence and Prediction Level Estimates. This will generate the prediction for the dependent variable.

Step 9: Enter 95% for Confidence and Prediction Level estimates

Step 10: Click OK to finish.

Figure 12.12 illustrates the template that is created within PHStat in order to complete this problem. Notice that the user should input specific values for the independent variables. At this time, please input the values for the independent variables in the template.

Figure 12.12

Confidence and Prediction Estimate Intervals				
Confidence Level	95%			
	1			
Land Value x1 given value	0			
Improvements x2 given value	0			
Area x3 given value	0			
XX	20	184260	706226	27667
	184260	2.25E+09	7.65E+09	2.88E+08
	706226	7.65E+09	2.94E+10	1.08E+09
	27667	2.88E+08	1.08E+09	42394501
Inverse of XX	0.526486	7.05E-06	-8.1E-07	-0.00037
	7.05E-06	4.18E-09	-7.2E-10	-1.5E-08
	-8.1E-07	-7.2E-10	7.11E-10	-1.3E-08
	-0.00037	-1.5E-08	-1.3E-08	6.92E-07
X'G times Inverse of XX	0.526486	7.05E-06	-8.1E-07	-0.00037
[X'G times Inverse of XX] times XG	0.526486			
t Statistic	2.119905			
Predicted Y (YHat)	1470.276			
For Average Predicted Y (Yhat)				
Interval Half Width	12181.66			
Confidence Interval Lower Limit	-10711.4			
Confidence Interval Upper Limit	13651.94			
For Individual Response Y				
Interval Half Width	20742.43			
Prediction Interval Lower Limit	-19272.2			
Prediction Interval Upper Limit	22212.7			

PHStat User Note:

Enter the values for the given X's in the shaded range B5:B7. (You can change these values interactively at any time.)

To delete this note: Select this note and then select Edit | Cut.

Figure 12.13 illustrates the new template in PHStat with the values for the three independent variables included in the spreadsheet.

a. Notice that when the land value for a given property is $15,000, the improvement value is $50,000, and the property has 1,800 square feet, the estimate of the mean sale price for the residential property is $79,061.43 which is the same in the textbook.

b. In addition, the corresponding 95% confidence interval for the estimated mean sale price is ($73,380.71, $84,742.15).

c. The corresponding 95% prediction interval for the predicted sale price is ($61,337.83, $96,785.03)

173

Figure 12.13

Confidence and Prediction Estimate Intervals				
Confidence Level	95%			
	1			
Land Value x1 given value	15000			
Improvements x2 given value	50000			
Area x3 given value	1800			
X'X	20	184260	706226	27667
	184260	2.25E+09	7.65E+09	2.88E+08
	706226	7.65E+09	2.94E+10	1.08E+09
	27667	2.88E+08	1.08E+09	42394501
Inverse of X'X	0.526486	7.05E-06	-8.1E-07	-0.00037
	7.05E-06	4.18E-09	-7.2E-10	-1.5E-08
	-8.1E-07	-7.2E-10	7.11E-10	-1.3E-08
	-0.00037	-1.5E-08	-1.3E-08	6.92E-07
X'G times Inverse of X'X	-0.07555	7.56E-06	9.61E-07	1.59E-05

		PHStat User Note:
[X'G times Inverse of X'X] times XG	0.114493	
t Statistic	2.119905	
Predicted Y (YHat)	79061.43	Enter the values for the given X's in the shaded range B5:B7. (You can change these values interactively at any time.)
For Average Predicted Y (Yhat)		
Interval Half Width	5680.719	
Confidence Interval Lower Limit	73380.71	
Confidence Interval Upper Limit	84742.15	
		To delete this note:
For Individual Response Y		Select this note and then
Interval Half Width	17723.6	select Edit | Cut.
Prediction Interval Lower Limit	61337.83	
Prediction Interval Upper Limit	96785.03	

12.6   Model Building: Interaction Models

In this section of the text, the relationship between the mean of y and the independent variable, x, does depend on the values of the remaining x's held fixed. PHStat can provide limited analysis of the cross products of two or more independent variables. Although multidimensional graphing is beyond the scope of Excel and PHStat, it is possible to use the previously developed tests in PHStat to successfully navigate through this section of the textbook. Please refer to the following steps.

Step 1: Open a new workbook in Excel and be sure the cursor is located in the first cell of the spreadsheet.

Step 2: Enter the data manually or open the file containing the data.

Step 3: Once the data is opened, create a new column for the cross product of two or more x's.

Step 4: Do this by entering in a multiplication equation and pasting it down the column where the variable is located.

Step 5: PHStat-Regression-Multiple Regression

Step 6: Enter the y variable and x variable cell range

Step 7: Click First cell in both ranges contain labels

Step 8: Click Regression Statistics Table and ANOVA and Coefficients Table

Step 9: Click OK to finish.

Note that PHStat will not replicate the graphs but will replicate the regression statistics and ANOVA statistics for analysis. Please refer to the following example.

Example 12.5 Please refer to Example 12.5 (page 685) from the textbook.

Suppose the collector of grandfather clocks, having observed many auctions, believes that the rate of increase of the auction price with age will be driven upward by a large number of bidders. Consequently, the interaction model is proposed:
$$E(y) = \beta_0 + \beta_1 x_1 + \beta_2 x_2 + \beta_3 x_1 x_2 + \epsilon$$
The 32 data points listed in Table 12.2 of the text were used to fit the model with interaction.

a. Test the overall utility of the model using the global F test at $\alpha=0.05$.
b. Test the hypothesis that the price-age slope increases as the number of bidders increases.
c. Estimate the change in auction price of a 150-year-old grandfather clock, y, for each additional bidder.

Solution:

Please refer to the following steps in order to complete this problem.

Step 1: Before answering this question, the data presented in Table 12.2 needed to be modified. Please refer to Table 12.3. Notice that a new column, A1, was created. In this column, the product of x_1 and x_2 was created. Now the multiple regression may be run as normal.

Table 12.3

	A	B	C	D
1	x1*x2	Age x1	# of bidders x2	Auction price,y
2	1651	127	13	1235
3	1380	115	12	1080
4	889	127	7	845
5	1350	150	9	1522
6	936	156	6	1047

Step 2: Select PHStat-Regression-Multiple Regression

Step 3: Y variable cell range is D1:D33

Step 4: X variable cell range is A1:C33

Step 5: Click First cell in both ranges contain label

Step 6: Select Regression Statistics and ANOVA and Coefficients Table

Step 7: Select Confidence and Prediction Interval Estimates

Step 8: Select a 95% interval as specified by the problem.

Step 9: Click OK to finish.

The data, once entered, should appear as in Figure 12.14. Once the multiple regression is run, the output generated in Figure 12.15 resembles the MINITAB printout located within the textbook.

Figure 12.14

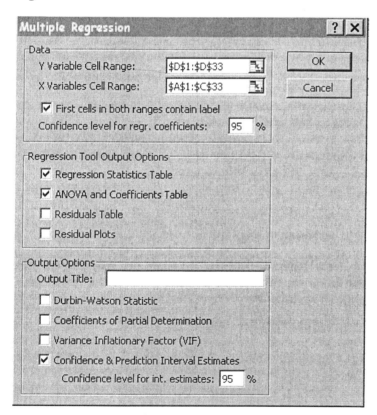

Using this output, the problem can now be completed.
   a. The F statistic is 193.0410958 and significance F is zero. Thus, the model is a statistically useful predictor of auction price.

176

b. The t-statistic for the interaction parameter, β_3, is 6.11 with a p-value of zero. Thus, as the number of bidders increase, the rate of change of the mean price with age increases.
c. The estimated rate of change of y for a unit increase in x_2 is y = -93.26 + 1.30(150) = 101.74. These values for the coefficient of the intercept and the x1*x2 are rounded to reflect the results in the text.

Figure 12.15

Regression Analysis

Regression Statistics	
Multiple R	0.9766682
R Square	0.9538809
Adjusted R Square	0.9489395
Standard Error	88.914512
Observations	32

ANOVA

	df	SS	MS	F	Significance F
Regression	3	4578427.367	1526142	193.0411	8.34956E-19
Residual	28	221362.1332	7905.79		
Total	31	4799789.5			

	Coefficients	Standard Error	t Stat	P-value	Lower 95%	Upper 95%
Intercept	320.45799	295.141285	1.08578	0.286837	-284.112202	925.028188
x1*x2	1.2978458	0.212332598	6.11232	1.353E-06	0.862901724	1.73278992
Age x1	0.8781425	2.03215593	0.43212	0.6689613	-3.28454492	5.04082987
# of bidders x2	-93.264824	29.89161615	-3.1201	0.0041646	-154.495093	-32.034556

12.7   Model Building: Quadratic and Other Higher-Order Models

In this section, models that allow for curvature in relationships will be developed. PHStat and Excel may be used to construct second-order models through the creation of additional columns to represent squared or cubed terms of x, the independent variable. Please refer to the first chapter to remember how to create columns in Excel and how to write formulas for squaring or cubing terms. The following example will be used how to build higher-order models.

Example 12.7 Please refer to Example 12.7 (page 698) in the textbook

In all-electric homes, the amount of electricity expended is of interest to consumers, builders, and groups involved with energy conservation. Suppose we wish to investigate the monthly electrical usage, y, in all-electric homes and its relationship to the size, x, of the home. Moreover, suppose we think that monthly electrical usage in all-electric homes is related to the size of the home by the quadratic model

$$y = \beta_0 + \beta_1 x_1 + \beta_2 x^2 + \epsilon$$

To fit the model, the values of y and x are collected for 10 homes during a particular month. The data are shown in Table 12.4. Answer parts a through f in the text by generating output in PHStat.

177

Table 12.4

	A2	▾	=	=(B2)^2

	A	B	C
1	x-square	Size of Home x	Monthly Usage y
2	1664100	1290	1182
3	1822500	1350	1172
4	2160900	1470	1264
5	2560000	1600	1493
6	2924100	1710	1571
7	3385600	1840	1711
8	3920400	1980	1804
9	4972900	2230	1840
10	5760000	2400	1956
11	8584900	2930	1954

Solution:

Notice that Table 12.4 displays the data with a new column inserted before the independent variable to represent the squared term of x. For example, the square of 1,290 square feet is 1,664,100 square feet. Please refer to the following steps to complete Example 12.6.

Step 1: Open a new workbook and be sure the cursor is located in the first cell of the spreadsheet.

Step 2: Input the data or open the correct file where the data is located.

Step 3: Insert a column before the location of the independent variable. In this case, **highlight column A** and then **select Insert-Column** from the top of the tool bar.

Step 4: Label column A with a descriptive name. In this case, the label is x-square to represent the quadratic term.

Step 5: Place the cursor in the second cell of column A (A2), click the = sign on the equation bar and input "=(B1)^2. Hit Enter and notice that the square of 1,290 (1,664,100) is located in A2. Please refer to Table 12.4 to check the equation.

Step 6: Drag the equation down throughout the column.

Step 7: Select PHStat-Regression-Multiple Regression

Step 8: Y-variable cell range is C1:C11

Step 9: X-variable cell range is A1:B11

Step 10: Select first cell in both ranges contain a label.

Step 11: Select a confidence interval of 99% since the problem specifies interpreting the model when $\alpha = 0.01$.

Step 12: Select Regression Statistics Table and ANOVA and Coefficients Table

178

Step 13: Enter an output title if desired. In this case the output title is "**Solution for Example 12.6 from the text.**"

Step 14: Click OK to finish.

Figure 12.16

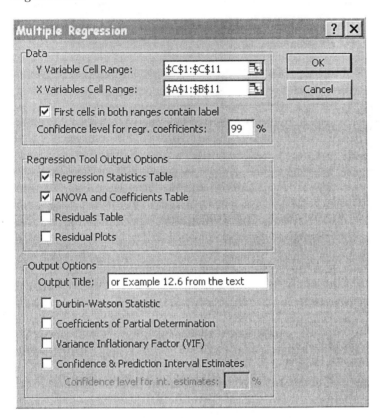

The output generated in Figure 12.17 replicates the output found in figure 12.16 of the text. Notice that the coefficients, R^2, and ANOVA statistics are the same. This data may be used to answer the specific questions outlined by the problem.

Figure 12.17

Solution for Example 12.6 from the text

Regression Statistics	
Multiple R	0.9909011
R Square	0.981885
Adjusted R Square	0.9767093
Standard Error	46.801333
Observations	10

ANOVA

	df	SS	MS	F	Significance F
Regression	2	831069.546	415534.8	189.7103	8.0008E-07
Residual	7	15332.5536	2190.365		
Total	9	846402.1			

	Coefficients	Standard Error	t Stat	P-value	Lower 95%	Upper 95%	Lower 99.0%	Upper 99.0%
Intercept	-1216.1439	242.806369	-5.008698	0.00155	-1790.2893	-641.99847	-2065.84	-366.448
x-square	-0.00045	5.9077E-05	-7.617907	0.000124	-0.0005897	-0.0003103	-0.00066	-0.00024
Size of Home x	2.3989302	0.2458356	9.75827	2.51E-05	1.81762177	2.9802386	1.538633	3.259227

12.8   Sample Exercises

The following exercises were selected from the textbook to test the reader's understanding of PHStat and its usage within the chapter. Try to replicate the output in PHStat for the following problems.

Problem 12.19, Incomes of Mexican Street Vendors, page 680

Detailed interviews were conducted with over 1000 street vendors in the city of Pueblo, Mexico in order to study the factors influencing vendors' incomes (*World Development*, Feb. 1998). Vendors were defined as individuals working in the street, and included vendors with carts and stands on wheels and excluded beggars, drug dealers, and prostitutes. The researchers collected data on gender, age, hours worked per day, and annual earnings, and education level. A subset of these data appears in the table in the textbook. Write a first-order model for the mean annual earnings, $E(y)$, as a function of age (x_1) and hours worked (x_2).

180

Figure 2.18, Solution to Problem 12.19

Regression Analysis

Regression Statistics	
Multiple R	0.763064634
R Square	0.582267636
Adjusted R Square	0.512645575
Standard Error	547.737482
Observations	15

ANOVA

	df	SS	MS	F	Significance F
Regression	2	5018231.543	2509115.772	8.363263464	0.005313599
Residual	12	3600196.19	300016.3492		
Total	14	8618427.733			

	Coefficients	Standard Error	t Stat	P-value	Lower 95%	Upper 95%
Intercept	-20.35201243	652.7453202	-0.031179101	0.975639286	-1442.561866	1401.857841
Age, x1	13.35044655	7.671676501	1.740225432	0.107375655	-3.364700348	30.06559345
Hours worked per Day, x2	243.7144645	63.51173661	3.837313819	0.002363965	105.3342803	382.0946487

Problem 12.30, Incomes of Mexican Street Vendors, page 688

Refer to the World Development (Feb. 1998) study of street vendors' earnings (y). Find the 95% prediction interval for y and a 95% confidence interval for E(y) for a 45-year-old vendor who works 10 hours a day.

Figure 2.19, Solution to Problem 12.30

Confidence and Prediction Estimate Intervals			
Confidence Level	95%		
	1		
Age, x1 given value	45		
Hours worked per Day, x2 given value	10		
X'X	15	527	131
	527	23651	4549
	131	4549	1219
Inverse of X'X	1.420177	-0.00811	-0.12234
	-0.00811	0.000196	0.00014
	-0.12234	0.00014	0.013445
X'G times Inverse of X'X	-0.16836	0.002113	0.018411
[X'G times Inverse of X'X] times XG	0.110835		
t Statistic	2.178813		
Predicted Y (YHat)	3017.563		
For Average Predicted Y (Yhat)			
Interval Half Width	397.3108		
Confidence Interval Lower Limit	2620.252		
Confidence Interval Upper Limit	3414.873		
For Individual Response Y			
Interval Half Width	1257.816		
Prediction Interval Lower Limit	1759.747		
Prediction Interval Upper Limit	4275.379		

PHStat User Note:

Enter the values for the given X's in the shaded range B5:B6. (You can change these values interactively at any time.)

To delete this note: Select this note and then select Edit | Cut.

Problem 12.144, Carp experiment, page 773

Fisheries Science (Feb. 1995) reported on a study of the variables that affect endogenous nitrogen excretion (ENE) in carp raised in Japan. Carp were divided into groups of 2 to 15 fish each according to body weight and each group placed in a separate tank. The carp were then fed a protein-free diet three times daily for a period of 20 days. One day after terminating the feeding experiment, the amount of ENE in each tank was measured. The next table gives the mean body weight (in grams) and ENE amount (in milligrams per 100 grams of body weight per day) for each carp group. The quadratic model $E(y) = \beta_0 + \beta_1 x_1 + \beta_2 x^2$ was used to fit to the data using MINITAB. Use the printout to test $H_0: \beta_2 = 0$ against $H_a: \beta_2 \neq 0$ using $\alpha = 0.10$.

182

Table 12.5

	A	B	C	D
1	Tank	Body Weight,x	x squared	ENE, y
2	1	11.7	136.9	15.3
3	2	25.3	640.1	9.3
4	3	90.2	8136.0	6.5
5	4	213.0	45369.0	6.0
6	5	10.2	104.0	15.7
7	6	17.6	309.8	10.0
8	7	32.6	1062.8	8.6
9	8	81.3	6609.7	6.4
10	9	141.5	20022.3	5.6
11	10	285.7	81624.5	6.0

Figure 12.20, Solution to Problem 12.144, Carp Experiment, page 773

Regression Analysis

Regression Statistics

Multiple R	0.858733141
R Square	0.737422607
Adjusted R Square	0.662400495
Standard Error	2.194326595
Observations	10

ANOVA

	df	SS	MS	F	Significance F
Regression	2	94.65851555	47.32925778	9.82940343	0.009276853
Residual	7	33.70548445	4.815069207		
Total	9	128.364			

	Coefficients	Standard Error	t Stat	P-value	Lower 95%	Upper 95%
Intercept	13.71273728	1.306249375	10.49779433	1.55178E-05	10.62395054	16.80152402
Body Weight,x	-0.101839013	0.028810911	-3.534737733	0.009536966	-0.169965942	-0.033712084
x squared	0.000273478	0.000101599	2.691740332	0.031007513	3.32348E-05	0.000513721

13.1   Introduction

In this chapter, various methods are presented for analyzing categorical data which can be classified according to two factors in contingency tables. Within a multinomial experiment, qualitative variables which result in one of two factors may be analyzed using the binomial property distribution Similarly, the Chi-squared distribution (a type of nonparametric test) may also be used to analyze categorical probabilities for data classified according to two qualitative variables.

The following examples will be used to illustrate techniques in PHStat and Excel.

> Discussion from Section 13.3, Table 13.3, Journal of Marketing Study, page 795
> Example 13.3, Table 13.8, Survey Results regarding Marital Status & Religious Affiliation, page 800
> Sample Exercise #1: Problem 13.28, Seatbelt Study Use, page 807
> Sample Exercise #2: Problem 13.26, JAMA study of heart patients, page 806
> Sample Exercise #3: Problem 13.31, Politics and religion, page 808

13.2   Testing Categorical Probabilities

PHStat provides a method for testing probabilities using a contingency table only and allows the user to specify the number of columns and rows corresponding to that table. Please refer to the following steps in order to calculate these probabilities.

Step 1: Open a new workbook and be sure the cursor is located within the first cell in the upper left-hand corner (A1).

Step 2: Select PHStat- c-Sample tests- Chi Square Test.

Step 3: Within the display menu, **input the level of significance** for the problem. Please refer to Figure 13.1 below.

Step 4: Enter the number of rows of data.

Step 5: Enter the number of columns of data.

Step 6: Select an output title, if desired.

Step 7: Click OK to finish. Refer to Figure 13.2.

Step 8: Enter the data in the observed frequency table and the template will calculate the Chi-squared statistic automatically as well as the expected frequencies.

Figure 13.1

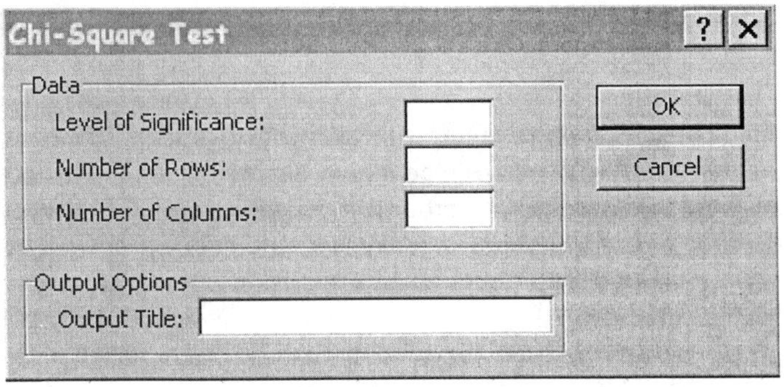

Notice that PHStat provides a template for inserting multinomial count data classified on two scales. From this template, the user should input the **observed frequencies** for the **2x2 contingency table**. The template then calculates the necessary information for the problem. Please refer to Figure 13.2 to review the template. Note the similarity in design to those derived in Chapter 3.

Figure 13.2

Chi-Square Test					
Observed Frequencies:			**Column Variable**		
		Row Variable	C1	C2	Total
		R1	0	0	0
		R2	0	0	0
		Total	0	0	0
Expected Frequencies:			**Column Variable**		
		Row Variable	C1	C2	Total
		R1	#DIV/0!	#DIV/0!	#DIV/0!
		R2	#DIV/0!	#DIV/0!	#DIV/0!
		Total	#DIV/0!	#DIV/0!	#DIV/0!
#DIV/0!					
Level of Significance	0.05				
Number of Rows	2				
Number of Columns	2				
Degrees of Freedom	1				
Critical Value	3.841455				
Chi-Square Test Statistic	#DIV/0!				
p-Value	#DIV/0!				
#DIV/0!					

PHStat User Note:
Enter replacement labels for the row and column variables and categories as well as the observed frequencies in the shaded cells in rows 3, 4, and 5 above.

Note: The #DIV/0! and #NUM! error messages will disappear after the observed frequencies have been entered.

Select this note and then select Edit | Cut to delete this note from the worksheet.

NOTE:
As p-value approaches zero, the Chi-square test statistic may appear as #NUM! due to a Microsoft Excel bug.

Please refer to the following example illustrate how to generate categorical probabilities.

Example 13.1 Please refer to Section 13.3 and Table 13.3 (page 795) and the example that follows from this data

Consider a study in the Journal of Marketing (Fall 1992) on the impact of using celebrities in television advertisements. The researchers investigated the relationship between gender of a viewer and the viewer's brand awareness. Three hundred TV viewers were asked to identify products advertised by male celebrity spokespersons. Test whether the two classifications, gender and brand awareness, are independent.

Table 13.1   Contingency Table for Marketing Example

	Gender of TV Viewer		
	Male	Female	Totals
Identify Product	95	41	136
Could not Identify Product	55	109	164
Totals	150	150	300

Solution:

Please refer to the following steps in order to answer this question.

Step 1: Open a new workbook and be sure the cursor is located within the first cell in the upper left-hand corner (A1).

Step 2: Select PHStat- c-Sample tests- Chi Square Test.

Step 3: Within the display menu, **input the level of significance** for the problem. In this case, select **0.05**.

Step 4: Enter 2 as the number of rows of data.

Step 5: Enter 2 as the number of columns of data.

Step 6: Select an output title, if desired. Refer to Figure 13.3

Figure 13.3

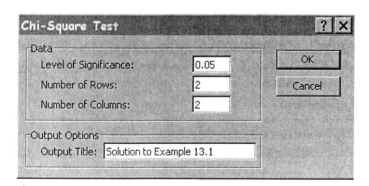

186

Step 7: Click OK to finish.
Step 8: Enter the data in the observed frequency table.

Figure 13.4

Solution for Example 13.1					
			Column Variable		
Observed Frequencies:		Row Variable	Male	Female	Total
		Identify Prod	95	41	136
		Not Indentify Prod	55	109	164
		Total	150	150	300
Expected Frequencies:			**Column Variable**		
		Row Variable	Male	Female	Total
		Identify Prod	68	68	136
		Not Indentify Prod	82	82	164
		Total	150	150	300

Level of Significance	0.05	**PHStat User Note:**
Number of Rows	2	Enter replacement labels for the row and
Number of Columns	2	column variables and categories as well as the
Degrees of Freedom	1	observed frequencies in the shaded cells in
Critical Value	3.841455	rows 3, 4, and 5 above.
Chi-Square Test Statistic	#NUM!	
p-Value	3.78E-10	Note: The #DIV/0! and #NUM! error messages
Reject the null hypothesis		will disappear after the observed frequencies have been entered.
NOTE:		
As p-value approaches zero,		Select this note and then select Edit | Cut to
the Chi-square test statistic		delete this note from the worksheet.
may appear as #NUM! due to		
a Microsoft Excel bug.		

Please compare the answer derived in Table 13.5 of the text with that derived in PHStat.

1. From the expected frequency table generated in PHStat, it is expected that 68 males and 68 females identify products advertised by male celebrity spokespersons. It is expected that 82 males and 82 females do not identify with male celebrity spokespersons.
2. The critical value located in cell B20, for the Chi-squared statistic is 3.84146, which is identical to the critical value, derived in the text.
3. Notice that the Chi-squared statistic, which is 39.22 in the text, is NOT calculated. If the p-value is zero, then PHStat does not calculate the Chi-squared statistics. In this example, the p-value is 3.78E-10 or 0.000000000378 which is essentially zero. In order to calculate the Chi-squared statistic in this case, the formula presented in the textbook would need to be manually calculated within Excel. See Figure 13.5 to derive that formula.
4. Thus, the solution is to reject the null! The conclusion is that the viewer gender and brand awareness are dependent events!

Figure 13.5

	B14	▼	=	=(B4-B10)^2/B10 + (C4-C10)^2/C10 + (B5-B11)^2/B11 + (C5-C11)^2/C11

	A	B	C	D	E	F	G	H
1								
2		**Column Variable**						
3	**Row Variable**	**Male**	**Female**	**Total**				
4	**Identify Prod**	95	41	136				
5	**Not Indentify Prod**	55	109	164				
6	**Total**	150	150	300				
7								
8		**Column Variable**						
9	**Row Variable**	**Male**	**Female**	**Total**				
10	**Identify Prod**	68	68	136				
11	**Not Indentify Prod**	82	82	164				
12	**Total**	150	150	300				
13								
14	Chi-squared statistic	39.22166						

Example 13.3   Please refer to Example 13.3 (page 800) in the text

A social scientist wants to determine whether the marital status (divorced or not divorced) of U.S. men is independent of their religious affiliation (or lack thereof). A sample of 500 U.S. men is surveyed and the results are tabulated in Table 13.8 of the text (Table 13.2 here).
  a.   Test to see whether there is sufficient evidence to indicate that the marital status of men who have been or are currently married is dependent on religious affiliation? Test at $\alpha = 0.01$.
  b.   Plot the data and describe the patterns revealed. Is the result of the text supported by the plot?

Table 13.2   Survey Results (Observed Counts), Example 13.2

		Religious Affiliation					
		A	B	C	D	None	Totals
Marital Status	Divorced	39	19	12	28	18	116
	Never Divorced	172	61	44	70	37	384
	Totals	211	80	56	98	35	500

Solution for part a:

Please refer to the following steps in order to answer this question.

Step 1: Open a new workbook and be sure the cursor is located within the first cell in the upper left-hand corner (A1).

Step 2: Select PHStat- c-Sample tests- Chi Square Test.

Step 3: Within the display menu, **input the level of significance** for the problem. In this case, select **0.01**.

Step 4: Enter 2 as the number of rows of data.

Step 5: Enter 5 as the number of columns of data.

188

Step 6: Select an output title, if desired. Please refer to Figure 13.6

Step 7: Click OK to finish.

Step 8: Change the observed frequency table to the numbers reflected in Table 13.2. Refer to Figure 13.6 for help. Notice that when the data are inputted into the observed frequency table, the template calculates the necessary expected frequencies and Chi-squared statistics.

Figure 13.6

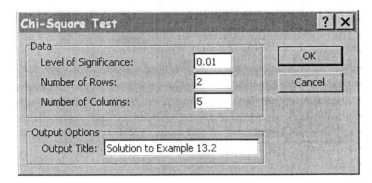

Please refer to the data displayed in Figure 13.7 and compare these results to Figure 13.4 of the text.

1. The expected frequencies generated by PHStat are comparable to those generated in Figure 13.4 in SAS.
2. The Chi-squared statistic is **7.13546 and the critical value is 13.2767**.
3. **Therefore, the null hypothesis cannot be rejected and marital status of U.S. men does not depend on their religious affiliation.**

Figure 13.7

Solution for Example 13.2 from text

Observed Frequencies:

Row variable	Column variable					
	A	B	C	D	None	Total
Divorced	39	19	12	28	18	116
Never Divorced	172	61	44	70	37	384
Total	211	80	56	98	55	500

Expected Frequencies:

Row variable	Column variable					
	A	B	C	D	None	Total
Divorced	48.952	18.56	12.992	22.736	12.76	116
Never Divorced	162.048	61.44	43.008	75.264	42.24	384
Total	211	80	56	98	55	500

Level of Significance	0.01
Number of Rows	2
Number of Columns	5
Degrees of Freedom	4
Critical Value	13.2767
Chi-Square Test Statistic	7.13546
p-Value	0.1289
Do not reject the null hypothesis	

NOTE:
As p-value approaches zero,
the Chi-square test statistic
may appear as #NUM! due to
a Microsoft Excel bug.

PHStat User Note:
Enter replacement labels for the row and
column variables and categories as well as
the observed frequencies in the shaded
cells in the Observed Frequencies table
that starts in row 3.

Note: The #DIV/0! and #NUM! error
messages will disappear after the
observed frequencies have been entered.

Select this note and then select Edit | Cut
to delete this note from the worksheet.

Solution for part b:

To create a bar chart in Excel for the data below refer to the following steps.

Step 1: Recopy the observed frequency table into a clean worksheet.

Step 2: Calculate the percentages of the number of men in each religious affiliation category. For example, in category A: 39/211 or 18.48% is the percentage of men in category A that are divorced. Please refer to Figure 13.8 to derive the equations.

Figure 13.8

B9		=	=B3/B5			
Name Box	**B**	**C**	**D**	**E**	**F**	**G**
1			Column variable			
2 Row variable	**A**	**B**	**C**	**D**	**None**	**Total**
3 Divorced	39	19	12	28	18	116
4 **Never Divorced**	172	61	44	70	37	384
5 Total	211	80	56	98	55	500
6						
7						
8 Row variable	A	B	C	D	None	
9 Divorced	0.184834	0.2375	0.214286	0.285714	0.327273	
10 Never Divorced	0.815166	0.7625	0.785714	0.714286	0.672727	
11 Total						

Step 3: Select Insert-Chart-Column

Step 4: In Chart Wizard, Step 1 of 4, select the default chart as in Figure 13.9. Then **click Next**.

Figure 13.9

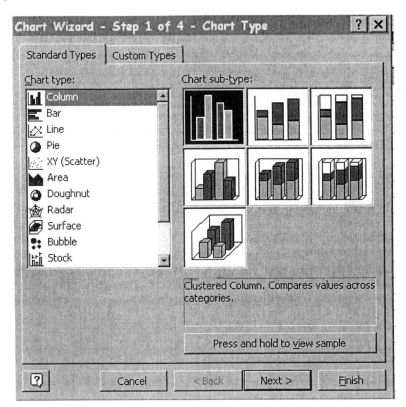

Step 5: In Chart Wizard, Step 2 of 4, select the data range, including the labels A, B, C, D, and None from the table. Be sure the data are **grouped by columns. Click Next.** See Figure 13.10 here.

191

Figure 13.10

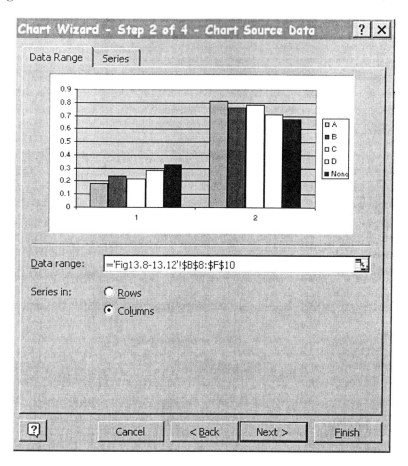

Step 6: In Chart Wizard, Step 3 of 4, type in a title such as Bar Graph for Example 13.2 in the text.

Step 7: Label the x axis religious affiliation

Step 8: Label the y axis: percentage of divorced and non divorced men. Click next.

Figure 13.11

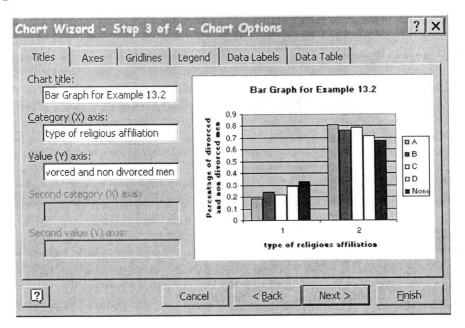

Step 9: In Chart Wizard, Step 4 of 4, insert in the current worksheet.

Figure 13.12

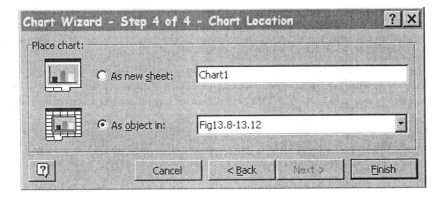

Step 10: Click Finish.

Please notice how the graph differs in appearance than the one presented in Figure 13.5 of the text. The legend in Excel defines, by colors, the religious affiliations of divorced and never divorced men. The first five bars cluster the divorced statistics in one chart; the last five bar clusters represent the never divorced statistics.

Figure 13.13

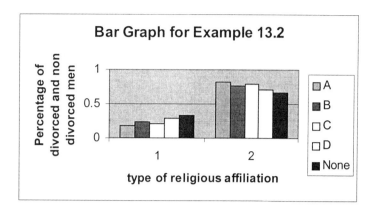

Bar Graph for Example 13.2

13.3　Sample Exercises

The following exercises were selected from the textbook to test the reader's understanding of PHStat and its usage within the chapter. Try to replicate the output in PHStat for the following problems.

Problem 13.28, Seatbelt Study Use, page 807

The *American Journal of Public Health* (July 1995) reported on a population-based study of trauma in Hispanic children. One of the objectives of the study was to compare the use of protective devices in motor vehicles used to transport Hispanic and non-Hispanic white children. On the basis of data collected from the San Diego County Regionalized Trauma System, 792 children treated for injuries sustained in vehicular accidents were classified according to ethnic status (Hispanic or non-Hispanic white) and seatbelt usage (worn or not worn). The data are summarized in the table below. Compare the two sample proportions and determine if the true population proportions differ. Conduct a test to determine whether seatbelt usage in motor vehicle accidents depends on ethnic status in the San Diego County Regionalized Trauma System. Use $\alpha = 0.01$

	Hispanic	Non-Hispanic	Whites
Seatbelts worn	31	148	179
Seatbelts not worn	283	330	613
Totals	314	478	792

Figure 13.14

Solution for Problem 13.24 from text				
Observed Frequencies:			**Column variable**	
	Row variable	Hispanic	on-Hispan	Total
	Seatbelts worn	31	148	179
	Seatbelts not worn	283	330	613
	Total	314	478	792
Expected Frequencies:			**Column variable**	
	Row variable	Hispanic	on-Hispan	Total
	Seatbelts worn	70.96717	108.0328	179
	Seatbelts not worn	243.0328	369.9672	613
	Total	314	478	792

Level of Significance	0.01	**PHStat User Note:**
Number of Rows	2	Enter replacement labels for the row and
Number of Columns	2	column variables and categories as well as
Degrees of Freedom	1	the observed frequencies in the shaded
Critical Value	6.634891	cells in the Observed Frequencies table
Chi-Square Test Statistic	#NUM!	that starts in row 3.
p-Value	3.88E-12	
Reject the null hypothesis		Note: The #DIV/0! and #NUM! error
		messages will disappear after the
NOTE:		observed frequencies have been entered.
As p-value approaches zero,		
the Chi-square test statistic		Select this note and then select Edit | Cut
may appear as #NUM! due to		to delete this note from the worksheet.
a Microsoft Excel bug.		

Problem 13.26, JAMA study of heart patients, page 806

The *Journal of the American Medical Association* (Apr. 18, 2001) published the results of a study of alcohol consumption in-patients suffering from acute myocardial infarction (AMI). The patients were classified according to average number of alcoholic drinks per week and whether or not they had congestive heart failure. A summary of the results for 1,913 AMI patients is shown in the table. Find the sample proportion of abstainers with congestive heart failure, moderate drinkers with heart failure, and heavy drinkers with heart failure. Does it appear that the proportion of AMI patients with congestive heart failure depends on alcohol consumption?

Congestive Heart Failure	Abstainers	Less than 7 per week	7 or more per week
Yes	146	106	29
No	750	590	292
Totals	896	696	321

195

Figure 13.15

Solution for Problem 13.22 from text					
Observed Frequencies:			Column variable		
	Row variable	Abstainers	Less than 7 per week	7 or more drinks	Total
	Yes	146	106	29	281
	No	750	590	292	1632
	Total	896	696	321	1913
Expected Frequencies:			Column variable		
	Row variable	Abstainers	Less than 7 per week	7 or more drinks	Total
	Yes	131.61317	102.2352326	47.15159435	281
	No	764.38683	593.7647674	273.8484056	1632
	Total	896	696	321	1913

Level of Significance	0.05
Number of Rows	2
Number of Columns	3
Degrees of Freedom	2
Critical Value	5.991476
Chi-Square Test Statistic	10.19667
p-Value	0.006107
Reject the null hypothesis	

NOTE:
As p-value approaches zero, the Chi-square test statistic may appear as #NUM! due to a Microsoft Excel bug.

PHStat User Note:
Enter replacement labels for the row and column variables and categories as well as the observed frequencies in the shaded cells in the Observed Frequencies table that starts in row 3.

Note: The #DIV/0! and #NUM! error messages will disappear after the observed frequencies have been entered.

Select this note and then select Edit | Cut to delete this note from the worksheet.

Problem 13.31, Politics and religion, page 808.

University of Maryland professor Ted R. Gurr examined the political strategies used by ethnic groups worldwide in their fight for minority rights (*Political Science & Politics*, June 2000). Each in a sample of 275 ethnic groups was classified according to world region and highest level of political action reported. The data are summarized in the accompanying contingency table. Conduct a test at $\alpha = 0.10$ to determine whether political strategy of ethnic groups depends on world region. Support your answer with a graph.

World Region	No Political Action	Mobilization	Terrorism, Rebellion
Latin American	24	31	7
Post-Communism	32	23	4
South, SE, and East Asia	11	22	26
Africa/Middle East	39	36	20

Figure 13.16

Solution for Problem 13.25					
Observed Frequencies:			Column variable		
	Row variable	No politic	lobilizatio	rrorism, e	Total
	Latin Am	24	31	7	62
	Post-Communism	32	23	4	59
	South, SE, East Asia	11	22	26	59
	Africa, Middle East	39	36	20	95
	Total	106	112	57	275
Expected Frequencies:			Column variable		
	Row variable	No politic	lobilizatio	rrorism, e	Total
	Latin Am	23.89818	25.25091	12.85091	62
	Post-Communism	22.74182	24.02909	12.22909	59
	South, SE, East Asia	22.74182	24.02909	12.22909	59
	Africa, Middle East	36.61818	38.69091	19.69091	95
	Total	106	112	57	275

Level of Significance	0.1
Number of Rows	4
Number of Columns	3
Degrees of Freedom	6
Critical Value	10.64464
Chi-Square Test Statistic	35.30602
p-Value	3.59E-06
Reject the null hypothesis	

NOTE:
As p-value approaches zero,
the Chi-square test statistic
may appear as #NUM! due to
a Microsoft Excel bug.

PHStat User Note:

Enter replacement labels for the row and column variables and categories as well as the observed frequencies in the shaded cells in the Observed Frequencies table that starts in row 3.

Note: The #DIV/0! and #NUM! error messages will disappear after the observed frequencies have been entered.

Select this note and then select Edit | Cut to delete this note from the worksheet.

197

Figure 13.17

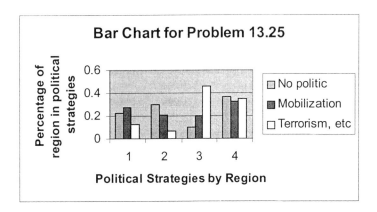

14.1   Introduction

In Chapter 14 of the text, McClave and Sincich present techniques for analyzing data in which there are no underlying assumptions about the probability distribution of the population. In addition, nonparametric tests will be used for data that are nonnumerical and can be ranked. PHStat can be used to run only two of the tests presented in the chapter:

1. Wilcoxon Signed Rank Test for Paired Difference Experiments
2. Kruskal-Wallis H-Test for a Completely Randomized Design

All other techniques presented in the chapter may be calculated manually in Excel by referring to the equations delineated within the text.

The following examples will be used to illustrate techniques in PHStat and Excel.

> Example 14.1, Manufacturer of CD Players, page 825
> Example 14.2, Drug Reaction Times, page 833
> Discussion from Section 14.5, Table 14.7, Number of Available Beds, page 848
> Sample Exercise #1: Problem 14.97, Spending on IS Technology, page 872
> Sample Exercise #2: Problem 14.93, Methods of Teaching Reading, page 871

14.2   Single Population Inferences: The Sign Test

The z and the t statistics for testing hypotheses about a population mean or population proportion may now be used to test hypotheses about the central tendency of a nonnormal distribution. In this case, the population median may be used a proxy for the population mean since it is less susceptible to skewness. Excel 97/2000 may be used to approximate the z or t statistics necessary to solve these problems.

Example 14.1   Please refer to Example 14.1(page 825) in the text.

A manufacturer of compact disk (CD) players has established that the median time to failure for its players is 5,250 hours of utilization. A sample of 20 CDs from a competitor is obtained, and they are continuously tested until each fails. The 20 failure times range from 5 hours (a "defective" player) to 6,575 hours, and 14 of the 20 exceed 5,250 hours. Is there evidence that the median failure time of the competitor differs from 5,250 hours? Use $\alpha=0.10$.

Solution:

Please refer to the following steps in order to solve Example 14.1

Step 1: Open a new worksheet and be sure that the cursor is located in the first cell in the upper left-hand corner of the spreadsheet.

Step 2: Click the "=" sign within the equation editor.

Step 3: Type in the following equation "=((14-.5) -.5*20)/(.5*SQRT(20))

Step 4: Equation editor provides an answer of 1.565

Step 5: Compare this to the critical value of 1.645 derived in the textbook

Step 6: Since 1.565 < 1.645, the null hypothesis cannot be rejected. Thus, the competitor's CDs have a median failure time that differs from 5,250 hours.

Figure 14.1

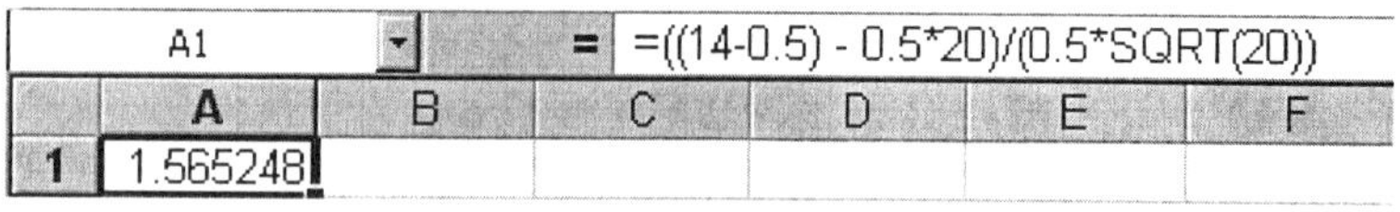

A1	▼	**=**	=((14-0.5) - 0.5*20)/(0.5*SQRT(20))		
A	B	C	D	E	F
1 1.565248					

14.3 Comparing Two Populations: The Wilcoxon Rank Sum Test for Independent Samples

The Wilcoxon rank sum test is used to test the hypothesis that the probability distribution associated with two independent populations are equivalent. In this case, the researcher may not be sure if the population distribution is normal or not; therefore, the t-test discussed in previous chapters may not be used. Remember from the text that in order to use the Wilcoxon rank sum test the distributions are assumed to be continuous; however, there is no specification on the shape or type of probability distribution.

Using PHStat, refer to the following steps regarding the use of the Wilcoxon rank sum test.

Step 1: Open a new worksheet and be sure the cursor is located in the upper left-hand cell of the spreadsheet.

Step 2: Open the data file for the specific problem or enter the data manually.

Step 3: Select PHStat-Two-Sample Tests - Wilcoxon Rank Sum Test

Step 4: Enter the level of significance. By default, this is set at 0.05.

Step 5: Enter the Population 1 Sample Cell Range. This may be done manually or by dragging and clicking on the relevant range.

Step 6: Enter the Population 2 Sample Cell Range

Step 7: Select first cells in both range contain labels if applicable.

Step 8: Select the Test Option: Two-tailed, Upper-Tail, or Lower-Tail test.

Step 9: Enter an Output Title, if desired.

Step 10: Click OK to finish.

Figure 14.2

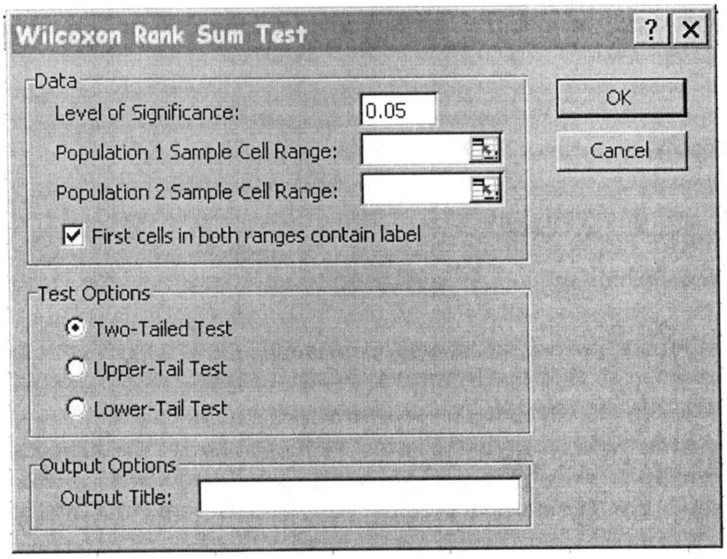

Figure 14.2 presents the Wilcoxon Rank Sum Test menu in PHStat. Notice that only one test option may be selected at one time unlike SAS or other statistical packages. Please refer to the following example, which illustrates the use of the Wilcoxon Rank Sum Test.

Example 14.2 Please refer to Example 14.2 (page 833).

Do the data given in Table 14.2 of the text (Table 14.1 here) provide sufficient evidence to indicate a shift in the probability distributions for drugs A and B? That is, does the probability distribution corresponding to drug A lie either to the right or left of the probability distribution corresponding to drug B? Test at the 0.05 level of significance.

Table 14.1

	A	B	C	D	E	F
1	Reaction Times of Subjects Under the Influence of Drug A or B					
2	Drug A	Drug B				
3	1.96	2.11				
4	2.24	2.43				
5	1.71	2.07				
6	2.41	2.71				
7	1.62	2.5				
8	1.93	2.84				
9		2.88				

Solution:

Please refer to the following steps in order to answer Example 14.2.
Step 1: Open a new worksheet and be sure the cursor is located in the upper left-hand cell of the spreadsheet.

Step 2: Open the data file for the specific problem or enter the data manually.

Step 3: Select PHStat-Two-Sample Tests - Wilcoxon Rank Sum Test

Step 4: Enter the level of significance as 0.05

Step 5: Enter the Population 1 Sample Cell Range as A2:A8.

Step 6: Enter the Population 2 Sample Cell Range as B2:B9.

Step 7: Select first cells in both range contain labels.

Step 8: Select the Test Option: Lower Tail.

Step 9: Enter an Output Title, if desired.

Step 10: Click OK to finish.

Figure 14.3

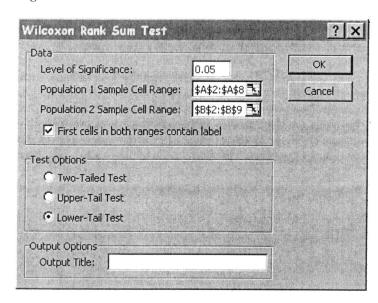

Figure 14.4a, Figure 14.4b, and Figure 14.4c present the output generated by Example 14.2 of the text. PHStat cannot automatically run a test with respect to the direction of the rejection region. Therefore, three individual tests must be computed for
1.   the lower tailed test (Figure 14.4a).

202

Figure 14.4a

Solution for Example 14.2: Lower Tail						
Level of Significance	0.05					
Population 1 Sample						
Sample Size	6					
Sum of Ranks	25					
Population 2 Sample						
Sample Size	7					
Sum of Ranks	66					
Warning: Large-scale approximation formula not designed for small sample sizes.						
Total Sample Size n	13					
$T1$ Test Statistic	25					
$T1$ Mean	42					
Standard Error of $T1$	7					
Z Test Statistic	-2.428571429					
Lower-Tail Test						
Lower Critical Value	-1.644853					
p-value	0.007579219					
Reject the null hypothesis						

2. the upper tailed test (Figure 14.4b)

Figure 14.4b

Solution for Example 14.2: Upper Tail					
Level of Significance	0.05				
Population 1 Sample					
Sample Size	6				
Sum of Ranks	25				
Population 2 Sample					
Sample Size	7				
Sum of Ranks	66				
Warning: Large-scale approximation formula not designed for small sample sizes.					
Total Sample Size n	13				
$T1$ Test Statistic	25				
$T1$ Mean	42				
Standard Error of $T1$	7				
Z Test Statistic	-2.42857				
Upper-Tail Test					
Upper Critical Value	1.644853				
p-value	0.992421				
Do not reject the null hypothesis					

3. the two tailed test (Figure 14.4c)

Figure 14.4c

Solution for Example 14.2: Two Tailed						
Level of Significance	0.05					
Population 1 Sample						
Sample Size	6					
Sum of Ranks	25					
Population 2 Sample						
Sample Size	7					
Sum of Ranks	66					
Warning: Large-scale approximation formula not designed for small sample sizes.						
Total Sample Size n	13					
$T1$ Test Statistic	25					
$T1$ Mean	42					
Standard Error of $T1$	7					
Z Test Statistic	-2.42857					
Two-Tailed Test						
Lower Critical Value	-1.95996					
Upper Critical Value	1.959961					
p-value	0.015158					
Reject the null hypothesis						

In this case, from Figure 14.4a, it can be concluded that the probability distributions for drug A and drug B are not identical. The test statistic (t=25) and the p-value (0.007) are highlighted in order to derive this conclusion.

Notice that a warning message is generated explaining that the sample utilized are not large enough to yield complete statistics to the user. Please refer to Figure 14.6 in the text which provides the SAS printout for the example. The z-statistic and p-value for the normal approximation as well as the test statistic are similar. Please refer to the end of the chapter exercises for an example in this procedure.

14.4 The Kruskal-Wallis H-Test for a Completely Randomized Design

According to the text, the Kruskal-Wallis test is used to compare the populations where no assumptions are made concerning the population probability distributions. Please refer to the following steps to use the Kruskal-Wallis test in PHStat.

Step 1: Open a new worksheet and be sure the cursor is located in the upper left-hand cell of the spreadsheet.

Step 2: Enter the data manually or open the appropriate data file.

Step 3: Select PHStat-c-Sample Tests-Kruskal-Wallis Rank Test

Step 4: Enter the level of significance.

Step 5: Select the appropriate sample data cell range.

Step 6: Click first cells contain label if appropriate.

Step 7: Enter an Output Title if desired.

Step 8: Click OK to finish.

Figure 14.5 presents the Kruskal-Wallis Rank Test menu. The following example from the text will illustrate this technique.

Figure 14.5

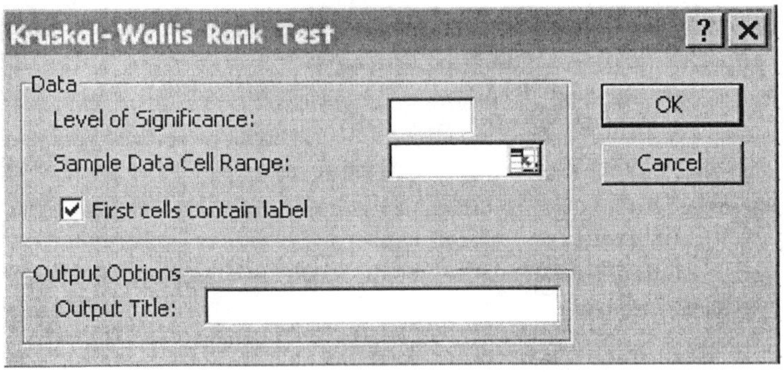

Discussion from Section 14.5 Please refer to Table 14.7 (page 848).

Suppose a health administrator wants to compare the unoccupied bed space for three hospitals located in the same city. She randomly selects 10 different days from the records of each hospital and lists the number of unoccupied beds for each day in Table 14.7 in the text (Table 14.2 here). Use the Kruskal-Wallis H-test to determine if the distributions for the number of beds available at the three hospitals differ. Use a 0.05 level of significance.

Table 14.2

	A	B	C
1	Hospital #1	Hospital #2	Hospital #3
2	6	34	13
3	38	28	35
4	3	42	19
5	17	13	4
6	11	40	29
7	30	31	0
8	15	9	7
9	16	32	33
10	25	39	18
11	5	27	24

Solution:

Please refer to the following steps in order to answer the preceding question.

205

Step 1: Open a new worksheet and be sure the cursor is located in the upper left-hand cell of the spreadsheet.

Step 2: Enter the data manually or open the appropriate data file.

Step 3: Select PHStat-c-Sample Tests-Kruskal-Wallis Rank Test

Step 4: Enter the level of significance as 0.05

Step 5: Select the appropriate sample data cell range: A1:C11

Step 6: Click first cells contain label if appropriate.

Step 7: Enter an Output Title if desired.

Step 8: Click OK to finish.

Figure 14.6

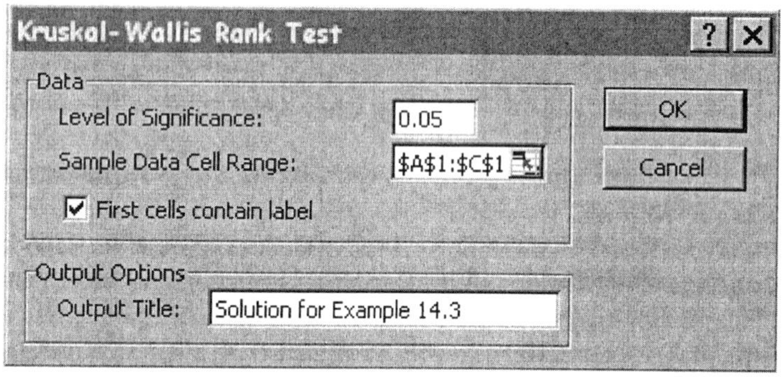

Figure 14.7 presents the output generated using the Kruskal-Wallis test in PHStat. Please compare it to the SPSS printout (Figure 14.11) in the text. The test statistic generated is 6.097419 (this is greater than the critical value of 5.991476). Notice that the test statistic generated here in Figure 14.7 is slightly different than the one presented in Figure 14.11 of the textbook. In addition, the p-value of 0.047 is less than the level of significance of 0.05. Therefore, the null is rejected. Therefore, at least one of the three hospitals tends to have a larger number of unoccupied beds than the others.

Figure 14.7

Solution for Example 14.3	
Level of Significance	0.05
Group 1	
Sum of Ranks	120
Sample Size	10
Group 2	
Sum of Ranks	210.5
Sample Size	10
Group 3	
Sum of Ranks	134.5
Sample Size	10
Sum of Squared Ranks/Sample Size	7680.05
Sum of Sample Sizes	30
Number of groups	3
H Test Statistic	6.097419
Critical Value	5.991476
p-Value	0.04742
Reject the null hypothesis	

14.5   Sample Exercises

The following exercises were selected from the textbook to test the reader's understanding of PHStat and its usage within the chapter. Try to replicate the output in PHStat for the following problems.

Problem 14.97, Spending on IS Technology, page 872

University of Queensland researchers sampled private sector and public sector organizations in Australia to study the planning undertaken by their information systems departments (*Management Science*, July 1996). As part of the process they asked each sample organization how much it had spend on information systems and technology in the previous fiscal year as a percentage of the organization's total revenues. The results are reported in the table. Do the two sampled populations have identical probability distributions or is the distribution for the public sector located to the right of Australia's private sector firm? Test using $\alpha = 0.05$.

Table 14.3

	A	B
1	Private Sector	Public Sector
2	2.58	5.40
3	5.05	2.55
4	0.05	9.00
5	2.10	10.55
6	4.30	1.02
7	2.25	5.11
8	2.50	24.42
9	1.94	1.67
10	2.33	3.33

Figure 14.8

Solution for Problem 14.21: Upper Tail					
Level of Significance	0.05				
Population 1 Sample					
Sample Size	9				
Sum of Ranks	66				
Population 2 Sample					
Sample Size	9				
Sum of Ranks	105				
Warning: Large-scale approximation formula not designed for small sample sizes.					
Total Sample Size n	18				
$T1$ Test Statistic	66				
$T1$ Mean	85.5				
Standard Error of $T1$	11.32475				
Z Test Statistic	-1.72189				
Upper-Tail Test					
Upper Critical Value	1.644853				
p-value	0.957456				
Do not reject the null hypothesis					

Figure 14.9

Solution for Problem 14.21: Lower Tail					
Level of Significance	0.05				
Population 1 Sample					
Sample Size	9				
Sum of Ranks	66				
Population 2 Sample					
Sample Size	9				
Sum of Ranks	105				
Warning: Large-scale approximation formula not designed for small sample sizes.					
Total Sample Size n	18				
$T1$ Test Statistic	66				
$T1$ Mean	85.5				
Standard Error of $T1$	11.32475				
Z Test Statistic	-1.72189				
Lower-Tail Test					
Lower Critical Value	-1.64485				
p-value	0.042544				
Reject the null hypothesis					

Figure 14.10

Soltuion for Problem 14.21: Two Tailed Test					
Level of Significance	0.05				
Population 1 Sample					
Sample Size	9				
Sum of Ranks	66				
Population 2 Sample					
Sample Size	9				
Sum of Ranks	105				
Warning: Large-scale approximation formula not designed for small sample sizes.					
Total Sample Size n	18				
$T1$ Test Statistic	66				
$T1$ Mean	85.5				
Standard Error of $T1$	11.32475				
Z Test Statistic	-1.72189				
Two-Tailed Test					
Lower Critical Value	-1.95996				
Upper Critical Value	1.959961				
p-value	0.085089				
Do not reject the null hypothesis					

Problem 14.93, Methods of Teaching Reading, page 871.

Sixth graders in an elementary school are taught reading by three different methods. Students were randomly assigned to three classes. One class used programmed instruction, a second used standard memorization techniques, and the third used an open classroom approach. The increases in reading levels attained by five students randomly selected from each of the three classes are shown below. Do the data provide sufficient evidence to indicate that the probability distributions of increases in reading level differ for at least two of the methods? Use $\alpha = 0.05$.

Table 14.4

	A	B	C
1	Programmed	Standard	Open
2	0.9	1.0	1.7
3	1.5	0.8	0.5
4	0.7	0.9	1.6
5	1.1	1.2	1.4
6	0.5	1.4	1.0

Figure 14.11

Solution for Problem 14.41 in text	
Level of Significance	0.05
Group 1	
Sum of Ranks	32
Sample Size	5
Group 2	
Sum of Ranks	38.5
Sample Size	5
Group 3	
Sum of Ranks	49.5
Sample Size	5
Sum of Squared Ranks/Sample Size	991.3
Sum of Sample Sizes	15
Number of groups	3
H Test Statistic	1.565
Critical Value	5.991476
p-Value	0.457261
Do not reject the null hypothesis	

Minitab™ Manual

Augustin M. Vukov
University of Toronto

■

Statistics

Tenth Edition

JAMES T. McCLAVE & TERRY SINCICH

Preface

This Minitab Manual is designed to carefully integrate the statistical computing package MINITAB™ into an introductory Statistics course, using Statistics by McClave and Sincich, 10th edition. It provides worked out illustrative examples using data from examples and exercises in the text (data available at www.prenhall.com/mcclave and on the text CD-ROM), which are suitable for solution using Minitab. I have presented a gradual one step at a time approach, teaching just what you need to know for the current topic, rather than bombarding you with too much information too quickly. More ambitious students can easily use Minitab's Help menu to delve more deeply.

There are many statistical software packages available, but Minitab is ideal for an introductory course, due to its ease of use. With Minitab, you are freed from the tedium of long calculations, and instead can focus on what is important – learning the concepts, determining which procedures are relevant, checking their applicability, interpreting the results of the analysis

With Minitab, you issue commands that tell Minitab to do certain calculations or to produce graphs. But, instead of writing down commands, and the consequent need to learn syntax, you will use the *menu driven* approach, i.e. you select the appropriate procedure from the menu selections, and then fill in some required information in dialog boxes. If you make an error in your entries, Minitab will give you direct feedback, with an error message, pointing out the problem or omission.

Minitab is proprietary software, and is not free. You have to purchase it, or use it on computers at your school where it has been purchased or leased. Minitab software is available for various operating systems, but I will only deal with Minitab for Windows. There is a full *Professional version* and a less expensive abbreviated *Student version*. The latter is more than adequate for this course, and will also enable you to perform powerful statistical analyses of data for other courses. If you wish to work at home, you have to purchase (or rent) and install Minitab on your PC or Mac (for use on a Mac, you need Windows emulation software). I recommend that you purchase the Student version, which is available bundled together with the textbook at a *greatly discounted* price. For further information about Minitab, go to www.minitab.com, where you can download a 30-day free demo, or find information about renting Minitab for a semester.

Minitab is continually being updated and improved. Hence, there are different release numbers. This manual is based on Minitab Release 14. The screen captures in this manual were produced with the Student version rather than the Professional version, but any differences are purely superficial.

Augustin Vukov
Toronto, Ontario, Canada
Dec. 29, 2004

Table of Contents

[M/S = Statistics, 10th ed., by McClave and Sincich]

Getting Started

Before doing any of the examples or exercises in the text, let's start with a brief overview and simple first session.

Installing the software on your PC

If you have purchased the student version of Minitab, you have to install it on your computer. Close down all programs on your computer. Insert the CD into your computer's CD-ROM drive. An installation wizard should appear to guide you through the installation process. Keep hitting 'next' until the installation is completed. Choose the default 'Complete' setup rather than the 'Custom' setup (for advanced users), and if you wish, choose to install the Minitab icon to your desktop.

If you have installed the student version of Minitab on your computer, the first step is to open it up, as follows:

Start>Programs>MINITAB 14 Student>MINITAB 14 STUDENT

where the symbol '>' above leads you through the appropriate sequence of menu selections. Or if you installed the Minitab icon on your desktop, just double-click it.

If you are working at a computer at your school, follow log-on instructions given to you at your school for accessing Minitab.

The first view that you will have of Minitab will look like:

There are three key sections of the screen above:

(1) The main menu bar, at the top:

File Edit Data Calc Stat Graph Editor Tools Window Help

If you click on any of these, a submenu will drop down, displaying the available options. Below, I clicked on **Stat** from the main menu bar first, and from the **Stat** pull-down menu, I slid the cursor to **Nonparametrics,** which has it own sub-menu immediately appearing to the side, and from the **Nonparametrics** menu, I slid the cursor over to **Kruskal-Wallace**. Try this yourself.

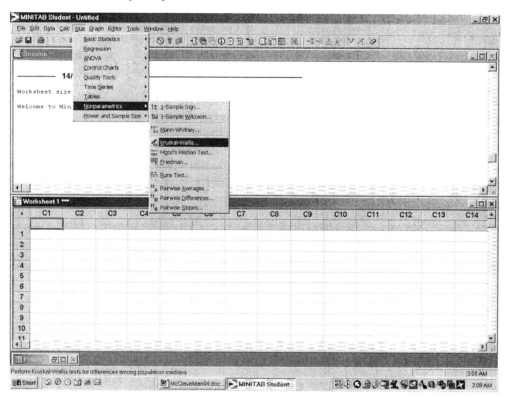

Click on **Kruskal-Wallace**, and up pops the **Kruskal-Wallace** dialog box, shown below.

2

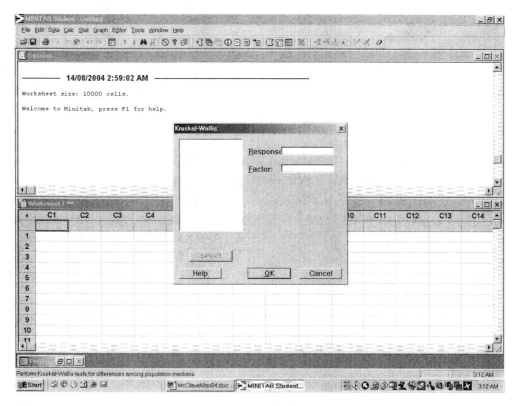

If you wanted to apply the Kruskal-Wallace statistical procedure to some data, you would now fill in the required information in this dialog box (**Response**, **Factor**), and hit **OK**. Since we have not entered any data yet, just click on **Cancel**.

You will learn gradually about the procedures available from this menu, but in brief, the key functions of each main menu item are:

   File: Open and save various types of files. Print windows
   Edit: Copy, cut, paste data and text
   Data: Manipulate the data in the data window
   Calc: Perform basic calculations and transformations on the data. Calculations and random sampling for common probability distributions
   Stat: Execute common statistical procedures
   Graph. Construct various plots of the data
   Editor: Graph, data, and session window editing tools. Enable use of the command language.
   Tools: Calculator, notepad, toolbar display options, …
   Window: Manage the display of the various windows
   Help: Get help on Minitab procedures

(2) The Data (Worksheet) Window, occupying the lower half of the screen
This is where the data sit. At the start, you have a new blank worksheet, named **Worksheet 1**:

3

Data obtained from subjects in a study are entered in a very specific way. The columns represent the variables, and the rows represent the subjects. Suppose that you have weight and height measurements on three subjects. Start by typing in an appropriate name for each column (Weight, Height), directly underneath the column number, but above row **1** as shown. Stick to 16 letters or less, since in some output, longer column names may be truncated to the first 16 letters. Next, type in the data shown below. If you hit *Enter* on your keyboard, after entering 120, you will be thrown into the cell directly below - because of that downward arrow in the upper left corner of the worksheet. Hit *Ctrl + Enter* (i.e. both together) on your keyboard, to move to the top of the next column. You may also just point and click on any cell, to get you there. If you click on the arrow, it will point to the right, and *Enter* will now move you over one column instead of down one row.

(3) The Session Window, occupying the upper half of the screen
This is where the output of your statistical procedures will appear, except for graphs, which pop up in separate new graphical windows. If you select (click, drag, click) from the menu as follows: **Stat>Basic Statistics>Display Descriptive Statistics**, you will see the following dialog box:

4

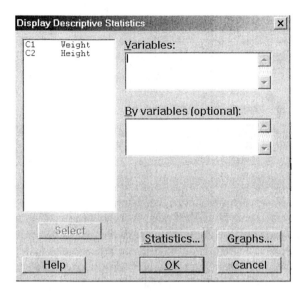

We want to display descriptives for the subjects' weights and heights, so we need to type in the column names under **Variables**. We can type either *C1* or *Weight* to indicate column 1, and then *C2* or *Height* to indicate the second column or variable. But instead of typing, try double-clicking (left mouse button) on 'C1 Weight' at the left. It will be copied over to where the active cursor is blinking, which should be under **Variables**. Repeat for C2. Alternatively, delete your two listed **Variables**, and instead try the following: left-click on C1, hold down and drag to C2, so that both are highlighted. Hit **Select**, and both should appear under **Variables**, if your cursor was blinking there. You now should see:

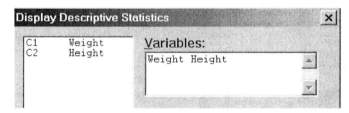

We will indicate the above series of selections, in this manual, by writing:

Stat>Basic Statistics>Display Descriptive Statistics: Variables: *Weight Height*

where bold entries indicate fixed menu and dialog items on display within Minitab, and italicized entries indicate items that you have to fill in, depending on the particular study.

Now, left-click on **OK**, and you get the following output in the Session Window occupying the upper half of your screen:

5

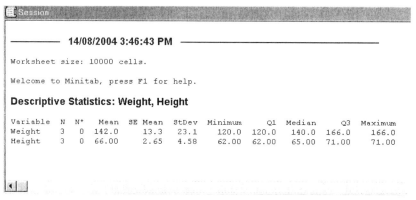

To print this window, click on **File>Print Session Window**. If you do not see this selection, it is because the session window is not currently the *active window*. Clicking on a window makes it *active*. Click on the data window. You will see **File>Print Worksheet**. Click on the session window, and instead you will see **File>Print Session Window**.

Now would also be a good time to check out the printer settings. Click on **File>Print Setup** and check that the selections are as you desire.

Graphical windows: Click on **Graph>Plot** and you will see:

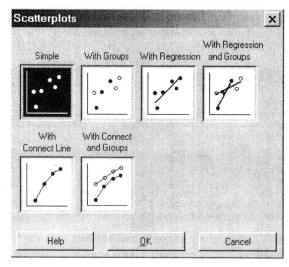

Click **OK** since all we want here is the simple scatterplot. You get another dialog box:

6

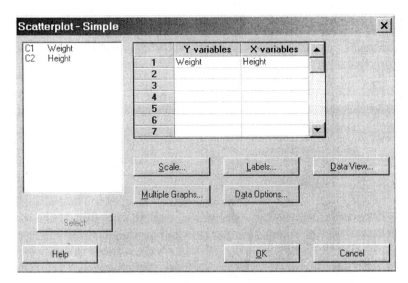

Above, we entered *Weight* and *Height* under **Y** and **X** respectively, then hit **OK**. A new *graphical* window pops up on top of the other windows:

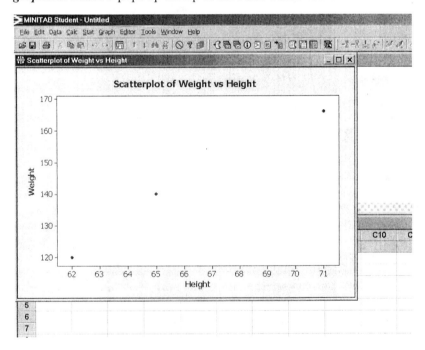

To print out a hard copy of the graph, use **File>Print Graph**.

You may choose to add labels, such as titles, to a graph if you wish. In the dialog box for the graph, look for **Labels...** and click it:

7

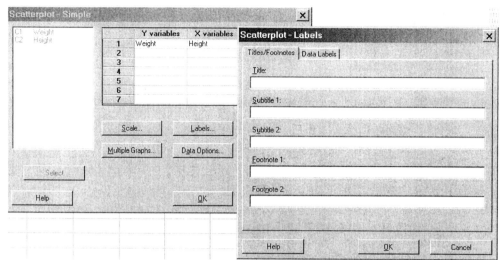

Fill in your desired **Title**, e.g. *Exercise 3a* and perhaps a **Subtitle** *Plot of Weight vs. Height for College Students*. Add additional labels if you wish. You can also label data in the plot, if you click on **Data Labels**. **OK, OK**, to get the plot again, with your labels.

Click anywhere on the data window, and it comes to the top (becomes the active window). Click now on the session window, and the graph appears to be completely gone. It is just hiding! Click on **Window** and you see:

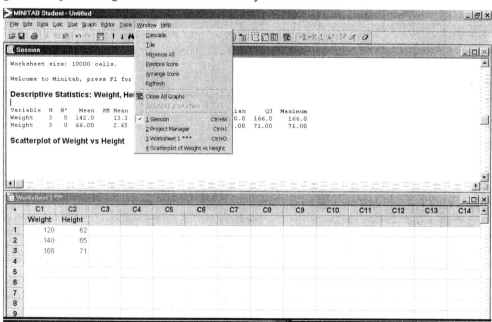

Select **4 Scatterplot of Weight vs Height**, and the window with this graph will be back on top.

Project Manager Window: This is minimized at startup. Look to the bottom left, and you will see it lurking below the data window:

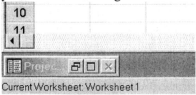

Click on the left button, to restore the window, and you see:

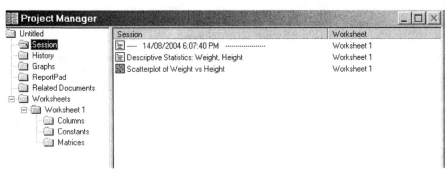

There are various folders displayed to the left:

Session folder which lists the session output you have produced in the session window as well as the graphs (as in screen capture above); useful for managing session window output, e.g. jump to output, delete output, append output to ReportPad.

History folder which lists commands you have used in this session;

Graphs folder for managing graphs;

ReportPad folder, a basic word processing tool for editing reports of your work;

Worksheets folder containing one folder for each open worksheet, where each worksheet folder provides basic info about that worksheet; e.g. click on **Columns** above and you will see the name, count, and type (numeric/text) of each column in worksheet 1.

You may enter data into several different worksheets, but only one worksheet can be the *current* or *active* one (indicated by 3 asterisks). If you wish to switch to another worksheet, you may use **Window** from the main menu bar, to select it. Or you can also proceed using window operations described below.

Window operations: In the upper right corner of the graphical, session, and worksheet windows, you will likely see three square buttons:

These three buttons perform certain operations on the window. To hide (minimize) the window, click on the left one, to maximize its size (or to restore its size), click on the middle one, and to be completely rid of the window, click on the right one. You may also use the icon in the upper *left corner* of a window, to execute any of these operations. Left-click on this icon and a menu of operations will appear.

9

If you wish to move a window, left-click on the upper (title) bar of that window, and drag the window.

For windows that are hidden, you can find and restore them using **Window** from the main menu. Or you can look for the hidden window's title bar, which will be lurking somewhere at the bottom of your screen, possibly hidden behind some windows. After hiding all the windows, I found the following four shortened title bars at the bottom of my screen:

The icon at the left tells you what type of window it is. To restore any one of these, click on the left button, or click anywhere on the bar, for a pop-up menu.

Stored Constants: In the course of doing certain calculations, you may wish to calculate and store for later use, *one single number*, rather than an entire column of data. Do this by writing k1 (k2, etc.) rather than a column designator, C1, C2, etc. for the place of storage requested in the dialog box being used. The **Project Manager** window will list all such stored constants.

Missing Data: Occasionally, some data are missing or known to be erroneous. Suppose that the height of the second listed subject was incorrectly recorded, but the weight measurement is fine. Click on the second row in C2, and hit *Backspace* on your keyboard. An '*' will appear to indicate the missing value (or type '*' from your keyboard).

↓	C1 Weight	C2 Height
1	120	62
2	140	*
3	166	71

Note that if you hit *Delete* instead, you mess up the data set, since the third height measurement of 71 moves up a cell to become the second subject's height.

Saved Data Files

In the example above, we typed the data directly into a blank worksheet. While you can always fall back on this approach, if necessary, the data sets used in the text have already been entered and saved for you. However, there are different formats in which data may be saved. The format used is indicated by attaching an appropriate suffix to the file name. Three common formats, with corresponding suffixes (shown in parentheses) are:

(i) **Minitab Worksheet file (*.mtw)**

This is the standard and most efficient Minitab format for saving data. However, it is not transportable to other operating systems running Minitab, nor is it usable in other statistical software packages. And there may be problems going back and forth between newer and much older versions of Minitab. To open up such a file, just use **File>Open Worksheet: Files of Type: Minitab (*.mtw)**. Select this format when you wish to save data for later analysis using Minitab on your computer (via **File>Save Worksheet**). If copying it for a friend, it may be safer to use one of the other formats listed below.

The data sets included on your text CD-ROM (and at www.prenhall.com/mcclave) were saved in this most convenient format, as well as in *.mtp and *.dat formats described below.

(ii) **Minitab Portable Worksheet file (*.mtp)**
This 'portable' format may be read correctly by any version of Minitab, regardless of the operating system running Minitab. To open up these data files, use **File>Open Worksheet: Files of Type: Minitab Portable (*.mtp)**.

(iii) **Text file (*.dat or *.txt)**
This is a very general format that allows you to use the data file for analysis by any statistical software, but you will have to give Minitab some help in interpreting the information in the file. When you open up such a file, you have to check that the **Options** are set properly in the **File>Open Worksheet: Options** dialog box. If the **Options** are not set properly, you will get a confusing jumble of data, perhaps with multiple entries appearing in each cell, or with data in the row where the column names should be. Often, names for the columns are not included.

Data files used in this manual
The examples and sample exercises in this manual use data sets that have already been saved on the text CD-ROM, in various formats. An icon with the name of the data file appears in your textbook next to the corresponding example/exercise. Retrieve the data file from your CD using **File>Open Worksheet**. In the dialog box, select **Files of Type: Minitab (*.mtw, *.mpj)**. At the top, locate the folder on your CD with the example/exercise data files in Minitab format.

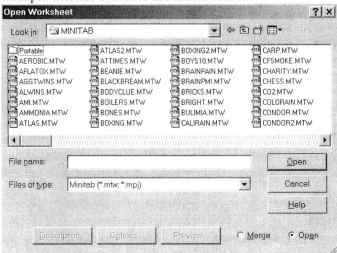

Double-click on the desired data set. Hit **OK** in response to Minitab's query about adding this to the current project. The data will appear in a new worksheet (but in the same *project file*– discussed later).

If opening files saved in the portable Minitab format, proceed similarly, but switch to **Files of type: Minitab Portable (*.mtp)** in the box. If not sure which format was used for saving the data, try **Files of type: All**.

Note that if the files are of one type, and you specify a different type in the dialog box above, no files will be visible!

While there should be no need to open up any saved *text files* during this course, you will almost surely need to use such files in the future, so let's briefly run through this procedure here as well. Once again, you would start with **File>Open Worksheet**. In the dialog box, select **Files of Type: Data (*.dat)**. Find the desired *.dat file and select it, but don't open it just yet! First you need to make sure that Minitab knows how to interpret the file. Click on **Preview** at the bottom of the **Open Worksheet** dialog box. If things don't look right, click on **Options**, change the settings, then **Preview** again. When the **Preview** looks right, you can proceed to **Open** the selected file. The following **Options** selections, e.g., would be appropriate for the *.dat files on your CD:

Above, I chose **Variable Names: None**, since the *.dat files on your CD do not include column names. In this case, after opening up a file, you should write in appropriate names for each column. You can then save the file as a Minitab Worksheet (**File>Save Current Worksheet As**) so that you will not have to bother repeating these worksheet changes the next time you require this data set. Whenever you open up Minitab, you will have to re-set the **File>Open Worksheet: Options**, which go back to the default settings. But once you set the options as above, they will stay unchanged until you exit Minitab (or open a new project – see below).

If using Minitab at facilities at your school, be sure to check in which format the data files were saved. In all likelihood, they will be the *.mtw or *mtp files, rather than the *.dat files.

12

Exiting from Minitab

When finished using Minitab, click on **File>Exit** and you get

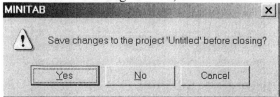

Generally, you will select **No**, which results in none of your work being saved. If you want to save some or all of your work, click on **Yes** and you will see:

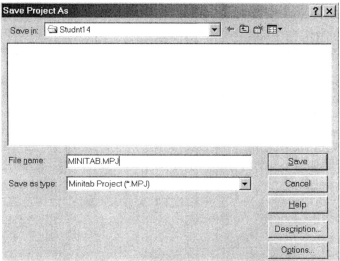

The default name of 'MINITAB' should be replaced by a more useful name, and decide where to save the file (**Save in**). Hit **Save**, and all your work (session, worksheets, graphs) will be saved in a Minitab Project file, designated by the suffix *.mpj. To open it up again, use **File>Open Project**. However, at any time during your session, you can save the content of any of your windows, by clicking on that window, then on **File>Save...**, or save all your work up to that point by using **File>Save Project**. If you want to eliminate some of your graphs, just close each graph window directly or via the **Project Manager Window**.

Are you ready to write your report? If so, before exiting, go to the **Project Manager Window**, and use the **ReportPad**. E.g. click on the **Session folder** or the **Graphs folder**. Right-click an item, and choose **Append to Report**. Right-click on the **ReportPad** folder, and from the selections, you may choose to print your report, or to save it (rich text format or web page), or to edit it in your word processor.

Chapter 2 – Methods for Describing Sets of Data

In this chapter, we will construct various data plots using the commands:
Graph>Pie Chart
Graph>Bar Chart
Graph>Histogram
Graph>Stem-and-Leaf
Graph>Dotplot
Graph>Boxplot
Graph>Scatterplot

Do some simple calculations on the data, using
Calc>Column Statistics
Calc>Calculator
Calc>Standardize

Open saved worksheets, and new worksheets, using
File>Open Worksheet
File>New>Minitab Worksheet

Organize and summarize data, using
Stat>Tables>Tally Individual Variables
Stat>Basic Statistics>Display Descriptive Statistics

Manipulate data in the worksheets, using
Data>Sort
Data>Code

Qualitative data

Qualitative data can be summarized in a table using **Stat>Tables>Tally Individual Variables** or with a bar chart or pie chart using **Graph>Bar Chart** or **Graph>Pie Chart**

Figures 2.1-2.4 - M/S pages 31-32

We want to summarize and display the data on 22 adult aphasiacs shown in Table 2.1. The data in this example have been saved in a Minitab worksheet file named APHASIA on your text CD-ROM. Insert the CD into the appropriate drive on your computer (or if working at computer facilities at your school, find out where the data sets have been saved, and in what format). Open up Minitab. Click on **File>Open Worksheet**. The dialog box below will appear. Next to **Files of type:**, choose **Minitab (*.mtw)**. At the top, next to **Look in**, your current folder is displayed. You need to find the folder with this data file. Click the little pointer at the right to see a list of available folders/drives and select the drive with your CD (or use the yellow folder icon at the top, with an up arrow, to go up one folder level at a time). Once you find the drive with the text CD,

14

select the folder with data files for textbook examples. Then choose the MINITAB folder. You should now see a file named APHASIA.

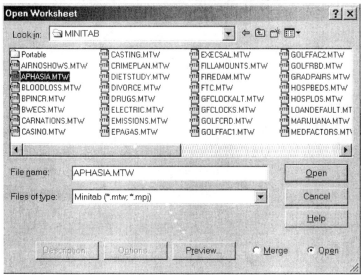

Double-click on APHASIA or select it and click on **Open**. Now you will see the saved worksheet in the bottom half of your screen:

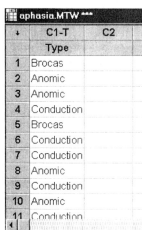

Note: Why do we see C1-T, rather than just C1 at the top? The '-T' indicates that this is a text column. A column must be either text or numerical, and cannot be a combination of the two. Your first entry determines the format, unless you choose to format the column before data entry, which is done by right-clicking the column, then from the pop-up menu, choosing Format Column>Text. If you use numbers in a text column, they will be interpreted as text labels or names, without numerical meaning.

Let's summarize the data. First, produce the frequency distribution table, via **Stat>Tables>Tally Individual Variables**. Fill in as follows:

15

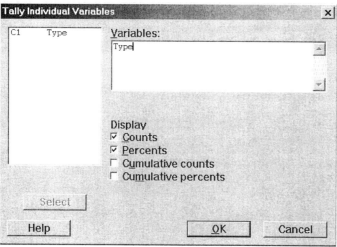

Hit **OK**. You will then see in the Session Window in the upper half of your screen:

Tally for Discrete Variables: Type

```
      Type   Count   Percent
Anomic           10    45.45
  Brocas          5    22.73
Conduction        7    31.82
       N=         22
```

Counts are the same as frequencies, and Percents are the same as relative frequencies except for a factor of 100 (compare with Figure 2.1 in text). This neatly tells us how this variable is distributed, so e.g. we see that nearly half have the Anomic type of aphasia.

Now for some pictures: We can produce a Pie Chart with **Graph>Pie Chart**

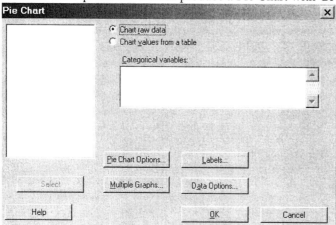

Everything is blank! But just click in the box under **Categorical variables**, and 'C1 Type' will appear in the box on the left side (where Minitab produces a list of candidate variables). Double click on 'C1 TYPE' now (or click it, then **Select**) to copy this column over to the box on the right side, i.e. to where your cursor is now blinking:

16

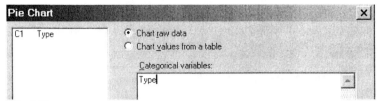

Hit **OK** now to get

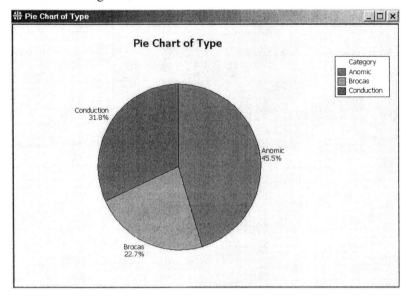

You will be seeing this in some nice bright colors! In order to display the names and percents alongside the pie, I also selected in the main dialog box: **Labels>Slice Labels** and ticked off **Category name** and **Percent**. Comparing areas, we readily see that the Anomic type dominates at nearly 50% of cases, followed by Conduction, then Broca's.

To produce the same information in a Bar Chart, just click on **Graph>Bar Chart**. In the first box you see, click **OK** for a **Simple** bar chart, and then a second box appears:

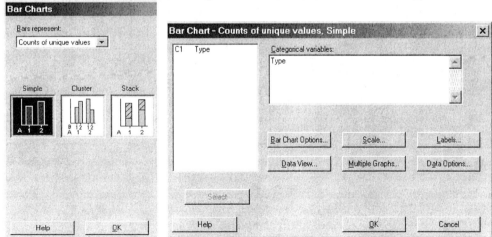

17

In the second box, we entered *Type* under **Categorical variables**. Hit **OK**, to get:

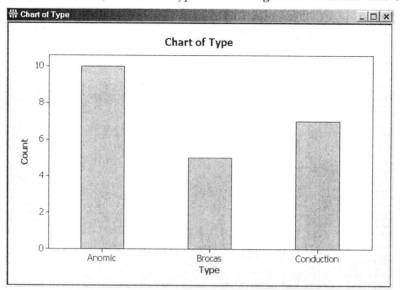

In this display, we easily see the relative frequency of occurrence of the various types of aphasia, by a comparison of the heights of the respective bars (rather than the distinct areas of the Pie Chart). If you prefer them ordered, *Pareto* style, from high to low or from low to high, use the **Bar Chart Options** in the dialog box shown above.

To print the session window, or one of the graphical windows, just click on it to make it the active window, the use **File>Print Session Window** or **File> Print Graph**, as the case may be (after making sure that **File>Print Setup** has the right specs).

Example 2.1 - M/S page 33

The data in this example have been saved in a Minitab worksheet file named BLOODLOSS on your text CD. Click on **File>Open Worksheet**. Specify **Files of Type: Minitab (*mtw).** Find the drive with your text CD, and select the folder with text examples, in Minitab format. Double-click on the file BLOODLOSS, to open it. You will now see the saved worksheet. The first 6 rows are:

bloodloss.MTW *

↓	C1-T	C2-T	C3-T	C4
	PATIENT	DRUG	COMP	
1	1A	NO	REDO	
2	2A	NO	NONE	
3	3A	NO	NONE	
4	4A	NO	NONE	
5	5A	NO	NONE	
6	6A	NO	NONE	

Ignore C1 (patient designation). In C2, we see whether or not the patient in that row used the drug, and in C3 we have the type of complication, if any.

18

We want to compare the distribution of complications for those taking the drug (drug = yes) and those not taking the drug (drug = no). For side-by-side bar charts of the two distributions, there are two possible ways to proceed:

(i) **Graph>Bar Chart>Simple>Categorical Variables:** *Comp*

   Multiple Graphs: By Variables (with groups in separate panels): *Drug*

   Bar Chart Options: Show Y as Percent

                             OR

(ii) **Graph>Bar Chart>Cluster>Categorical Variables:** *Drug Comp*

   Bar Chart Options: Show Y as Percent and **Take Percent Within categories at level 1 (outermost)**

(i)   (ii)

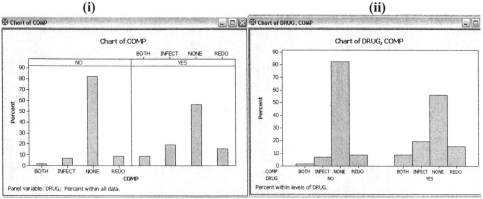

We see that complications are more frequent when using the drug. Note the importance of using *percents, not counts* - which would make for a difficult comparison, if there were more subjects in one group than in the other.

For the exact counts/percents, use **Stat>Tables>Cross Tabulation and Chi-Square**

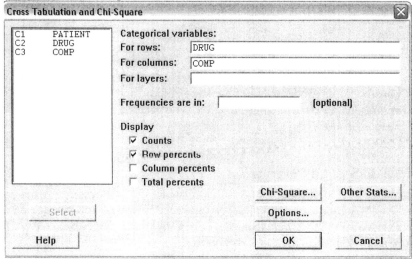

which produces:

19

Tabulated statistics: DRUG, COMP

```
Rows: DRUG   Columns: COMP

        NONE   BOTH  INFECT   REDO     All

NO        47      1       4      5      57
        82.46   1.75    7.02   8.77  100.00

YES       32      5      11      9      57
        56.14   8.77   19.30  15.79  100.00

All       79      6      15     14     114
        69.30   5.26   13.16  12.28  100.00

Cell Contents:       Count
                     % of Row
```

You can also produce the above summary via: **Stat>Tables>Descriptive Statistics**

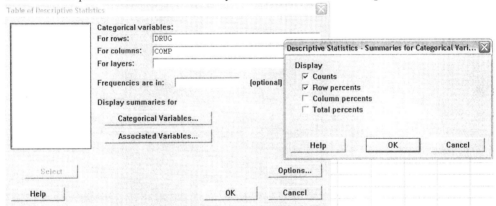

where you have to click **Categorical Variables** in the main box.

Ordering the categories:
Note the default alphabetical ordering of categories in the summary above and in the earlier displays. If you wish to change the ordering, click anywhere on the column, and select **Editor>Column>Value Order**, or right-click the column and from the pop-up menu, choose **Column>Value Order**. Then choose **User-specified order** or **Order of occurrence in worksheet.**

Exercise 2.14 - M/S page 39

Use **File>Open Worksheet** to open up the worksheet HEARLOSS. Here we have *raw data* on types of hearing loss in C1. To summarize, graphically, use **Graph>Pie Chart (Chart raw data)** or **Graph>Bar Chart (Simple)** with **Categorical Variables**: *C1*, **Options: Show Y as Percent**. Hit **OK**, and you get

20

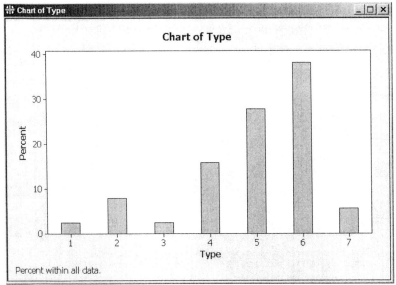

You might wish to use more informative labels for the categories. To do this, you can use **Data>Code>Numeric to Text**, and enter meaningful names or abbreviations.

Exercise 2.19 - M/S page 40

Click on **File>Open Worksheet**, find the data file DOLPHIN on your text CD, and open it up. We want to graphically summarize the types (categories) of dolphin whistles. Note that 'o' does not indicate a type but rather 'other types'.

NOTE that this is not *raw data*, which would consist of a list of 185 whistle classifications, but rather a *summary table* of the raw data, showing categories and counts. For a pie chart, use **Graph>Pie Chart**, and tick off **Chart values from a table** rather than **Chart raw data**. Fill in as below, and be sure to include some sort of **Labels** for the **Slice**s in your pie.

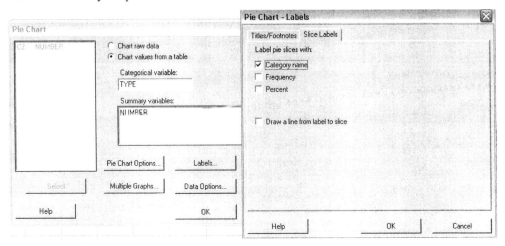

Click on **OK, OK,** and we get

21

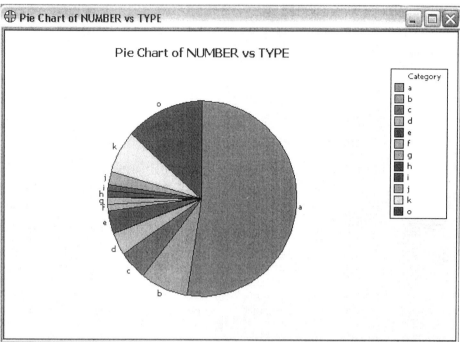

You can also click on **Percent** in the sub-dialog box for **Labels**, but I chose to omit this information here because the resulting display is just too messy in some parts. We easily see e.g. that more than half of whistles are type a.

For the bar chart, there are two possible ways to proceed with summarized data. Select **Graph>Bar Chart**. Note that there is a default setting at the top, indicating what the bars in your display will represent:

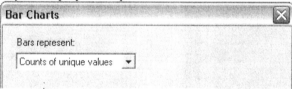

Click on the arrow and you will see 3 choices. Switch to the choice below:

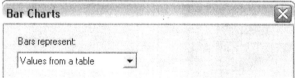

We still want a **Simple** bar chart. Hit **OK**, and then fill in:

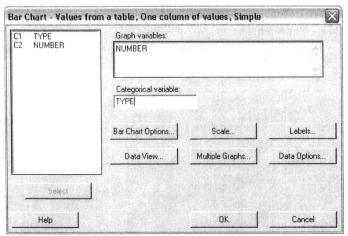

OK and we get:

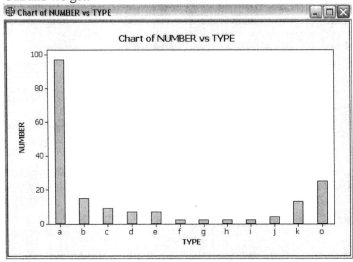

Again, we see the prevalence of type a whistles, though the percentage is not as clear as with the pie chart. To switch from counts to percents above, use the **Bar Chart Options: Show Y as Percent**.

Alternatively, you could stick with **Bars represent**: *Counts of unique values*, **Simple** chart, **Categorical variables**: *TYPE*, **Data Options**: **Frequency**: **Frequency variable(s)**: *NUMBER*

Exercise 2.17 - M/S page 40

Here, you can easily just type in the data yourself:

↓	C1-T	C2	C3	C4
	SITE	1999%	2001%	
1	Inside	7	4	
2	Outside	38	47	
3	Both	41	22	
4	Don'tKnow	14	26	

23

These are *summarized* data, not *raw* data. Use **Graph>Bar Chart (Simple)**: **Bars represent**: *Values from a table*, **Graph Variables**: *1999%, 2001%*, **Categorical Variables: SITE**, **Multiple Graphs>Multiple Variables**: **Show Graph Variables**: *In separate panels of the same graph.*

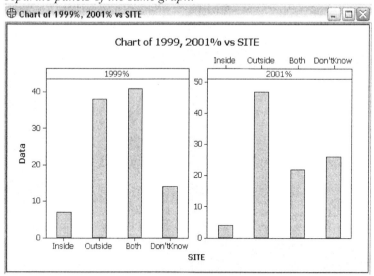

If instead, you were interested in comparing pie charts, side-by-side, you would use: **Graph>Pie Chart**: **Chart values from a table**, **Categorical Variable**: *SITE*, **Summary Variables**: *1999%, 2001%*, **Multiple Graphs>Multiple Variables**: **Show Pie Charts from Different Variables**: *On the same graph.*

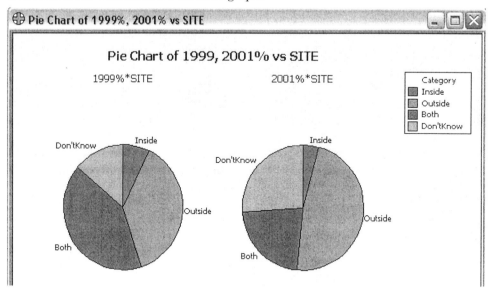

Quantitative Data: Graphical Displays

Dotplots, histograms, stem & leaf plots, and boxplots (discussed later) are the customary methods for displaying quantitative data. You will find them listed under **Graph**.

Example 2.2 - M/S page 45

The 50 default rates in Table 2.4 are saved in a file named LOANDEFAULT. Click **File>Open Worksheet**. Locate the file and open it. You will see in C1 the 50 default rates:

↓	C1	C2
	Default	
1	12.0	
2	19.7	
3	12.1	
4	12.9	

For a histogram, click on **Graph>Histogram (Simple)** and plug in 'Default' under **Graph Variables**, as below.

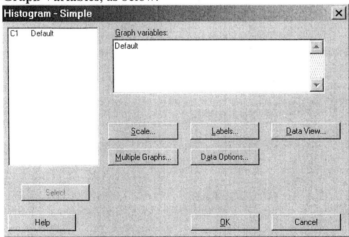

There are many sub-dialog boxes displayed, which allow you to change various characteristics of the graph – you will not need to use any of these unless you want to compare histograms, in which case, the **Multiple Graphs** option will be needed. Keep in mind that if you do change some of the *default settings*, the changes will persist through your session, until you change them again, or exit.

Just hit **OK** and Minitab will do a good job of producing a reasonable histogram, with the classes and labeling decided by Minitab. We will show how you can change the class intervals later.

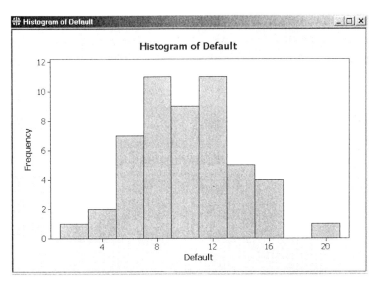

If you want to produce a histogram exactly like the one in Figure 2.12 in the text, with midpoints of 1.5, 4.5, 7.5, … , 19.5, then you need to tell Minitab to make some changes to the display. Double-click on any of the bars in the histogram, or on any of the numbers on the x-axis, and you will see the **Edit Bars** dialog box. Or click once on the bars, then right-click or go to the main menu and select **Editor>Edit Bars** (but be sure that all the bars are selected, *not just one bar*, in which case you will only be able to *edit the characteristics of that single bar*). Classes are also called 'bins', so choose **Binning** from the four choices shown, select **Interval Type:** *Midpoint*, and fill in as below:

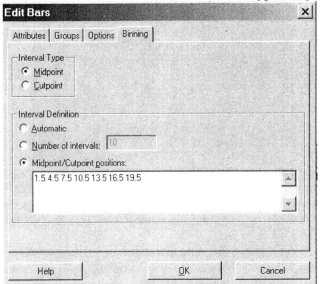

Note that you can specify either the midpoints of the classes or the limits (cutpoints) of the classes, or you can just specify the number of classes that you would like. To specify the midpoints above, you may also write 1.5:19.5/3, which is a short way to say: run from 1.5 to 19.5 in steps of 3.

Hit OK, and now you get exactly the histogram in the text display:

26

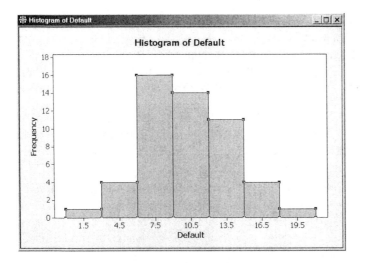

For a stem and leaf plot, select **Graph>Stem-and-Leaf**

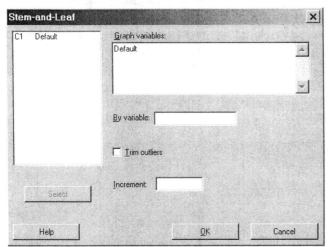

Fill in the relevant **Graph variable**, hit **OK**, and you will see the plot in the Session Window (a separate graphical window is not needed for this type of display):

Stem-and-Leaf Display: Default

```
Stem-and-leaf of Default   N = 51
Leaf Unit = 1.0


    1    0  2
    5    0  4455
   14    0  666667777
 (12)    0  888888899999
   25    1  000011111
   16    1  2222223
    9    1  4444555
    2    1  6
    1    1  9
```

27

The Leaf Unit is shown at the top as '1.0', so we know that the data are 02, 04, 04, ... 16 (and not say 0.2, 0.4, 0.5, ... , 1,6 , which would be indicated by 'Leaf Unit = 0.1') . The last figure (tenths of percent) for each default % gets chopped off (truncated), as computer programs prefer to use only one-digit leaves. So the smallest entry in the plot, the '2', could actually have been anything from 2.0% to 2.9%.

What are those numbers in the left column? To find out, try Minitab's **Help** function. Go back to the **Stem and Leaf** dialog box and select **Help** (or from the main menu, select **Help>Help**, then **Search**). You will see an example and some information about this plot, including the following remark:

- *Counts (left) – If the median value for the sample is included in a row, the count for that row is enclosed in parentheses. The values for rows above and below the median are cumulative. The count for a row above the median represents the total count for that row and the rows above it. The value for a row below the median represents the total count for that row and the rows below it.*

Increment: You can alter the display by your choice of **Increment**, in the dialog box. The increment is the difference between the lowest possible value in one row and the lowest possible value in either adjacent row (or the stem unit divided by the number of splits). In the display above, this equals $(4 - 2)$ or $(16 - 14) = 2$. Choose '1' for the increment, and the number of rows will double. Choose '5' and there will be fewer rows. The only possible values of **Increment** are .01, .02, .05, .1, .2, .5, 1, 2, 5, 10, 20, 50, etc. since you can only repeat the same stem once, twice, or five times, and still maintain equal width classes (rows).

We can also produce a dotplot using **Graph>Dotplot (One Y, Simple)**, then enter the variable 'Default' under **Graph Variables**:

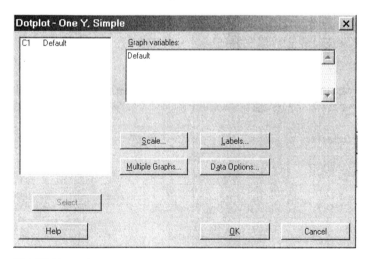

Hit **OK** to get

28

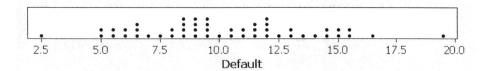

Default

Figure SIA2.5 - M/S page 49

To produce these side by side histograms, open up the EYECUE data set, then select
Graph>Histogram (Simple): Graph Variables: *Deviation*,
   Multiple Graphs: By Variables (with groups in separate panels): *Gender*,
   Scale: Y-Scale Type: *Percent*

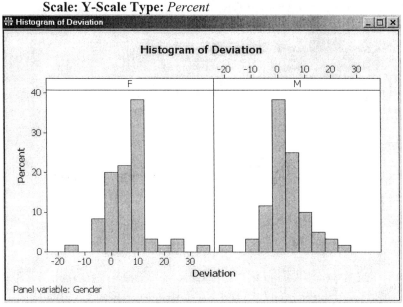

Exercise 2.27 - M/S page 50

The actual data are not given here, just a grouped distribution with classes and relative
frequencies. So, start with a new worksheet via **File>New: Minitab Worksheet** unless
you have just opened up Minitab, in which case a blank worksheet is already sitting there.
Enter the measurements' midpoints in C1, relative frequencies in C2, and name your
columns:

29

	C1	C2	C3
	Measurement	RelFreq	Counts
1	1.5	0.10	50
2	3.5	0.15	75
3	5.5	0.25	125
4	7.5	0.20	100
5	9.5	0.05	25
6	11.5	0.10	50
7	13.5	0.10	50
8	15.5	0.05	25

Since we need counts, multiply C2 by 500, and place these counts into C3 by using **Calc>Calculator.** Type 'Counts' next to **Store result in variable**, and under **Expression**, enter '500 * C2', as below. Hit **OK**. No column is yet named 'Counts', so Minitab assigns this name to the first free column in the worksheet.

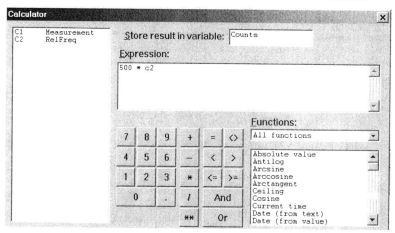

Now, in your worksheet, you see the Counts in C3. Let's try **Graph>Histogram (Simple), Graph Variables:** *Measurement*; **Data Options: Frequency: Frequency Variables:** *Count*. You see the display on the left below.

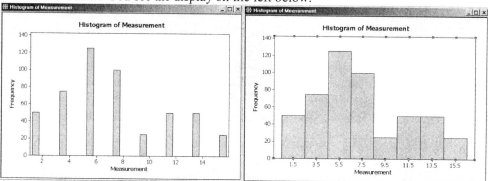

Whoops, it seems Minitab assumed that the frequencies (counts) were attached exactly to each midpoint value, and not spread out over the entire class. Double-click on the x-axis, then **Edit Scale: Binning: Interval Type:** *Midpoints*, **Midpoint positions:** *1.5:15.5/2*. Now, you get the proper histogram on the right.

30

Alternatively, try **Graph>Bar Chart: Simple, Bars represent:** *Values from a table*; **Graph Variables:** *Count*, **Categorical Variable**: *Measurement*. To fatten the bars, double click on the x-axis values, then **Edit Scale: Scale: Gap between clusters:** *0*

Exercise 2.37 - M/S page 52

Click **File>Open Worksheet**. Find and open BRAINPMI on your text CD. The data are in C1. Use **Graph>Dotplot (One Y, Simple)** with **Graph Variables:** *C1*. You get:

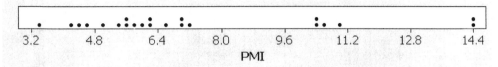

Note the distinct groups.

Exercise 2.41 - M/S page 53

(b) Use **File>Open Worksheet**, find and open LISTEN. Note the structure of the data in the worksheet, with three columns:

↓	C1-T	C2	C3
	Group	Response Rate	Percent Correct
1	Control	250	23.6
2	Control	230	26.0
3	Control	197	26.0
4	Control	238	26.7

This is a very convenient structure, referred to as *stacked* data, where each subject takes up one row, in which we see that subject's Group Membership, Response Rate and Percent Correct. All the Response Rates are 'stacked' in one column. Likewise, the Percentages Correct are 'stacked' in a column. The Group Membership of each individual (one per row) can be found in the first column. The data would have been *unstacked* if we had used 6 columns in total: C1: Response Rates for Controls, C2: %Correct for Controls, C3: Response Rates for Treatment Group, etc. - similar to the data display in the textbook. [To convert from one structure to the other, use **Data>Unstack Columns** or **Data>Stack**]

If you wish to use dotplots to compare several groups, select: **Graph>Dotplot: One Y, With Groups.** You have to switch from the default **Simple** dotplot choice. Note how the picture helps you choose the correct type of dotplot.

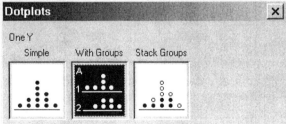

31

Fill in as follows:

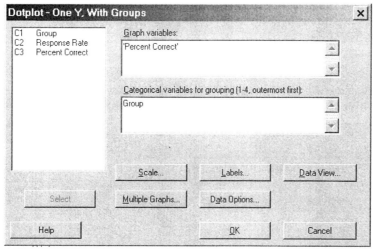

Note the quotes around the variable: 'Percent Correct'. They are needed (in the actual Minitab command language) whenever the variable name has a space in it.

Minitab now knows it should separate the data in C3 based on the distinct group settings it finds in the column C1 = Group. We get:

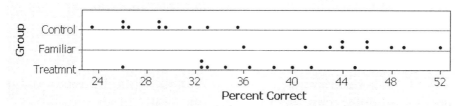

[If the data had been *unstacked* like in the textbook display, you would select **Graph>Dotplot (Simple, One Y),** list all three variables under **Graph Variables,** and select **Multiple Graphs: Multiple Variables: In separate panels…, Same Scales…**]

Likewise, we can produce *separate stem-and-leaf plots by Group*, except that *Minitab's Stem-and-Leaf procedure requires Group to be a numerical variable.* So recode the three group names in C1 into the numbers 1,2,3 in a new column (C4) named 'GroupNumber' using **Data>Code>Text to Numeric** as follows:

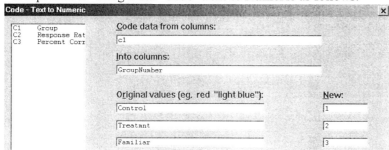

Now, use **Graph>Stem and Leaf** with **Graph Variables:** *C3*, and **By Variable:** *GroupNumber*, and you get, in the session window:

Stem-and-Leaf Display: Percent Correct

```
Stem-and-leaf of Percent Correct   GroupNumber = 1    N  = 10
Leaf Unit = 1.0

  1    2   3
 (6)   2   666999
  3    3   12
  1    3   5

Stem-and-leaf of Percent Correct   GroupNumber = 2    N  = 10
Leaf Unit = 1.0

  1    2   6
  5    3   2234
  5    3   68
  3    4   014

Stem-and-leaf of Percent Correct   GroupNumber = 3    N  = 10
Leaf Unit = 1.0

  1    3   6
  5    4   1344
  5    4   6689
  1    5   2
```

Minitab uses the same *increment*, in each plot, in order to maintain comparability. [For *unstacked data*, you would list all three variables under **Graph Variables** and specify one common increment]

For comparative histograms, use **Graph>Histogram (Simple): Graph Variables:** *Percent Correct*, **Multiple Graphs>By Variables: By Variables with groups in separate panels:** *Group*.   You get

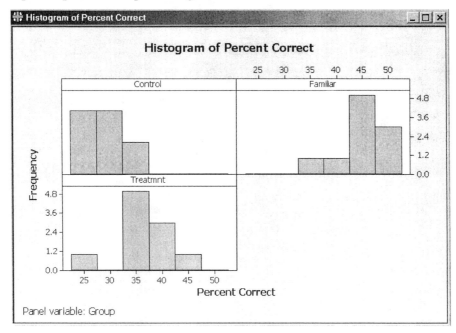

We readily see the shift up in accuracy as we go from the Control group to the Treatment group to the Familiarization group.   [For *unstacked data*, you would use **Multiple Graphs: Multiple Variables,** rather than **Multiple Graphs: By Variables]**

Quantitative Data: Measures of Center and Spread

We can use Minitab to calculate individual measures such as mean or standard deviation with **Calc>Column Statistics** or get a batch of useful measures with **Stat>Basic Statistics>Display Descriptive Statistics**.

Example 2.4 - M/S page 56

Open up the file EPAGAS, where the 100 EPA mileage ratings of Table 2.2 have been stored in C1. To find the mean rating, use **Calc>Column Statistics**, tick the statistic needed, and input the variable:

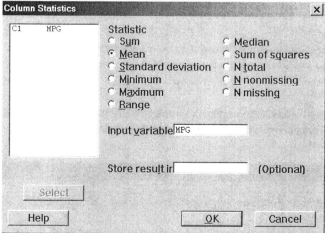

Note that there are many other statistics you can choose to calculate for a column of numbers. The output is:

Mean of MPG

```
Mean of MPG = 36.994
```

You also have the option, in the dialog box, to store the result as a saved constant $(k1, k2, \ldots)$. Storing it as a constant may come in handy if you are going to use this single number as part of some future calculation.

Example 2.6 - M/S page 59

To get the median mileage rating, we can proceed as in Example 2.4, choosing **Median** in the dialog box, or we can proceed to ask Minitab for a good selection of useful summary statistics, via **Stat>Basic Statistics>Display Descriptive Statistics**

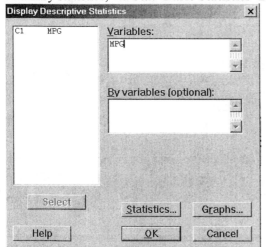

Note the '**By variables**' option, which we would use if e.g. there were a categorical variable, like gender in C2, and we wished separate stats for males and for females. You can also produce a histogram or boxplot (studied later) for the data via the **Graphs...** selection in the dialog box.

In the session window, we see:

Descriptive Statistics: MPG

Variable	N	N*	Mean	SE Mean	StDev	Minimum	Q1	Median	Q3
MPG	100	0	36.994	0.242	2.418	30.000	35.625	37.000	38.375

Variable	Maximum
MPG	44.900

Under Median, we see 37.0, telling us that half of the mileages are in excess of 37.0. You can choose additional statistics or delete some of those displayed above, by clicking on **Statistics** in the dialog box.

In the output, **StDev** stands for 'standard deviation', which will be discussed in section 2.5. **Q1** and **Q3** stand for first (lower) and third (upper) quartile, discussed in section 2.8. **Minimum** and **Maximum** are self-explanatory. **N** is the number of EPA ratings. N* indicates the number of missing observations in the column.

Example 2.7 - M/S page 60

Open up a fresh worksheet, via **File>New>Minitab Worksheet** and enter the 10 numbers in column 1. To find the mode, use **Stat>Tables>Tally Individual Values: Variables:** *C1*, and you get

Tally for Discrete Variables: C1

```
C1    Count
 5      1
 6      1
 7      2
 8      2
 9      3
10      1
N=     10
```

Since this command lists all distinct outcomes and their frequencies, you can readily see which outcome is the most common (i.e. the mode).

Exercise 2.65 - M/S page 65

Open up the data file SLI. The DIQ scores are all stacked into one column, with a group indicator in another column, so we can use **Stat>Basic Statistics>Display Descriptive Statistics** with **Variables:** *DIQ* and **By variables:** *Group*, and we get:

Descriptive Statistics: DIQ

Variable	Group	N	N*	Mean	SE Mean	StDev	Minimum	Q1	Median	Q3
DIQ	OND	10	0	101.90	3.06	9.69	87.00	93.50	103.00	110.75
	SLI	10	0	93.60	2.87	9.08	84.00	86.00	91.50	100.25
	YND	10	0	95.30	2.38	7.51	86.00	90.00	92.00	101.25

Variable	Group	Maximum
DIQ	OND	113.00
	SLI	110.00
	YND	110.00

You can easily pick out the means or medians for each of the three groups from the above.

Example 2.11 - M/S page 74

Open up the RATMAZE file, containing the times for the 30 rats in C1. Use **Stat>Basic Statistics>Display Descriptive Statistics** with **Variables:** *Time*, and we get:

Descriptive Statistics: RUNTIME

Variable	N	N*	Mean	SE Mean	StDev	Minimum	Q1	Median	Q3	Maximum
RUNTIME	30	0	3.744	0.401	2.198	0.600	1.960	3.510	5.165	9.700

As in the text, the 2 standard deviation interval about the mean works out to be (-.66, 8.14), and we want to count how many observations fall in it.

We could use **Data>Sort**: **Sort column:** *C1*, **By column:** *C1*, **Store in:** *Original Column*, to sort the numbers in C1 in ascending (or descending) order, making a direct count easy.

Alternatively, we can use **Calc>Calculator** with the following entries:

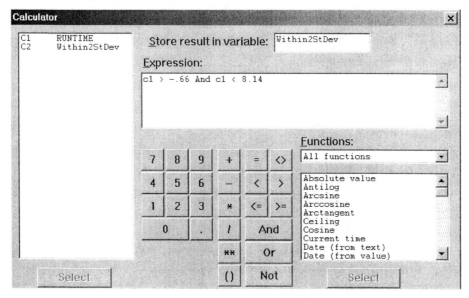

If you put a 'true or false' statement under **Expression**, Minitab checks it out for each row, and enters a '1' if true, '0' if false, into the storage column that you specify at the top. Such a true/false expression is called a *Logical Operation*. My **Expression** above says that the entry in C1 is greater than -.66 and also is less than 8.14, i.e. within the 2 standard deviation interval. You will see now 0's and 1's in C2, named 'Within2StDev'. Now, you can tally the 1's and 0's with say **Stat>Tables>Tally Individual Variables:** *C2*

Tally for Discrete Variables: Within2StDev

```
Within2StDev   Count
           0       2
           1      28
          N=      30
```

We see that 28 of 30, or about 93% lie within the interval, quite close to the 95% predicted by the empirical rule, even though the distribution is not very bell-shaped.

Exercise 2.176 - M/S page 110

Open up the PAF file. For a stem and leaf plot of the PAF Inhibition percentages shown in C1, use **Graph>Stem and Leaf: Graph variables:** *C1*, to get

Stem-and-Leaf Display: PAF Inhibition

```
Stem-and-leaf of PAF Inhibition  N  = 17
Leaf Unit = 1.0

   6    0  000009
   8    1  25
  (2)   2  45
   7    3  13
   5    4  0
   4    5
   4    6  2
   3    7  057
```

From the above, we can identify the mode as 0. For the other descriptive statistics, use **Stat>Basic Statistics>Display Descriptive Statistics**, with **Variables:** *C1*, and you see

Descriptive Statistics: PAF Inhibition

```
Variable         N   N*    Mean  SE Mean  StDev      Minimum              Q1   Median
PAF Inhibition  17   0    27.82    6.76   27.88  0.000000000  0.000000000   24.00

Variable          Q3  Maximum
PAF Inhibition  51.00    77.00
```

The *Mean*, *Median*, and *StDev* (standard deviation) are apparent. For the range, subtract the *Minimum* from the *Maximum*, and square the *StDev* to get the variance, or if you prefer, click on **Statistics** in the dialog box for this procedure, and tick off *Range* and *Variance* as additional choices.

Quantitative Data: The Boxplot

This useful display for quantitative data provides measures of center, spread, and shape, helps detect outliers, and is very useful for comparing distributions. Look for it under **Graph**.

Example 2.16 - M/S page 88

Open up the LOANDEFAULT file. To produce a boxplot, use **Graph>Boxplot: One Y, Simple,** and fill in the obvious entries:

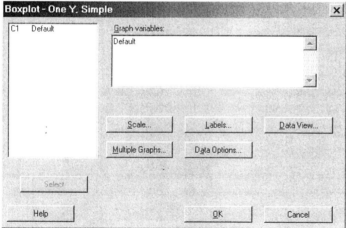

Hit **OK**, and we get

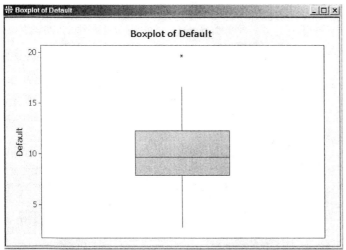

In the above, the asterisk points out one unusual observation, the nearly 20% default rate for Alaska.

In these Minitab boxplots, all unusual observations are denoted with an '*', *regardless* of how far beyond the inner fences they may lie, i.e. no 'outer fences' are computed by Minitab, as mentioned in the textbook.

To read values more precisely from the plot, you might wish to show more ticks on the Y-axis. Double-click on any of the numbers shown on the y-axis, to display the **Edit Scale** dialog box:

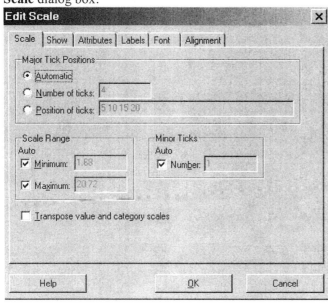

Using this, you can change the **Major Tick Positions** or their **Number**, as well as the **Number** of **Minor Ticks** (unclick box under **Auto**, then enter desired **Number** of ticks), or reverse (**Transpose**) the orientation of the plot, so that the Default axis stretches from left to right instead of vertically.

Example 2.17 - M/S page 89

Open up the REACTION data file. The reaction times are stacked in C2 = TIME, with the stimulus type in C1 = STIMULUS, where NT = non-threatening and T = threatening. To compare reaction times for the two types of stimulus, use
Graph>Boxplot - One Y, With Groups, then fill in:

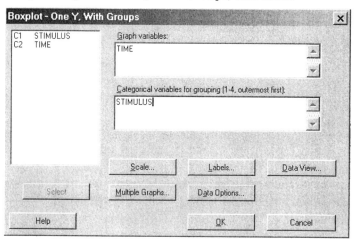

We get

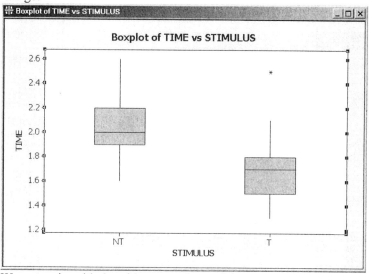

We get a nice side-by-side comparison of the two samples, showing generally higher, i.e. slower, reaction times for the non-threatening stimulus, similar variability in the two groups, and one possible reaction time outlier in the threatening stimulus group.

Exercise 2.126 - M/S page 94

Open up the worksheet LM2_126, showing the Sample (A or B) in C1 and the Data in C2. Since the data are all stacked in C2, we readily get side-by-side boxplots, via **Graph>Boxplot: One Y, With Groups: Graph Variables:** *Data*, **Categorical Variables**: *Sample.*

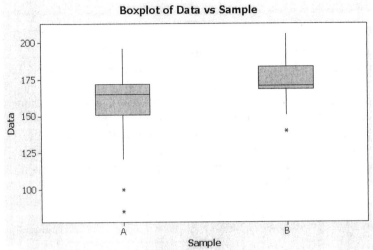

We see a lower center and greater variability for sample A, which exhibits some left (negative) skewness, while right (positive) skewness is exhibited by sample B. Using the scale on the left side, you can read off approximate numerical values for the two suspect outliers in sample A, and the single suspect outlier in sample B.

Exercise 2.177 - M/S page 110

Open up the DOWNTIME file. Downtime is in C2 and the customer number is in C1. For a boxplot of the downtimes, use **Graph>Boxplot (One Y, Simple): Graph Variables:** *Downtime*

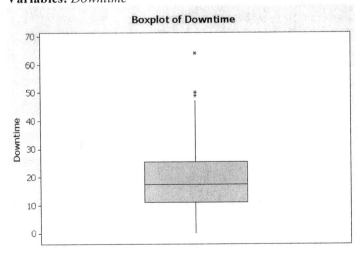

41

(a) You can easily determine the median, IQR, and skewness from the plot above.
(b) The 3 suspect outliers have downtimes around 50 and 65, so you can look for these values in C2, and then identify the customer numbers from C1. Or, to make things easier, sort the data first via **Data>Sort: Sort Columns:** *C1 C2*, **By column:** *C2,* **Store in:** *Original columns.* You will then see the customers with the three biggest downtimes listed last in the worksheet.

Shall we brush?

Aside from looking through the worksheet in order to identify the outliers in the boxplot, you could also proceed as follows: Click on the graph to make it the active window. Click **Editor>Brush**, which produces a special *brushing cursor* in the graphical window. Encircle the three outliers. In a little window off to the left, called the *brushing palette*, the row numbers 35, 39, 40 will appear. Look through the worksheet, and you will also see markers next to these three rows. Or select **Editor>Set ID Variables: Variables:** *C1 C2*, to display complete information about the highlighted observations right in the brushing palette. ***Brushing allows you to highlight extreme points on a graph in order to learn more about them.***

(c) To standardize all the downtimes, use **Calc>Standardize,** with entries as below

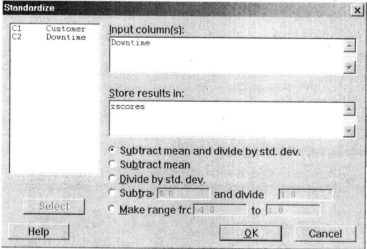

which produces, in part:

35	264	64	3.14095
36	265	19	-0.09900
37	266	18	-0.17100
38	267	24	0.26100
39	268	49	2.06097
40	269	50	2.13297

from which you can read off the z-score for each customer.

Bivariate Relationships – The Scattergram

With two quantitative variables, we may use **Graph> Scatterplot** to plot one variable versus the other in a scattergram (also known as a scatterplot), in order to assess the nature of the relationship between the variables.

Example 2.19 - M/S page 96

Open up the MEDFACTORS worksheet, which shows Length of Stay in C2 and #Factors in C1, for 50 coronary patients. To plot Factors on the y-axis versus LOS on the x-axis, use **Graph>Scatterplot: Simple: Y variables:** *Factors*, **X variables:** *LOS*.

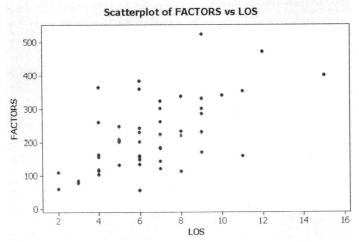

Hit **OK** to produce the scattergram:

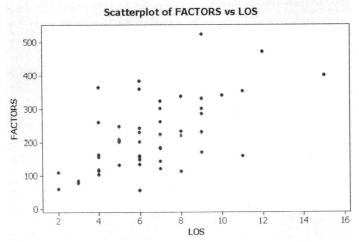

from which we see a roughly linear positive relationship.

Exercise 2.178 - M/S page 111

Open up the worksheet SLI worksheet. We can investigate the relationship between DIQ and Pronoun Errors for all children and also just for the SLI children, with a very useful scattergram variant. Select **Graph>Scatterplot: With Groups** and fill in as below, to produce a plot using *different symbols* for each GROUP.

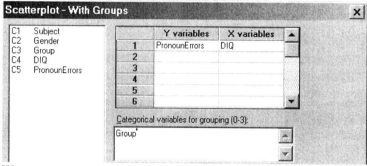

We get:

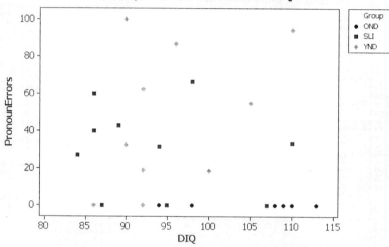

This plot allows us to see all the data, but also to focus on just the data from the SLI group, indicated by red square symbols (with the bright colors that you will be seeing, the differences between groups are much clearer). We see little relationship between the two variables here, overall, and for the SLI group.

If you want to plot only the 10 SLI observations, select and delete (using the *Delete* key on your keyboard) the other 20 rows in the worksheet. Or use **Data>Delete Rows:** *1:10 21:30* **From:** *c1-c5*, or **Data>Split Worksheet: By Variables:** *GROUP*, to isolate the SLI group in one new worksheet. Then **Graph>Scatterplot (Simple):** **Y:** *PronounErrors*, **X:** *DIQ*, produces:

44

Bivariate Relationships – The Scattergram

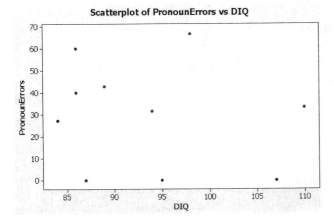

45

Chapter 3 – Probability

In this chapter, we will produce random data and draw random samples, using
Calc>Random Data>Bernoulli
Calc>Random Data>Integer
Calc>Set Base
Calc>Random Data>Sample From Columns

Summarize qualitative data using
Stat>Tables>Tally Individual Variables

Sort the data from low to high, with
Data>Sort

Enter a simple set of patterned numbers into a column, with
Calc>Make Patterned Data>Simple Set of Numbers

Display data in the session window, using
Data>Display Data

Do some calculations on the data in the worksheet using:
Calc>Calculator

Random Events

We can simulate various types of random experiments, like coin tossing, using
Calc>Random Data.

Figure 3.2 & the Law of Large Numbers - M/S page 123

If probability is just long run relative frequency, then the proportion of heads obtained when flipping a balanced coin repeatedly, in a truly random fashion, should get quite close to .5 after a large number of flips. Instead of actually flipping a real coin, we can simulate flipping a coin by using **Calc>Random Data>Bernoulli** as indicated below

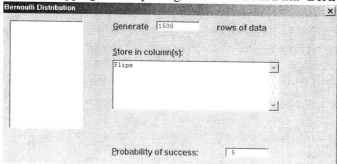

46

This will generate 1500 rows of random data, where in each row there is a .5 probability of '1' (termed a *success*) and .5 probability of '0'. Consider '1' to be a head, and '0' to be a tail. The individual flips are stored in *Flips* = C1.

Let's put the running or cumulative total of heads in C2, using a function called *Partial Sums (PARS)*, from the **Calc>Calculator**

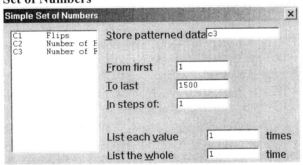

Enter the *Number of Flips*: 1:1500 in C3, using **Calc>Make Patterned Data> Simple Set of Numbers**

Divide c2 by c3, giving the current proportion of heads in c4 = *Proportion of Heads*:

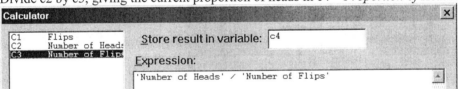

Here is some of the worksheet:

	C1	C2	C3	C4
	Flips	Number of Heads	Number of Flips	Proportion of Heads
1	0	0	1	0.000000
2	1	1	2	0.500000
3	0	1	3	0.333333
4	0	1	4	0.250000
5	1	2	5	0.400000
6	1	3	6	0.500000
7	1	4	7	0.571429
8	0	4	8	0.500000
9	1	5	9	0.555556
10	1	6	10	0.600000

Now, plot the proportion of heads versus flip number, with **Graph>Scatterplot (Simple), Data View: Data Display: Connect Line; Data Options: Subset: Include: Row Numbers:** *6:1500/30* (i.e. showing only every 30th point). Two different simulations of 1500 flips are shown below:

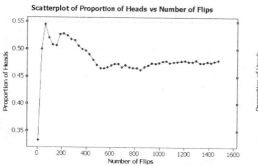 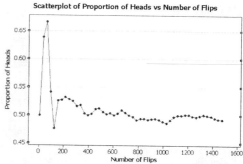

In each simulation above, the proportion of heads can be seen to be gradually converging toward 0.5. If you continue flipping, the proportion of heads must surely converge to the true probability of obtaining a head, in the long run. [This coin-flipping situation may be classified and studied further under a fascinating part of Probability theory called *Random Walk* theory]

Exercise 3.152 - M/S page 180

Open up the EVOS data set. Column C4 indicates oil contamination (yes/no) for each of the 96 transects. The probability that a randomly selected transect is contaminated with oil may be estimated by the proportion in the sample that are contaminated, which you can find using **Stat>Tables>Tally: Individual Variables:** *C4*, **Display:** *Percents*

Tally for Discrete Variables: Oil

```
Oil   Count   Percent
 no     36     37.50
yes     60     62.50
 N=     96
```

Random Samples

To make inferences about a population from a sample, we must use some type of *random* (*probability*) sample. To select a *simple random sample* (analogous to drawing names from a hat), you can use a random number table like the one in the back of the textbook, or you can use random number generators present in all statistical packages.

Example 3.22 - M/S page 161

The households have been numbered 00000-99999. To select fifty households, use **Calc>Random Data>Integer** as indicated below. The randomly generated data then appears in C1 in the worksheet, as shown to the left.

48

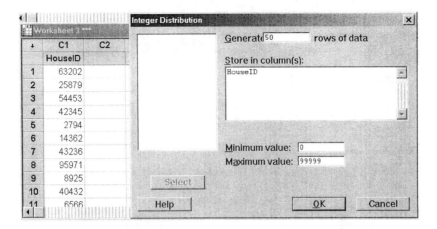

You might wish to print out the 50 selected households in order, so use **Data>Sort: Sort Columns:** *C1*, **By column:** *C1*, **Store in:** *Original column*, to sort the values in C1 into ascending order, and then use **Data>Display Data: Columns to display:** *C1*, to print the values out in your session window:

Data Display

```
HouseID
    144    2794    4136    6566    7163    7646    8925   13356   13744
  14362   14956   15671   17175   18147   18692   22548   25879   29100
  35129   36088   39839   40256   40261   40432   42345   42595   42772
  43236   45043   46323   51465   52914   53369   54453   56696   56723
  56850   57117   60458   63202   64185   68299   76966   81114   82184
  83230   87275   87545   89214   95971
```

The numbers that you get though will *not be the same* as these, as there is a random 'starting point' mechanism at work with Minitab. If you want your random selection to be repeatable, then before using **Calc>Random Data**, select **Calc>Set Base** and enter some number - this tells Minitab's random number generator where to start, in its sequence of pseudo-random numbers. Set the same base, before generating random data again, and you will get the same results.

If there are any repeated households in the random sample, just make additional random selections (or ask for a bit more than actually needed at the start).

Example 3.23 - M/S page 163

If you have a population listed explicitly in say C1, you can select a portion of them randomly, using **Calc>Random Data>Sample From Columns**. If this is an experiment, the ones selected will get the active treatment, and the rest will get the placebo.

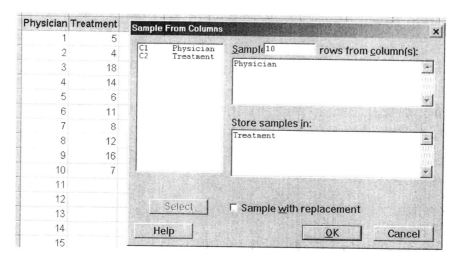

Physician	Treatment
1	5
2	4
3	18
4	14
5	6
6	11
7	8
8	12
9	16
10	7
11	
12	
13	
14	
15	

Exercise 3.99 - M/S page 164

(a) and (b): We want to generate 10 random numbers from the population of all possible phone numbers: 0000000 – 9999999, or if no phone numbers start with 0 or 1, as is often the case, from 2000000-9999999. Use **Calc>Random Data>Integer: Generate:** *10* **rows, Store in:** *Phone*, **Minimum:** *2000000*, **Maximum:** *9999999*. Display them with **Data>Display Data: Columns to display:** *Phone*

Data Display

```
Phone
   6270999     9868830     9243011     4784577     7187920     6837347
   2308035     5489204     8153160     2482625
```

(c) To generate 5 numbers starting with 373, use **Calc>Random Data>Integer: Generate:** *5* **rows, Minimum:** *3730000*, **Maximum:** *3739999*

Chapter 4 – Discrete Random Variables

In this chapter, we will calculate the mean and standard deviation of a probability distribution, using
Calc>Calculator

Do calculations using the Binomial and Poisson distributions, with
Calc>Probability Distributions>Binomial
Calc>Probability Distributions>Poisson

Plot a probability distribution, with
Graph>Bar Chart

Enter a simple set of patterned numbers into a column, with
Calc>Make Patterned Data>Simple Set of Numbers

Examine the contents of the worksheet, using
Window>Project Manager

Mean and Standard Deviation

To calculate the mean and standard deviation of a probability distribution, we need to enter the outcomes in one column, the corresponding probabilities in a second column, and then use **Calc>Calculator** to construct these measures.

Example 4.7 - M/S page 200

Let's calculate the mean and standard deviation for the specified distribution using Minitab. First, we enter the distribution itself into two columns, as follows:

	C1	C2	C
↓	Outcome	Prob	
1	0	0.002	
2	1	0.029	
3	2	0.132	
4	3	0.309	
5	4	0.360	
6	5	0.168	

Now, use **Calc>Calculator**. Find the function **Sum**, in the **Functions** list. Double-click it to make it appear under **Expression.** Double-click 'Outcome', click '*', double-click 'Prob', or just type in directly the following expression:

51

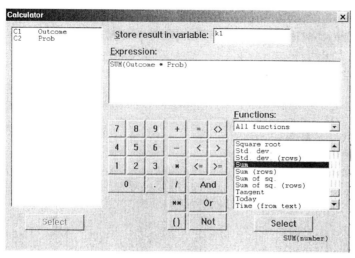

Note that I have stored the result in the form of a *stored constant*, denoted 'k1'. Use **Window>Project Manager** (or **Data>Display**) to see the value of k1:

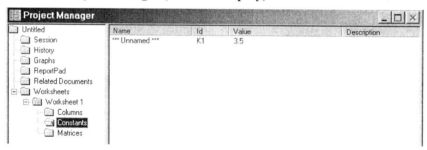

So, the mean is 3.5. To calculate the standard deviation, type the following expression (or select 'Square root', then 'Sum', etc.), being careful with all the parentheses:

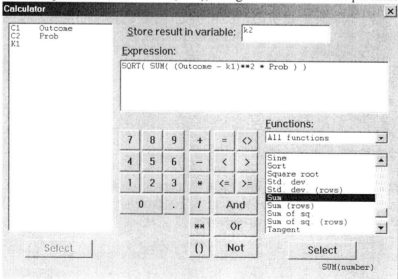

I chose to **Store result in:** k2. Then, **Window>Project Manager**

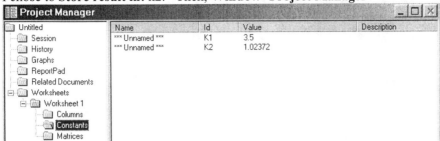

tells us that the standard deviation equals 1.024.   To remind yourself of what k1 actually represents, you can right-click k1 above, and then rename it say 'mean', or you can double-click on k1 to supply a description above. Likewise for k2.

You can graph p(x) via **Graph>Bar Chart: Bars represent: Values from a table; Simple:**

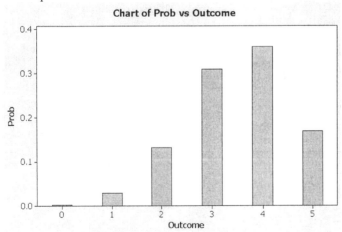

which produces

Chart of Prob vs Outcome

If you prefer a histogram look, you can reduce the space between bars to zero, by double-clicking on any of the x-values, producing the **Edit Scale: Scale** dialog box, in which you unclick the check mark next to **Gap between clusters**, and then type '*0*'.

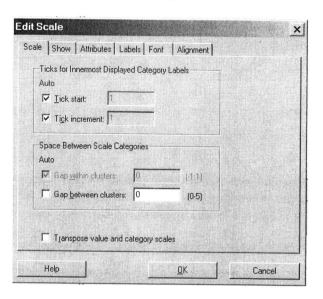

Exercise 4.36 - M/S page 203

Enter the number of homes in C1 = *Homes* and the probabilities in C2 = *Prob* as below

	C1	C2
↓	Homes	Prob
1	0	0.09
2	1	0.30
3	2	0.37
4	3	0.20
5	4	0.04

To calculate the mean, **Calc>Calculator** and enter the expression

Expression:
```
SUM( Homes * Prob )
```

Store the mean in 'k1'.

Then, for the standard deviation, use **Calc>Calculator** with expression

Expression:
```
SQRT( SUM( (Homes-k1)**2 * Prob ) )
```

Store the result in 'k2'. To see the values of the stored constants k1 and k2, use **Data>Display Data** or check your worksheet's contents via **Window>Project Manager**. You can then calculate $\mu \pm 2\sigma$ and determine the probability within this interval.

54

The Binomial Probability Distribution

Many real world situations give rise to random variables, whose behavior can be well modeled by the *Binomial distribution*. Probabilities can be easily calculated using statistical software.

Example 4.12 - M/S page 212

The number who favor the candidate is Binomially distributed with n = 20 and p = .6. To find probabilities of single outcomes, like P(x = 11), we may use
Calc>Probability Distributions>Binomial

```
Binomial Distribution                                        x
                        ⊙ Probability
                        ○ Cumulative probability
                        ○ Inverse cumulative probability

               Number of trials:            20
               Probability of success:      .6

               ○ Input column:
                 Optional storage:

               ⊙ Input constant:            11
                 Optional storage:

     Select
```

We have plugged in for n and p, and clicked on **Probability** among the three choices at the top. We **input** the **constant** *11*. If we wanted to find the probabilities for 11, 12 and 13 successes, we would type all of these numbers into say C1, and then use **Input column** with entry *C1*.

  The output is:

Probability Density Function

```
Binomial with n = 20 and p = 0.6

  x   P( X = x )
 11    0.159738
```

Minitab uses the term 'Probability Density Function' in the output, though strictly speaking this term applies only to the case of continuous random variables, to be discussed later on.

To find the probability that x falls in some range, it is easier if we use the selection **Cumulative Probability** rather than finding a bunch of individual probabilities and summing them. So, for P(x≤10), we use **Calc>Probability Distributions>Binomial** with

Binomial Distribution

- ○ Probability
- ● Cumulative probability
- ○ Inverse cumulative probability

Number of trials: `20`
Probability of success: `.6`

- ○ Input column: ` `
  Optional storage: ` `
- ● Input constant: `10`
  Optional storage: ` `

Select

which produces:

Cumulative Distribution Function

```
Binomial with n = 20 and p = 0.6

  x   P( X <= x )
 10      0.244663
```

The output gives the probability up to and including the number entered. If you wanted P(x<10), you would input '9', since < 10 is equivalent to ≤ 9.

To find P(x>12), note that this equals $1 - P(x \leq 11)$, and then find the **Cumulative Probability** for the input constant '11'.

If you need the probability for a range of values like P(8≤ x ≤16), find the **Cumulative Probability** for both '7' (not 8) and '16' and subtract.

To graph this function, enter values for x in say C1 (name it 'x') and then let Minitab compute the probabilities, which you would store in say C2 (name it 'prob'). Graph C2 versus C1. See the steps below.

Calc>Make Patterned Data>Simple Set of Numbers

Simple Set of Numbers

Store patterned data in: `x`

From first value: `0`
To last value: `20`
In steps of: `1`

List each value `1` times
List the whole sequence `1` times

Calc>Probability Distribution>Binomial

56

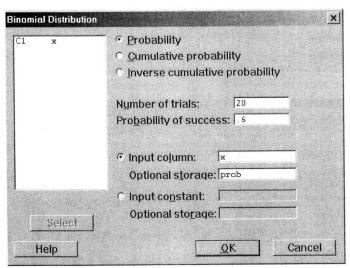

This gives you a worksheet where the first 5 rows are:

↓	C1	C2
	x	prob
1	0	0.000000
2	1	0.000000
3	2	0.000005
4	3	0.000042
5	4	0.000270

Now, use **Graph>Bar Chart: Simple; Bars represent: Values from a table: Graph variables:** *prob*, **Categorical variable:** *x*

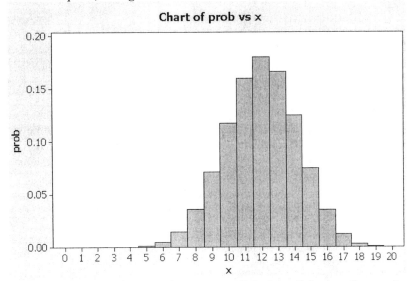

To reduce the space between bars to zero, I double-clicked on the x-values, producing the **Edit Scale** dialog box, then **Scale: Space Between ...: Gap between clusters**: *0*

You can also use **Graph>Scatterplot (Data View: Data Display: Project lines)**, to graph 'prob' vs 'x', with skinny bars projecting down.

$\mu \pm 2\sigma$ works out to be (7.6, 16.4) here (see text). Since the distribution is fairly bell-shaped, we would then expect about .95 in probability to lie between 8 and 16 (inclusive). Verify this by adding up nine probabilities from the graph or worksheet above; or you can evaluate $P(8 \le x \le 16)$ using **Calc>Probability Distributions>Binomial: Cumulative probability** (input '*7*', input '*16*', and subtract).

Exercise 4.125 - M/S page 232

(a) Here we have 25 students, with 80% chance of nasal allergies. Enter these parameters and calculate $P(x < 20) = P(x \le 19)$: **Calc>Probability Distributions> Binomial: Cumulative probability, Number of Trials:** *25*, **Probability of Success:** *.8*, **Input Constant:** *19*

Cumulative Distribution Function

```
Binomial with n = 25 and p = 0.8

  x   P( X <= x )
 19     0.383311
```

(b) **Calc>Probability Distributions>Binomial: Cumulative probability, Number of Trials:** *25*, **Probability of Success:** *.8*, **Input Constant:** *15* gives

Cumulative Distribution Function

```
Binomial with n = 25 and p = 0.8

  x   P( X <= x )
 15     0.0173319
```

Hence, 1 - .0173 is the requested probability of finding more than 15 with allergies, by the complementary rule.

Exercise 4.66 - M/S page 217

Instead of finding the cumulative probability up to a specified x_0, here we need to find the value of x_0, which possesses a certain cumulative probability. Rather than the Cumulative Probability function, we use the *INVERSE* Cumulative Probability function. The desired cumulative probability is .95 (5% passing means 95% failing), so input this constant using **Calc>Probability Distributions>Binomial: Inverse cumulative probability, Number of Trials:** *20*, **Probability of Success:** *.5*, **Input Constant:** *.95*

Inverse Cumulative Distribution Function

```
Binomial with n = 20 and p = 0.5

  x   P( X <= x )        x   P( X <= x )
 13     0.942341        14     0.979305
```

Minitab shows the outcomes that come close to our requirement. With 15 as the passing grade, 98% will fail (obtain 14 or less) and 2% will pass. But if 14 is the passing grade, 6% (1 - .9423) pass – too many.

Poisson Distribution

This distribution tends to arise, when we are counting the number of events of a certain type occurring over a region of time or space, where events occur independently of each other, and at a constant rate or density.

Example 4.13 - M/S page 219

To answer questions (b) – (d), we just need cumulative probabilities for x = 1, 5, 4, so enter these three numbers into C1, and then:

Calc>Probability Distributions>Poisson with the following entries:

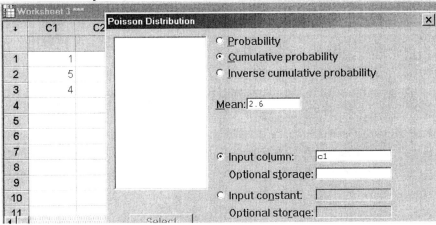

Hit **OK**, and we get

Cumulative Distribution Function

```
Poisson with mean = 2.6

x   P( X <= x )
1      0.267385
5      0.950963
4      0.877423
```

So, now you can easily answer (b) P(x<2) = .2674, (c) P(x>5)= 1 – P(x≤5) = 1 - .9510, (d) P(x=5) = (x≤5) – P(x≤4) = .9510 - .8774

Alternatively, for part (d), we could just select **Probability** rather than **Cumulative probability** in the **Poisson Distribution** dialog box, and input the constant '5'.

Example 4.132 - M/S page 232

Let x = number of admitted patients. We need $P(x \geq 5) = 1 - P(x \leq 4)$. Use **Calc>Probability Distributions>Poisson: Cumulative probability, Mean:** *2.5*, **Input Constant:** *4*

Cumulative Distribution Function

```
Poisson with mean = 2.5

x   P( X <= x )
4      0.891178
```

So, 1 - .8912 is the probability of at least 5 arrivals, hence too few beds.

Hypergeometric distribution

This distribution arises when we count the number of objects of type 'S' (Success) in a random sample of size n from a collection of N objects containing r elements of type S.

Example 4.14b. - M/S page 225

There are 10 applicants, 6 male, 4 female. We select at random 3 of the applicants. Find the probability that no females are selected. Select **Calc>Probability Distributions>Hypergeometric**:

We get: (M replaces the textbook symbol r)

```
Hypergeometric with N = 10, M = 4, and n = 3

x   P( X = x )
0     0.166667
```

Chapter 5 – Continuous Random Variables

In this chapter, we will do calculations using the normal and exponential distributions, using
Calc>Probability Distributions>Normal
Calc>Probability Distributions>Exponential

Plot and compare normal curves, with
Graph>Scatterplot: Multiple Graphs

Construct a normal probability plot, using
Graph>Probability Plot

Draw a histogram with a normal curve superimposed, using either
Stat>Basic Statistics>Display Descriptive Statistics: Graphs: Histogram with normal curve
     or
Graph>Histogram - With Fit

The Normal Distribution

Use **Calc>Probability Distributions>Normal** to find normal curve areas and percentiles, or even the height of the normal curve at any point.

Figure 5.6: Drawing normal curves - M/S page 244

Let's draw two of the three normal curves shown, i.e. Normal(0,1.5) and Normal(3,1)
Enter –5 to +6 in steps of 0.1 in C1, via **Calc>Make Patterned Data: Simple Set of Numbers**. Then use **Calc>Probability Distributions>Normal**, as indicated below.

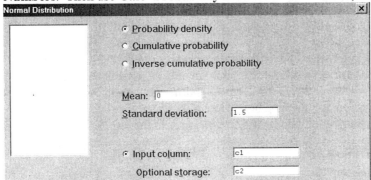

Repeat, changing the mean and standard deviation to 3 and 1, and the storage column to c3.

Then, **Graph>Scatterplot (Simple),** with **Multiple Graphs** as below:

61

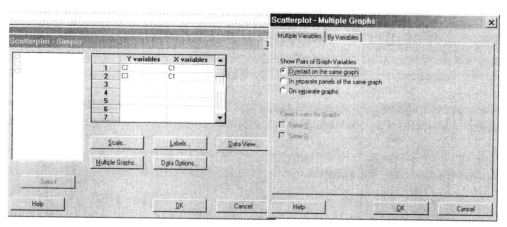

We get:

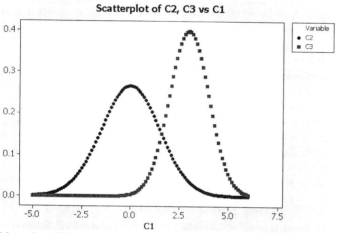

Note the difference in mean and spread, and also how the height is greater for the less dispersed curve, in order to ensure the same total area under each curve.

Example 5.6 - M/S page 249

The normal curve table in the text gives areas under a *standard* normal curve only, so when using that table, you need to standardize your values first. But when using Minitab for normal curve calculations, there is *no need to standardize*. Just specify the correct mean and standard deviation in the dialog box. Here, $\mu = 10$ and $\sigma = 1.5$. We want to find $P(8 \leq x \leq 12)$. To find the area under the normal curve to the left of 8, we must select **Cumulative Probability** from the choices listed at the top of the dialog box for **Calc>Probability Distributions>Normal**. First input the constant '8':

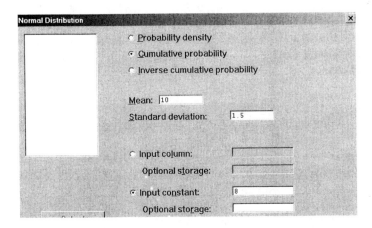

Repeat changing the input constant from '8' to '12'. You will see:

Cumulative Distribution Function

```
Normal with mean = 10 and standard deviation = 1.5

x   P( X <= x )
8     0.0912112
```

Cumulative Distribution Function

```
Normal with mean = 10 and standard deviation = 1.5

 x   P( X <= x )
12     0.908789
```

The desired probability $P(8 \leq x \leq 12)$ equals the difference of $P(x \leq 12)$ and $P(x \leq 8)$ or .9088 - .0912 = .8176. In general, $P(a \leq x \leq b) = \text{CumProb}(b) - \text{CumProb}(a)$, for a continuous distribution.

Example 5.10 - M/S page 255

Here, the problem is reversed from the example above. Given an area, we want to find the corresponding value of x_0. We want to find the 90th percentile, i.e. a value x_0 such that $P(x \leq x_0) = .90$. Instead of using the Cumulative Probability function, we use the **Inverse cumulative probability** selection from **Calc>Probability Distributions>Normal**

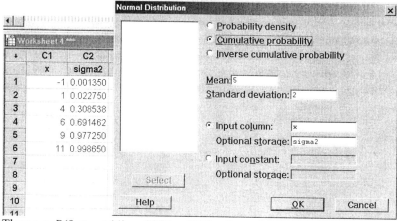

which gives us

Inverse Cumulative Distribution Function

```
Normal with mean = 550 and standard deviation = 100

P( X <= x )          x
       0.9      678.155
```

The answer is 678. This was simpler than the calculation in the text, because we did not have to start with a standard normal and then convert to the actual test score distribution.

Sometimes, the information given is not a left-side area. Then, you have to determine what is the area to the left of the unknown x_0 since this is the input constant required by Minitab. See exercise 5.37 below.

Figure SIA5.2 - M/S page 257

In this example, various cumulative probabilities have to be found for a $N(\mu=5, \sigma)$ distribution with varying choices for σ. For say $\sigma = 2$, enter the desired x-values in C1 (named 'x'), then use **Calc>Probability Distributions>Normal** as below, including a **storage** column.   **OK**, and the probabilities in C2 below will appear.

Then e.g. $P(9 \le x \le 11) = .9987 - .9773$.   Repeat for the other two values of σ.

64

Exercise 5.35 - M/S page 258

To find the three probabilities requested, we will need areas corresponding to 45, 55, 51, 52, so enter these numbers into C1. Then, **Calc>Probability Distributions>Normal: Cumulative Probability, Mean:** *50*, **Standard Deviation:** *3.2*, **Input Column:** *C1*

Cumulative Distribution Function

```
Normal with mean = 50 and standard deviation = 3.2

   x   P( X <= x )
  45     0.059085
  55     0.940915
  51     0.622670
  52     0.734014
```

(a) P(x > 45) = 1 - .0591
(b) P(x < 55) = .9409
(c) P(51 < x < 52) = .7340 - .6227

Exercise 5.43 - M/S page 260

We want to find the number *d* such that 30% of the tree diameters exceed *d*. This implies that 70% have diameters less than *d*, so we input *0.70* into **Calc>Probability Distributions>Normal: Inverse Cumulative Probability, Mean:** *50*, **Standard Deviation:** *12*, **Input Constant:** *.0.70*

Inverse Cumulative Distribution Function

```
Normal with mean = 50 and standard deviation = 12

P( X <= x )        x
       0.7   56.2928
```

So only 30% of trees will have diameters in excess of 56.3 cm.

Assessing Normality: The Normal Probability Plot

A common diagnostic tool for assessing normality of data is the *normal probability plot*. We plot the data versus the average values that would arise if sampling from a standard normal. The latter are called the 'normal scores' by Minitab and may be calculated using **Calc>Calculator** and choosing **Normal scores** from the **Functions** list. However, **Graph>Probability Plot** provides a quicker one-step approach.

Example 5.11 - M/S page 262

Open up the EPAGAS file, containing the EPA mileages first encountered in table 2.3 in chapter 2. We want to assess whether the EPA mileages may be roughly normal in distribution. To construct a normal probability plot select: **Graph>Probability Plot**, though the labeling in the plot differs, and some extra features are included. Select

Graph>Probability Plot (Single). Under **Graph variables,** enter *MPG*, then select **Distribution** and choose **Distribution:** *Normal (default choice)*, as shown below:

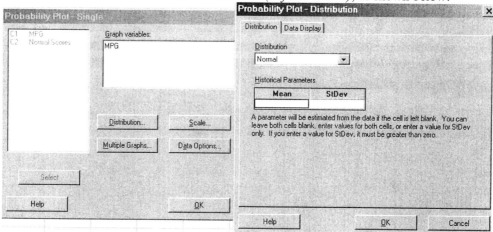

Also click on **Scale** in the dialog box above, then **Y-Scale Type**, and choose **Score** (the expected normal score), instead of the default **Percent** selection:

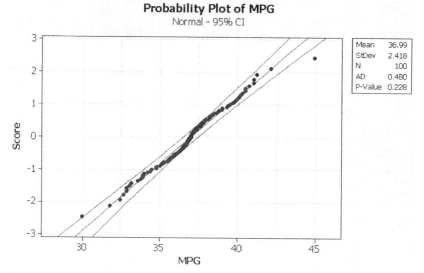

OK, **OK**, and you get:

Probability Plot of MPG
Normal - 95% CI

Mean	36.99
StDev	2.418
N	100
AD	0.480
P-Value	0.228

Since the points in our normal probability plot above fall reasonably close to a straight line, we may say that the distribution of the mileages appears to be close to normal.

Note the extra features in the plot above. A straight line, and two (95%) confidence interval bands have been drawn in for some extra guidance re linearity of the plot. If the

66

points lie within the bands, the normal distribution is a close fit to the data. While the mild S-shape here suggests a small departure from normality, the plot as a whole is quite close to being straight, with only one point outside the confidence bands. The distribution is nearly normal.

   Using the **Probability Plot: Distribution: Data Display** sub-dialog box (default settings shown below), you can remove, if you wish, the confidence interval bands, by un-clicking: *Show confidence interval*, and the straight line (= *distribution fit*), by selecting: *'Symbols only'*.

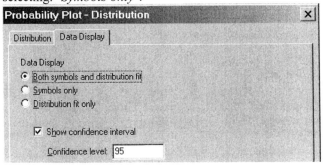

For alternative assessments of normality, you can use **Stat>Basic Statistics>Display Descriptive Statistics**: **Variables:** MPG, which produces:

Descriptive Statistics: MPG

Variable	N	N*	Mean	SE Mean	StDev	Minimum	Q1	Median	Q3
MPG	100	0	36.994	0.242	2.418	30.000	35.625	37.000	38.375

Variable	Maximum
MPG	44.900

Now, you can compute the IQR/*s* ratio, and compare with 1.3. Or you can calculate the 1, 2, 3 standard deviation intervals around the mean, for comparison with 68%, 95%, 100%, resp. [**Data>Sort** will put the data into ascending order for easier counting]

A histogram with superimposed normal curve can also be produced via:
Stat>Basic Statistics>Display Descriptive Statistics: Variables: *MPG*, **Graphs: Histogram with normal curve**.
We get the following graph (along with the descriptive stats displayed above):

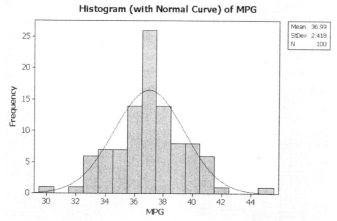

A little peaked in the middle, but quite close to a normal shape. You can also produce this plot using **Graph>Histogram - With Fit: Graph variables:** *MPG* (as illustrated in the next exercise).

Exercise 5.60 - M/S page 267

Open up the DARTS file on your text CD. We have **grouped data** here, with (rounded) distance from the target in c1, and corresponding frequencies in c2, so in any dialog box, you know you will have to specify, somewhere, where to locate these **frequencies** - otherwise, Minitab will have to assume that each listed distance is observed only once!

We want to assess whether the distances from the target may be normally distributed. For a normal probability plot, use:

Graph>Probability Plot (Single): Variables: *C1,*
        Distribution: Distribution: *Normal,*
        Data Options: Frequency: Frequency variable: *C2,*
        Scale: Y-Scale Type: *Score*

There is some clear curvature above - how do you interpret this?

68

Let's also examine the histogram via:
Graph>Histogram - With Fit: Graph variables: *C1,* **Data Options: Frequency: Frequency variable:** *C2.*

With **Histogram - With Fit**, there is a default setting of: **Data View: Distribution: Fit Distribution:** *Normal.* If instead you had selected the **Histogram - Simple**, you would then have to select and click your way through **Data View**, to have the *Normal* fit added to the histogram.

We get:

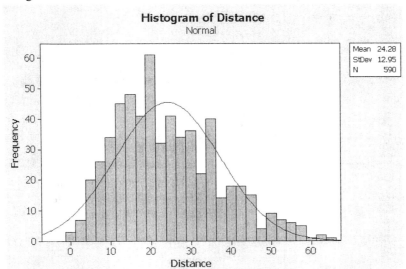

[Here, we cannot use **Stat>Basic Statistics>Display Descriptive Statistics: Graphs: Histogram with normal curve**, since it makes no allowance for a **frequency** column]

Exponential Distribution (optional)

This is another continuous distribution, often providing a good model for the time between occurrences of random events. It is one of the distributions listed under **Calc>Probability Distributions**.

Example 5.13 - M/S page 277

To find $P(x > 5) = 1 - P(x < 5)$, for an exponential distribution with mean equal to 2, we use **Calc>Probability Distributions>Exponential** as below:

69

Note that you should plug in the desired mean of *2* next to **Scale**, and keep the **Threshold** at its default value of *0.0* . Hit **OK**, and we get

Cumulative Distribution Function

```
Exponential with mean = 2

x   P( X <= x )
5      0.917915
```

So, 1 - .918 is our answer.

Exercise 5.96 - M/S page 280

The time to goal, x, is exponentially distributed with a mean of 9.15. We want P(x < 3) in (a) and P(x > 20) in (b), so enter '3' and '20' in C1, and then **Calc>Probability Distributions>Exponential: Cumulative Probability, Scale (= Mean):** *9.15*, **Input Column:** *C1*

Cumulative Distribution Function

```
Exponential with mean = 9.15

 x      P( X <= x )
 3      0.279543
20      0.887611
```

Hence, P(x < 3) = .2795, and P(x > 20) = 1 – P(x < 20) = 1 - .8876.

Chapter 6 – Sampling Distributions

In this chapter, we will simulate random samples, using
Calc>Random Data>Uniform
Calc>Random Data>Integer

Calculate statistics for random samples, stored in rows, using
Calc>Row Statistics

And calculate probabilities for a sample mean, using
Calc>Probability Distributions>Normal

[Also, you can learn much about sampling distributions by playing with the Java applet at http://www.ruf.rice.edu/~lane/stat_sim/sampling_dist/index.html]

Example 6.2 - M/S page 296

We want to approximate the sampling distribution of the sample mean and sample median for samples of size 11 from a uniform population. We will generate 700 (rather than 1000) samples of 11 observations each. For ease of computation, we put one sample of 11 observations in each row, rather than using the columns for the samples. Generate the random data using **Calc>Random Data>Uniform**

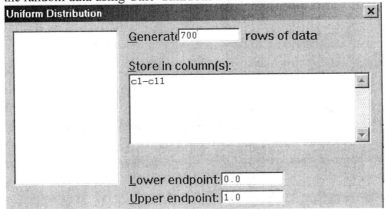

The maximal worksheet capacity with the student version of Minitab is only 10,000 cells, so if we generated 1000 rows × 11 columns, we would run over the capacity, and be forced to delete some cells in the worksheet. And we need space for some upcoming calculations too. But if you are using the full professional version of Minitab, you will have much greater capacity.

You will see entries ranging from 0 to 1, in each of c1-c11. Plot the data in any of the columns (**Graph>Histogram**), to confirm that the data is indeed uniformly distributed between 0 and 1, like in Fig 6.4 in the text. For each row, calculate the sample mean, and store it in a column named 'mean' (c12), using **Calc>Row Statistics**

71

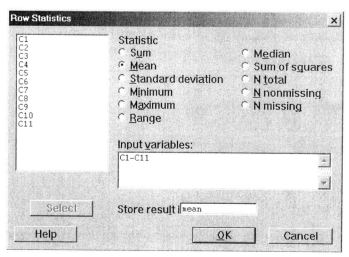

Repeat, changing the **Statistic** from **Mean** to **Median**, and store the results in a column named 'median', which will be c13. Now, the worksheet looks like:

↓	C1	C2	C3	C4	C5	C6	C7	C8	C9	C10	C11	C12	C13
												mean	median
1	0.759866	0.833816	0.936660	0.317832	0.724288	0.666838	0.332455	0.89263	0.916749	0.781813	0.942884	0.736894	0.781813
2	0.642128	0.467073	0.498788	0.939095	0.214278	0.133523	0.943245	0.83322	0.471283	0.939272	0.138774	0.565517	0.498788
3	0.352913	0.485840	0.094953	0.997388	0.271342	0.991927	0.605900	0.52499	0.511961	0.828799	0.494537	0.560050	0.511961
4	0.661971	0.508402	0.468794	0.814923	0.928570	0.791572	0.066269	0.34349	0.262218	0.033238	0.668704	0.504377	0.508402
5	0.824412	0.593455	0.881567	0.624515	0.854139	0.984412	0.375832	0.10210	0.082367	0.324977	0.471810	0.556326	0.593455

Check that the mean and median of the 11 numbers in row 1 are indeed .7369 and .7818, as displayed above.

Since we have selected a sample of size 11 many times, and recorded the mean of each sample in c12, a histogram for c12 will show us, approximately, the sampling distribution of the sample mean (for a sample of size 11 from a uniform population). Similarly, a histogram of c13 approximates the sampling distribution of the sample median. Let's have a look at these two distributions, side by side, with **Graph>Histogram - Simple: Graph variables:** *mean median*, **Multiple Graphs: Multiple Variables:** *In separate panels, Same Y, Same X,* **Scale: Y-Scale Type:** *Percent*

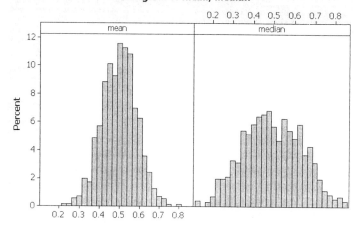

Histogram of mean, median

Both sampling distributions are mound-shaped (unlike the population), appear to center at 0.5 (the population center), and have less dispersion than the population. The sample mean is also less dispersed than the sample median, suggesting that the mean would be the superior estimator of the population center (if unknown, as is usually the case) for this type of population.

Exercise 6.9 - M/S page 298

We want to approximate the sampling distribution of the sample mean and also the sample variance for samples of size 25 from a discrete uniform population consisting of the integers 0 - 99. To accommodate space constraints in the student version of Minitab, generate 300 samples (rather than 500).

Calc>Random Data>Integers: *300* **rows of data, store in** *C1-C25*, **Min:** *0*, **Max:** *99*, gives you 300 rows of random data with 25 numbers per row. For each row (= sample), compute the mean and store in c26 ('mean'), via **Calc>Row Statistics: Statistic: Mean, Input Variables:** *C1-C25*, **Store result in:** *mean*. In c26, the row means will appear, as below. Repeat, but now click on **Standard deviation,** and store the result in c27. To convert the numbers in c27 into variances, you need to square them, using:

Calc>Calculator: Store result in: *C27*, **Expression:** *C27 * C27*. Name c27 'variance'.

↓	C14	C15	C16	C17	C18	C19	C20	C21	C22	C23	C24	C25	C26 mean	C27 variance
1	15	74	30	86	92	30	25	7	12	83	6	29	53.76	912.84
2	22	25	99	0	40	99	0	18	42	52	7	41	45.92	1048.44
3	47	42	67	45	62	52	79	79	65	69	91	86	54.60	782.42

Produce the histograms with **Graph>Histogram - Simple: Graph Variables:** *C26, C27*

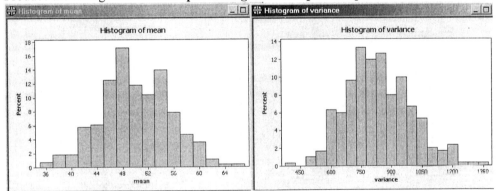

Both look roughly bell-shaped, while the variances show perhaps some slight positive skewness. If you use the professional version of Minitab with 500 or more samples (rows), the shapes will be smoother and closer to the 'true' limiting shape.

Example 6.7b. - M/S page 305

We want the probability that the sample mean is greater than 82, if we draw a random sample of size 25 from a population with mean and standard deviation of 80 and 5 resp. By the Central Limit Theorem, we know that the distribution of the sample mean can be approximated by a normal curve, where μ for the curve is 80, and σ for the curve is 5/√25 or 1. There is no need to standardize, just feed in these parameters to Minitab:

Calc>Probability Distributions>Normal

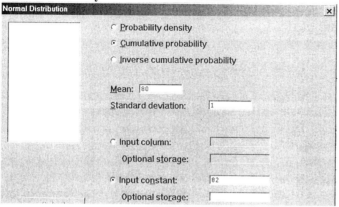

We get:

Cumulative Distribution Function

```
Normal with mean = 80 and standard deviation = 1

  x      P( X <= x )
 82      0.977250
```

By the complementary rule, the answer is 1 - .9772 .

Exercise 6.39 - M/S page 311

The population mean and standard deviation are 5.1 and 6.1 resp. We will draw a random sample of 150 victims. By the Central Limit Theorem, the distribution of the sample mean will be normal, with mean = 5.1, and standard deviation = 6.1/√150 = 0.50. For (c), we want the probability that the sample mean exceeds 5.5, and for (d), the probability that it lies between 4 and 5. So, we need areas corresponding to 5.5, 4, and 5. Enter these three numbers into C1. Select **Calc>Probability Distributions>Normal: Cumulative Probability, Mean:** *5.1*, **Standard Deviation:** *.5*, **Input Column:** *C1*

Cumulative Distribution Function

```
Normal with mean = 5.1 and standard deviation = 0.5

  x     P( X <= x )
 5.5      0.788145
 4.0      0.013903
 5.0      0.420740
```

Hence, the answers are: (c) 1 - .788, (d) .421 - .014

74

Chapter 7 - Confidence Intervals from Single Samples

In this chapter, we will draw a *t*-curve, using
Calc>Probability Distributions> t

Find a confidence interval for a population mean, using
Stat>Basic Statistics>1-Sample t
Stat>Basic Statistics>1-Sample Z   [if σ known]

Find a confidence interval for a population proportion, using
Stat>Basic Statistics>1 Proportion

Confidence Interval for a Population Mean

If σ is known (very unusual), use **Stat>Basic Statistics>1-Sample Z**, to calculate a
confidence interval for the mean (you will be prompted for the value for σ). Otherwise,
you should use **Stat>Basic Statistics>1-Sample t**, for which *s* will be calculated from
the sample and used in the C.I. For a large sample, the text uses a 'z' value in the C.I.,
whereas Minitab uses the more accurate 't' value, which differs little from the 'z' value in
this case. [If you *really* want to precisely imitate the text, for large samples, you would
first find *s* for the sample, then plug in this number for σ in **1-Sample Z**]

Example 7.1 - M/S page 326

Open up the AIRNOSHOWS data file. To find a 90% confidence interval for the
population mean, select **Stat>Basic Statistics>1-Sample t,** and tell Minitab where to find
the **Sample**, then click on **Options**, and change the default level of **Confidence** from *95*
to *90* percent, as below:

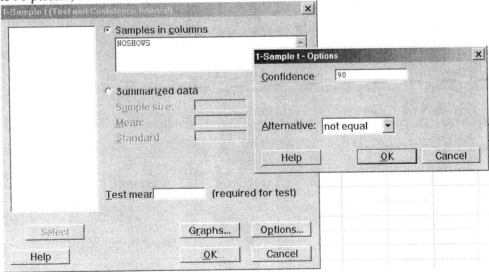

75

OK, OK, and we get some results:

One-Sample T: NOSHOWS

```
Variable    N    Mean    StDev  SE Mean       90% CI
NOSHOWS   225  11.5956  4.1026  0.2735  (11.1438, 12.0473)
```

Note that if we had the sample *summary stats*, rather than raw data, we would just click on **Summarized data** above, and enter the sample size, mean, and standard deviation.

Exercise 7.22 - M/S page 331

Open up the data file ATTIMES, where the attention times may be found in c1. To find a 90% confidence interval for the mean time, select **Stat>Basic Statistics>1-Sample t,** and fill in: **Sample in column:** *C1* and set **Options: Confidence:** *90* . **OK, OK,** and we get:

One-Sample T: Time

```
Variable    N    Mean    StDev   SE Mean       90% CI
Time       50  20.8480  13.4138  1.8970  (17.6676, 24.0284)
```

The 90% C.I. appears above. If you calculate the C.I. from the mean and stdev above, using a *z* value of 1.645, your result will differ slightly from the interval above, because Minitab uses the more accurate *t* value of 1.677 (see M/S section 7.3).

It would also be a good idea to have a look at the data via the **Graphs** selection in the main dialog box, where you can ask for a histogram or boxplot.

Figure 7.7: Compare *t* and *z* curves - M/S page 334

Let's draw this comparison of the standard normal curve and *t* curve (with 4 df), using Minitab. Enter -3 to +3 in steps of .1 in C1, via **Calc>Make Patterned Data>Simple Set of Numbers**. Then **Calc>Probability Distributions> t**

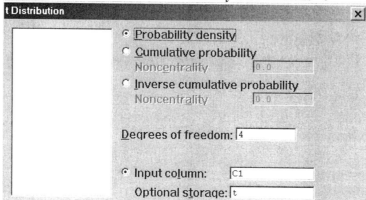

Above, we have calculated the probability density curve values for the numbers in C1, and stored them in a column named 't'.

Repeat this, but switch to **Normal** distribution, with **Mean:** *0*, **Standard Deviation:** *1*, and **Input column:** *C1*, **Optional storage:** *z*

Now plot 't' and 'z' vs. the C1 values, using **Graph>Scatterplot – With Groups,** as below; and in the dialog box, also select **Multiple Graphs: Multiple Variables:** *Overlaid on the same graph*

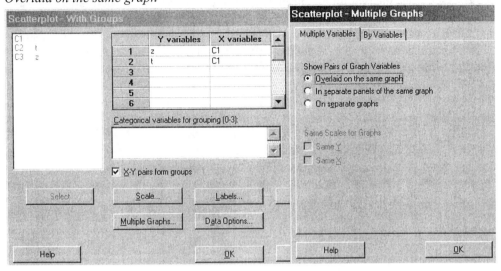

OK, **OK**, and we get

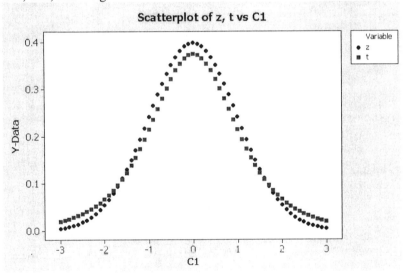

Different symbols and colors are used for each curve. We see that the shapes are similar, but the *t* curve is more dispersed (with fatter tails) than the standard normal curve.

77

Example 7.2 - M/S page 337

Open up the PRINTHEAD data file, showing the number of characters in c1. To produce a 99% confidence interval for the mean number of characters printed before failure, use **Stat>Basic Statistics>1-Sample t** as shown below, and don't forget to set the desired confidence level with the **Options**. To help check the distributional assumptions, also select **Graphs** in the dialog box, and tick off some of the choices.

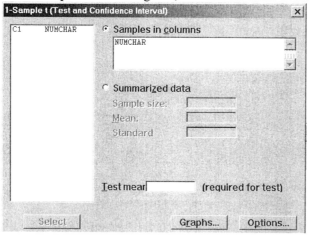

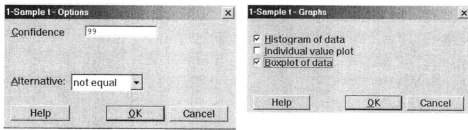

The resultant 99% C.I. is displayed below.

One-Sample T: NUMCHAR

```
Variable    N    Mean    StDev   SE Mean        99% CI
NUMCHAR    15  1.23867  0.19316  0.04987  (1.09020, 1.38714)
```

Note that 'SE Mean' in the output stands for *standard error of the mean*, and equals s/√n, which is .1932/√15 here.

To assess our underlying assumption of normality, have a look at the graphs displayed:

78

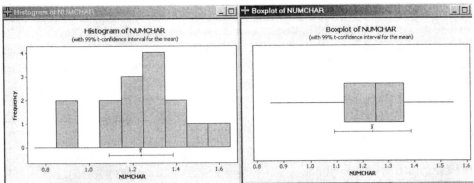

The data appear nearly normal, so our 99% level of confidence may be trusted. The C.I. is also displayed in each of these graphs, directly below the plot. You can also produce a stem and leaf plot, like in the text, using **Graph>Stem-and-Leaf**, or a normal probability plot, using **Graph>Probability Plot**, as discussed in earlier chapters.

Exercise 7.39 - M/S page 342

Open up BRAINPMI, containing the measurements in c1. For a 95% CI for the mean PMI, use **Stat>Basic Statistics>1-Sample t: Sample in column:** *C1*, **Options: Confidence:** *95*. Also, click on **Graphs** and tick off whichever ones you prefer. **OK, OK**, gives you

One-Sample T: PMI

```
Variable   N     Mean     StDev    SE Mean      95% CI
PMI        22   7.30000   3.18493  0.67903   (5.88788, 8.71212)
```

So, we can say, with 95% confidence that the true mean PMI of human brain specimens lies between 5.89 and 8.71. Or can we? Look at the accompanying data plots:

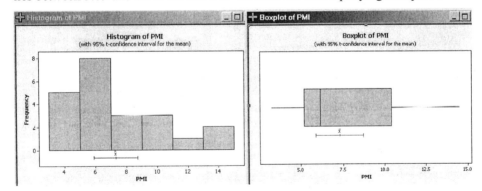

Confidence Interval for a Proportion

The text emphasizes use of the *large sample C.I.* for a proportion, which uses the normal approximation. Minitab will produce an *exact* C.I., based on the Binomial distribution *unless* you select **Options: Use test and interval based on normal distribution**. The

79

exact approach may be used if the normal approximation is not applicable or as an alternative to the 'Adjusted Confidence Interval' discussed in the text.

If you apply the normal approximation when the sample size is not adequate (Minitab's criterion is more lax than the text's), Minitab gives you the *following warning*:

```
* NOTE * The normal approximation may be inaccurate for small samples.
```

Example 7.4 - M/S page 345

We have 257 successes in 484 trials. We want a 90% C.I. for p. Enter this information in **Stat>Basic Statistics>1 Proportion,** under **Summarized data**, and then click on **Options**, enter **Confidence Level:** *90*, and tick **Use test and interval based on normal distribution,** as below. The items **Test Proportion** and **Alternative** relate to hypothesis testing - just ignore these for now.

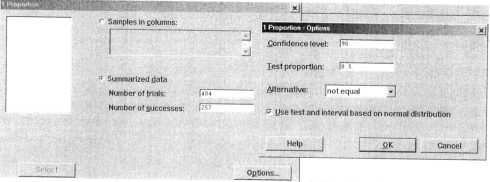

OK, OK, and you get:

Test and CI for One Proportion

```
Test of p = 0.5 vs p not = 0.5


Sample    X    N   Sample p          90% CI           Z-Value   P-Value
1       257   484  0.530992   (0.493681, 0.568303)     1.36      0.173
```

So, with 90% confidence, the proportion of all Florida consumers who are confident about the economy lies between .494 and .568. [Ignore the 'Z-value' and 'P-value' items, which relate to hypothesis testing - chapter 8]

Exercise 7.98 - M/S page 361

Open up the OILSPILL file (from back in chapter 2) showing causes of 50 oil spills in c2. Use **Stat>Tables>Tally Individual Variables: Variables:** *C2,* to find the count for hull failures. We see that 12 of 50 were from hull failure. Plug these numbers into **Stat>Basic Statistics>1 Proportion: Summarized Data:** *50* **trials,** *12* **successes,** with **Options: Confidence level:** *95,* **Use test and interval based on normal distribution**

Test and Confidence Interval for One Proportion

```
Test of p = 0.5 vs p not = 0.5

Sample      X      N    Sample p        95.0 % CI          Z-Value  P-Value
1          12     50    0.240000   (0.121621, 0.358379)    -3.68    0.000
```

We find that the proportion of oil spills caused by hull failure is between .12 and .36, with 95% confidence.

Exercise 7.101 - M/S page 361

(a) Open up SUICIDE, showing data on 37 suicides committed by inmates of correctional facilities. Y/N in C5 indicates whether or not the inmate was charged with murder/manslaughter. We want to estimate the proportion of suicides committed by inmates charged with murder/manslaughter. Here we have the *raw categorical data* in C5. While we could do a tally for C5, and enter the summary numbers, let's proceed instead as follows: Select **Stat>Basic Statistics>1 Proportion** but choose **Samples in Columns**, and enter column *C5*. Make sure **Options** show a *95%* confidence level, and use of the **normal distribution**. We get:

Test and CI for One Proportion: Murder?

```
Test of p = 0.5 vs p not = 0.5

Event = Y

Variable    X    N    Sample p        95% CI          Z-Value  P-Value
Murder?    14   37    0.378378   (0.222109, 0.534648)   -1.48    0.139
```

Minitab decided that the **Event** of interest (or **success**) = **Y** (Yes), and counted these up, reporting a total of X = 14 successes. Why was 'Y' and not 'N' deemed a success? By default, the entry which is last in alphabetical order, is deemed the 'success'. If desired, we can change the way Minitab orders the categories from 'N, Y' (alphabetical) to 'Y, N'. Right-click anywhere on C5, then choose **Column>Value Order** (or at the top: **Editor>Column>Value Order**). Click on **User-specified order**. Type *Y*, press Enter, then type *N*. Hit **OK**. Now, if you run the same procedure, the C.I. will be for the proportion of N's rather than Y's, since N now follows Y in the ordering.

(b) We want a C.I. for the proportion of suicides at Night. The raw data on time of suicide are in C6, showing three different times of day: D (day), A (afternoon), N (night). But this procedure only works for *dichotomous* columns (two categories only). We are counting N's, so it is our 'success' event. Convert everything that is not a success into one single category - e.g. convert all the A's into D's, directly or with **Data>Code>Text to Text**. 'N' comes last alphabetically, so Minitab deems it a 'Success', and there is no need to change the **Value Order** as explained in (a). Now, run **Stat>Basic Statistics>1 Proportion: Samples in columns:** *C6*, and we get:

Test and CI for One Proportion: Time

```
Test of p = 0.5 vs p not = 0.5

Event = N

Variable    X    N   Sample p        95% CI        Z-Value  P-Value
Time        26   37  0.702703  (0.555428, 0.849978)   2.47    0.014
```

From this, we see that the proportion of suicides committed at night is between .56 and .85, with 95% confidence.

(c) Use **Stat>Basic Statistics>1-Sample t** (as in example 7.2 above)

(d) Use **Stat>Basic Statistics>1 Proportion** (as in part a. above)

Chapter 8 – Test of Hypothesis for Single Samples

In this chapter, we will execute one sample tests for the mean of a population, using
Stat>Basic Statistics>1-Sample Z
Stat>Basic Statistics>1-Sample t

Test for the population proportion, using
Stat>Basic Statistics>1 Proportion

Determine power, or type II error probability, with
Stat>Power and Sample Size>1-sample Z

And execute a chi-square test for the population variance, using
Calc>Column Statistics
Calc>Calculator
Calc>Probability Distributions>Chi-Square

Testing for a hypothesized mean

To test a hypothesis about μ, we use the **1-Sample Z** test when σ is known (quite unlikely!), or the **1-Sample t** test when σ is unknown. When σ is unknown and the sample size n is large, the resulting statistic (using *s*) has a *t* distribution with large degrees of freedom. The text approximates this with the standard normal distribution. Minitab's *t* calculation is more accurate, but the difference is slight.

Example 8.4 - M/S page 386

Open up the HOSPLOS data file. We want to test for evidence that the true mean length of stay (LOS) is less than 5. Since we have an assumed value for σ equal to 3.68, we may use the z-test here: **Stat>Basic Statistics>1-Sample Z**

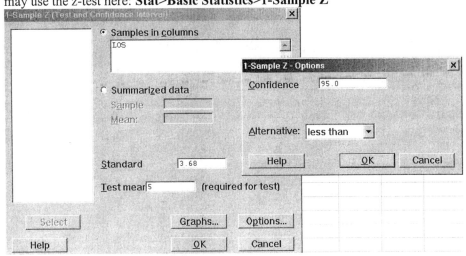

In the dialog box, we have entered the null **test** value for the **mean** and the assumed value of the **Standard deviation** (oops, the word *deviation* is missing from this expression in the dialog box), and used the **Options** to choose the appropriate one-sided **alternative**: *less than*. The output is:

One-Sample Z: LOS

```
Test of mu = 5 vs < 5
The assumed standard deviation = 3.68
                                                   95%
                                                  Upper
Variable    N     Mean    StDev   SE Mean   Bound      Z      P
LOS       100   4.53000  3.67755  0.36800  5.13531  -1.28  0.101
```

The value of the z-statistic is shown, but all we really need is the P-value. Since P = .101 is greater than α=.05, we do not have sufficient evidence to reject the null hypothesis.

Examples 8.5 & 8.6 - M/S page 392

Look for the EMISSION data file on your CD, or open a new worksheet (**File>New: Minitab Worksheet**), type the data *15.6, 16.2, 22.5, 20.5, 16.4, 19.4, 19.6, 17.9, 12.7, 14.9* into C1, and name this column *E-level*. To test the hypothesis that the true mean is *less than 20*, we perform a t-test via: **Stat>Basic Statistics>1-Sample t**

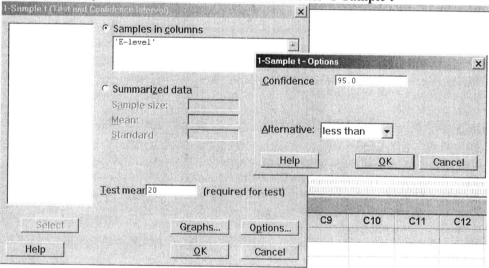

which produces:
One-Sample T: E-level

```
Test of mu = 20 vs < 20
                                                   95%
                                                  Upper
Variable    N     Mean    StDev   SE Mean   Bound      T      P
E-level    10   17.5700  2.9522   0.9336  19.2814  -2.60  0.014
```

There is some evidence to reject the null hypothesis, since the p-value is small, but the evidence is not quite sufficient at the requested 1% level (since p > α= .01). To check the accuracy of this p-value calculation, ***don't forget to examine the data***:

84

Use say **Graph>Stem-and-Leaf**, to produce:

Stem-and-Leaf Display: E-level

```
Stem-and-leaf of E-level   N  = 10
Leaf Unit = 1.0

    1    1  2
    3    1  45
   (3)   1  667
    4    1  99
    2    2  0
    1    2  2
```

No indication of non-normality, so we can trust the p-value reported. Or when you run the **One-Sample t**-test, just remember to click on **Graphs** in the dialog box, and then choose some:

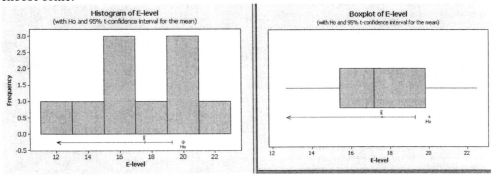

Or, you might have a look at the normal probability plot with **Graph>Probability Plot (Simple)**:

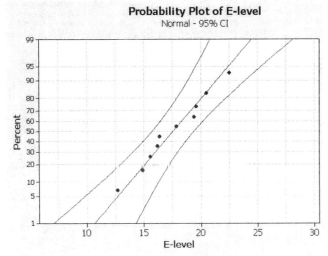

Pretty close to a straight-line pattern!

Exercise 8.64 - M/S page 395

Open up the IRONTEMP file. To test whether the true mean differs from the target of 2550 degrees, use **Stat>Basic Statistics>1-sample t: Samples in:** *C1*, **Test Mean:** *2550*, **Options: Alternative:** *not equal*, and we get

One-Sample T: Temp

```
Test of mu = 2550 vs not = 2550


Variable   N    Mean   StDev  SE Mean       95% CI           T      P
Temp       10  2558.70  22.75   7.19    (2542.43, 2574.97)  1.21  0.257
```

This shows no evidence of a problem, since P is large. But, is the p-value trustworthy? Using the **Graphs** option in the dialog box, we see e.g.:

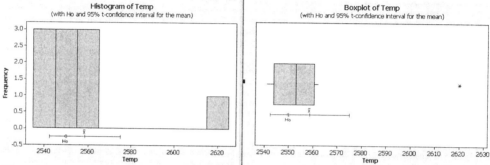

There is one extreme outlier! We should examine what went wrong for this particular production run. *We learn more from this plot* than from the t-test output above! If you remove the outlier, and redo the analysis, P will be even higher (.49).

Testing for a hypothesized proportion

To test a hypothesis about p, use **Stat>Basic Statistics>1 Proportion**. To utilize the normal approximation (large n), select the **Option: Use test and interval based on normal distribution**. If n is not sufficiently large, de-select this option and Minitab will perform an *exact* test. Minitab will issue a warning if you use the normal approximation inappropriately.

Examples 8.7 & 8.8 - M/S page 399

We have 10 defectives in a sample of 300 batteries. Does this constitute strong evidence that in the entire shipment, we have less than 5% defective? Use **Stat>Basic Statistics>1 Proportion,** along with the **Options** provided, as below:

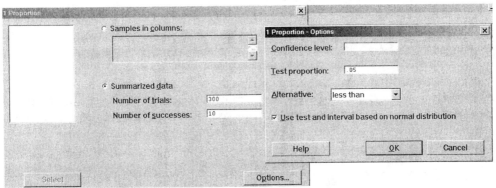

Note that here we have **Summarized data**, not raw data.   We get:

Test and CI for One Proportion

```
Test of p = 0.05 vs p < 0.05
                                    95%
                                  Upper
Sample    X    N   Sample p      Bound    Z-Value  P-Value
1        10  300   0.033333    0.050380    -1.32    0.093
```

Since the P-value of .093 is high (> α=.01), we do not have good support for the hypothesis that less than 5% of the entire lot are defective. The Z-value and hence P-value differ slightly from what you see in the text, due to rounding error.

Since no warning appears in the printout, the normal approximation should be adequate here. But, if you are not sure, de-select the **Option: Use test and interval based on normal approximation** and repeat.

Exercise 8.89 - M/S page 404

After taking a placebo, 70% of 7000 or 4900 patients reported improvement. If the placebo has no effect, then p = 0.5. Proceed with **Stat>Basic Stats>1 Proportion: Summarized data:** *7000* **trials,** *4900* **successes,** with **Options: Test proportion:** *0.5,* **Alternative:** *Not equal,* **Use test based on normal distribution**. I have chosen a two-sided alternative, but if you wish to test only for evidence of beneficial effect (rather than for evidence of any type of effect), select **Alternative:** *Greater than*. It won't change the clear conclusion below:

Test and CI for One Proportion

```
Test of p = 0.5 vs p not = 0.5

Sample     X     N   Sample p        95% CI          Z-Value  P-Value
1       4900  7000   0.700000  (0.689265, 0.710735)   33.47    0.000
```

Since P is 0.000, such a large outcome cannot be due to chance. We have strong evidence of a (beneficial) placebo effect.

87

Type II Error Probability (β) and Power of a Test

We usually need to know how well the test performs when the null is not true, i.e. how often we make the right decision [the power] or the wrong decision [β = probability of a type II error]. Minitab's **Stat>Power and Sample Size** helps you calculate power (or β = 1 - power) for a variety of tests. It also helps you find the sample size required in order to achieve some specified power, at alternatives deemed to be of practical importance.

Example 8.9 - M/S page 407

Here, we are testing H_0: $\mu = 1.2$ versus H_0: $\mu \neq 1.2$, with $\alpha = .05$ (and repeat for $\alpha = .01$). σ is assumed to equal .5. We want the power of the test (and $\beta = 1 -$ power) when $\mu = 1.1$. Enter this exact information, using **Stat>Power and Sample Size>1-sample Z:**

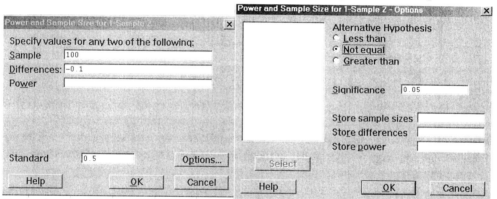

[Errors in dialog box: **Standard** should read **Standard deviation**, **Sample** should read **Sample sizes,** and **Significance** should read **Significance Level**]

Note that $\mu - \mu_0 = 1.1 - 1.2 = -0.1$, so we entered this under **Differences** above. We get

Power and Sample Size

```
1-Sample Z Test

Testing mean = null (versus not = null)
Calculating power for mean = null + difference
Alpha = 0.05   Assumed standard deviation = 0.5

                 Sample
Difference        Size      Power
     -0.1          100   0.516005
```

So, the power is .516 and hence the type II error probability, β, is 1 - .516 = .484. To find the power and β, if α is set at .01, repeat the above, but change the value of the **Significance level** in the **Options** box, to *.01*.

Exercise 8.98 - M/S page 411

We are testing H_0: $\mu=16$ vs H_a: $\mu<16$ with $\alpha=0.1$. We want the power of this test at $\mu=15.5, 15, 14.5, 14, 13.5$, assuming $\sigma=9.32$. Use **Stat>Power and Sample Size>1-sample Z: Differences:** *-.5 –1 –1.5 –2 –2.5*, **Sample Size:** *33*, **Standard deviation:** *9.32*, with **Options: Alternative Hypothesis:** *Less than*, **Significance level:** *0.10*, **Store differences in:** *Diff*, **Store power values in:** *Power*. We get:

Power and Sample Size

```
1-Sample Z Test

Testing mean = null (versus < null)
Calculating power for mean = null + difference
Alpha = 0.1  Assumed standard deviation = 9.32

               Sample
Difference     Size      Power
      -0.5       33   0.165185
      -1.0       33   0.252967
      -1.5       33   0.360547
      -2.0       33   0.480534
      -2.5       33   0.602326
```

Use **Graph>Scatterplot (Simple): Y:***Power*, **X:***Diff*, and we get

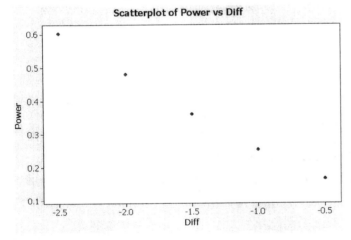

Scatterplot of Power vs Diff

(c) We can estimate the power when $\mu - 14.25$, or Diff = -1.75, at about .41 or .42 from the plot above, or find the exact value by repeating part (a.) as above but with **Stat>Power and Sample Size: 1-Sample Z: Sample size:** *33*, **Differences:** *-1.75*

Test for a Population Variance (optional)

Minitab does not explicitly perform this test, but you can still use Minitab to find the sample standard deviation with **Calc>Column Statistics**, and then compute the test statistic $(n-1)s^2/\sigma^2$, using **Calc>Calculator**. Next, use **Calc>Probability Distributions>Chi-Square: Cumulative Probability** to find the area to the left of your calculated statistic, from which the p-value can be determined.

Exercise 8.117 - M/S page 417

Enter the 4 measurements in C1, and name it CO2. The null hypothesis here is that $\sigma^2 = 1.0$. We calculate *s* via **Calc>Column Statistics**:

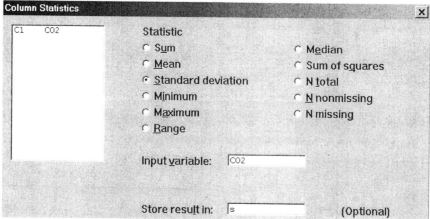

Next, calculate the chi-square statistic using **Calc>Calculator**

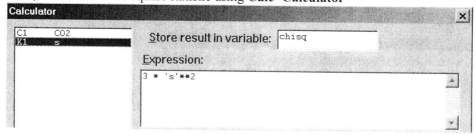

We stored the computed value in a column named *chisq*. Now find the probability to the left of the computed statistic using the chi-square distribution: **Calc>Probability Distributions>Chi-Square:**

Chi-Square Distribution ☒

| C1 | CO2 |
| C2 | chisq |

○ Probability density
◉ Cumulative probability
   Noncentrality   0.0

○ Inverse cumulative probability
   Noncentrality   0.0

Degrees of freedom: 3

◉ Input column: chisq
  Optional storage:

And we get:

Cumulative Distribution Function

```
Chi-Square with 3 DF

    x   P( X <= x )
 2.09      0.446062
```

Suppose we set $\alpha=.05$. The alternative is two sided ($\sigma^2 \neq 1$), so the rejection region has $.05/2 = .025$ in each tail. Does the chi-square statistic here (2.09) catch either the upper or lower .025 tail of the chi-square distribution? Since the probability above is not less than .025 or greater than .975, the answer is no. We may not reject the null hypothesis at the 5% level.

We may also choose to report the P-value here, which equals 2 x .4461 or .891, far exceeding any possible designated α.

If you want to find the exact critical values for a 5% level test, select **Inverse Cumulative Probability** from **Calc>Probability Distributions: Chi-Square**, and input the constants *.025* and *.975*. There is no need to do this here, since the p-value is more informative.

91

Chapter 9 – Two Sample Inference

In this chapter, we will compare two means from independent samples, using
Stat>Basic Statistics>2-Sample t

Check normality assumptions with
Graph>Probability Plot

We will compare means, from a paired-data design, using
Stat>Basic Statistics>Paired-t

Compare proportions, for independent samples, with
Stat>Basic Statistics>2 Proportions

And compare variances from two independent samples using
Stat>Basic Statistics> 2 Variances

Comparing Means - Independent samples

When comparing means from two independent samples, we may have:
(i) <u>Small samples</u>: If sample variances are similar, use Minitab's **2-Sample t** test procedure and click on **Assume equal variances**. If variances differ substantially, be sure that **Assume equal variances** is not selected, and then Minitab will execute the approximate Satterthwaite *t* test.
(ii) <u>Large samples</u>: The text uses a z-statistic with sample variances replacing the true variances. With Minitab, apply the *unequal variances* **2-Sample t** test, which uses the same statistic, but approximates its distribution with a *t* curve rather than a standard normal (negligible difference since the degrees of freedom will be large).
  Plot the data and look for severe departures from normality. If present, try a transformation (e.g. log) or non-parametric approach (see chap 14).

Examples 9.1-9.3 - M/S page 431

Open up the DIETSTUDY file. The data here are *stacked*, with all the weight losses in one column, C2, and the diet type (called the 'subscript' in the dialog box below) in another column, C1.

↓	C1-T	C2
	DIET	WTLOSS
1	LOWFAT	8
2	LOWFAT	10
3	LOWFAT	10
4	LOWFAT	12

  This is generally the preferred layout, as opposed to having the data *unstacked*, where we would have 100 weight losses from the low-fat diet in one column, and the 100 weight losses from the regular diet in another column.

  We wish to compare mean weight loss for the two diets. To execute the *large sample z-test*, we use Minitab's *unequal variances t-test* procedure, via **Stat>Basic Statistics>2-Sample t**, as follows:

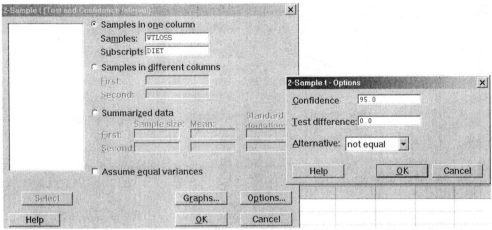

Since the data are stacked, we choose **Samples in one column,** and indicate where the **Samples** of weight losses are stacked, and also where the diet indicator - referred to as **Subscripts** – are to be found. We have chosen the two-sided **Alternative**: *not equal* to the null hypothesis **Test difference** of *0.0*, and a *95%* **confidence** level for the C.I. There is no reason to **Assume equal variances**, for the large sample test, so this is not ticked. We get

Two-Sample T-Test and CI: WTLOSS, DIET

```
Two-sample T for WTLOSS

DIET      N   Mean  StDev  SE Mean
LOWFAT   100  9.31   4.67   0.47
REGULAR  100  7.40   4.04   0.40

Difference = mu (LOWFAT) - mu (REGULAR)
Estimate for difference:  1.91000
95% CI for difference:  (0.69298, 3.12702)
T-Test of difference = 0 (vs not =): T-Value = 3.10   P-Value = 0.002   DF = 193
```

The P-value for the test is a very small .002, indicating very strong evidence (far beyond the 5% level) of a difference, with the low-fat diet producing bigger losses, on average.

Since the test is statistically significant, the next step is to determine the practical significance of our finding, and the C.I. displayed above helps you to do this. The displayed 95% C.I. is for $\mu_L - \mu_R$, rather than the other way around, due to alphabetical ordering (L before R). We see that the difference in average weight loss may be as small as 0.7 pounds or as much as 3.1 pounds, with 95% confidence.

Example 9.4 - M/S page 438

Open up the READING data file. The data are stacked with the scores in C1 = 'Scores' and the method in C2 = 'Method', where N = New method and S = Standard Method. We have two *small* samples to compare here. To execute the *equal variances* t-test or C.I., we use **Stat>Basic Statistics>2-Sample t** as follows:

93

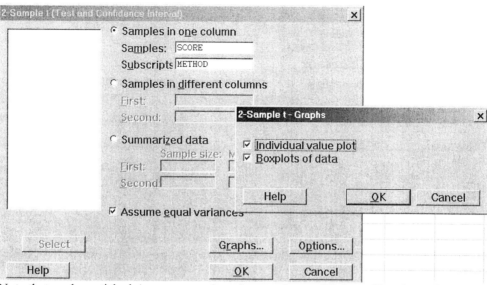

Note that we have ticked **Assume equal variances** above. Click on **Graphs** and select some, as shown above. Also click on **Options** and check the **confidence** level.

Even though we only want to construct a C.I. in this example, Minitab will execute both a test and C.I. whenever invoking this procedure. We get:

Two-Sample T-Test and CI: SCORE, METHOD

```
Two-sample T for SCORE

METHOD    N    Mean   StDev   SE Mean
NEW      10   76.40    5.83       1.8
STD      12   72.33    6.34       1.8


Difference = mu (NEW) - mu (STD)
Estimate for difference:  4.06667
95% CI for difference:  (-1.39937, 9.53270)
T-Test of difference = 0 (vs not =): T-Value = 1.55   P-Value = 0.136   DF = 20
Both use Pooled StDev = 6.1199
```

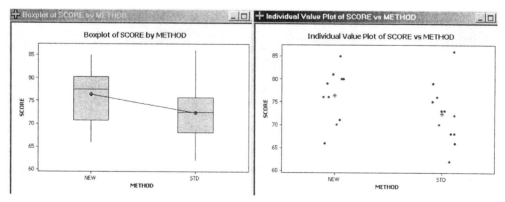

The confidence interval tells us, with 95% confidence, that the mean difference between the new method and the standard method may be anywhere from –1.4 to 9.5. Note that this interval contains '0', which indicates that there may well be no difference at all (at the 5% level of significance).

   The plots show no signs of severe skewness or other sharp departures from normality, and spreads are fairly similar (look at the box widths in the boxplot). So conclusions from the equal variances *t* procedure should be trustworthy here.

NOTE – Minitab has replaced ordinary dotplots in the **Graphs** choices here with *Individual value plots*, which are like dotplots except that the points are jiggled around randomly to avoid overlap of multiple points. I don't like them. Use **Graph>Dotplot** if you want to examine a standard dotplot comparison of the two samples here.

To assess normality, it is very useful to examine the normal probability plots:
Graph>Probability Plots - Simple: Graph Variables: *Score*, **Multiple Graphs: By Variables: By variables with groups in separate panels:** *Method*

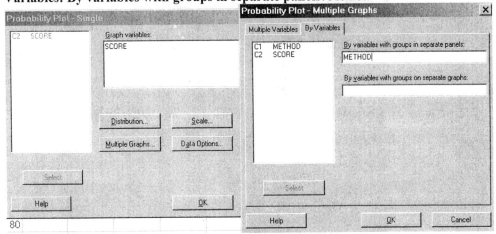

We need to ask for plots *by Method*, so that the data will be split up according to method. Otherwise, we would get one plot for all 200 scores.

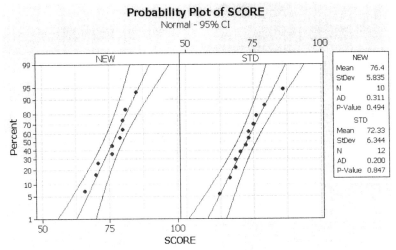

Probability Plot of SCORE
Normal - 95% CI

Panel variable: METHOD

No severe departures from normality are evident above.

Exercise 9.14 - M/S page 449

Here we have the ***summary statistics for each sample***, and not the raw data, so for a 90% confidence interval, proceed with:

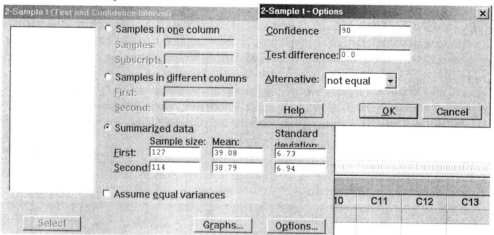

which produces:

Two-Sample T-Test and CI

```
Sample    N    Mean   StDev   SE Mean
1        127   39.08   6.73    0.60
2        114   38.79   6.94    0.65


Difference = mu (1) - mu (2)
Estimate for difference:  0.290000
90% CI for difference:   (-1.167651, 1.747651)
T-Test of difference = 0 (vs not =): T-Value = 0.33  P-Value = 0.743  DF = 234
```

Note that '0' is in the interval.

Exercise 9.27 - M/S page 452

Open up the SLI data file, where the DIQ scores are in C4 and the groups are in C3. To compare means for the two groups of children, use **Stat>Basic Statistics>2-Sample t: Samples in one column: Samples:** *C4*, **Subscripts:** *C3*, **Alternative:** *Not Equal*, **Confidence Level:** *95*, **Assume equal variances**. Also, click on **Graphs,** and select **Boxplot**.

Hit **OK**, and you get a message

```
* ERROR * There must be exactly two distinct subscripts.
```

The problem is that we have 3 instead of 2 distinct groups (subscripts) listed in C3. Select and delete the last 10 rows (OND group), rows 21 – 30. Repeat, and you get:

Two-Sample T-Test and CI: DIQ, Group

```
Two-sample T for DIQ

Group   N    Mean   StDev   SE Mean
SLI     10   93.60  9.08     2.9
YND     10   95.30  7.51     2.4


Difference = mu (SLI) - mu (YND)
Estimate for difference:  -1.70000
95% CI for difference:  (-9.53124, 6.13124)
T-Test of difference = 0 (vs not =): T-Value = -0.46   P-Value = 0.654   DF = 18
Both use Pooled StDev = 8.3350
```

Since the p-value is larger than $\alpha = .10$, we do not have evidence of a difference in mean DIQ, at the 10% level of significance.

Let's also check the assumptions required for the validity of this conclusion (though not explicitly requested here). The boxplots requested from **Graphs** in the dialog box above are below

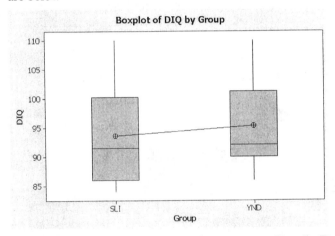

If you prefer dotplots, from the main menu, use **Graph>Dotplot – One Y, With Groups: Graph variables:** *DIQ*, **Categorical variables:** *Group*, which yields:

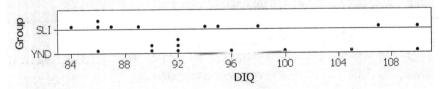

Spreads are similar, and while there is clear right-skewness, our results may still be tolerably accurate, considering the ***robustness*** of this test (also, see chapter 14 for some alternative nonparametric style tests).

We can also check normality by examining normal probability plots of the data, via **Graph>Probability Plot – Single: Graph:** *DIQ*, **Multiple Graphs: By Variables:** *Group*, as shown earlier in this section.

Comparing Means - Paired Data

When the observations in the two samples are paired (same subject used twice, paired subjects, etc.), we cannot use the two-sample t test discussed earlier, which assumes independent samples. Instead, we take the difference, for each set of paired observations, and perform a one-sample t test on the differences. In earlier versions of Minitab, you had to compute these differences (**Calc>Calculator: Expression:** $C1 - C2$), and then apply **Stat>Basic Statistics>1-Sample t**, but now, you have an explicit menu selection: **Stat>Basic Statistics>Paired-t**.

Example 9.5 - M/S page 458

Open up the GRADPAIRS data file.

↓	C1	C2	C3
	Pair	Male	Female
1	1	29300	28800
2	2	41500	41600
3	3	40400	39800
4	4	38500	38500

Each male has been paired with a female based on major and GPA, so the male and female salaries in each row of the table represent paired or correlated data.

Hence, we compare the males and female salaries, using the paired difference approach:
Stat>Basic Statistics>Paired t

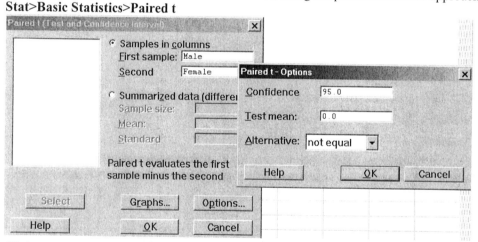

Click on **Options**. As you can see from the **Options** dialog box, this procedure executes both a test (default null mean is '0') and a C.I. Set the confidence level at '95', as requested here. Click on **Graphs** and select some of the choices, say the **Boxplot** and **Histogram**. We get

Paired T-Test and CI: Male, Female

```
Paired T for Male - Female

              N     Mean    StDev   SE Mean
Male         10  43930.0  11665.1    3688.8
Female       10  43530.0  11616.9    3673.6
Difference   10  400.000  434.613   137.437

95% CI for mean difference: (89.096, 710.904)
T-Test of mean difference = 0 (vs not = 0): T-Value = 2.91  P-Value = 0.017
```

The 95% C.I. is easily read off. It does not contain '0' - consistent with the test result indicating a significant difference in means (p=.017 < .05).

To confirm the accuracy of these conclusions, we should check the assumption of approximate normality (of the differences). Our **Graphs** choices yield:

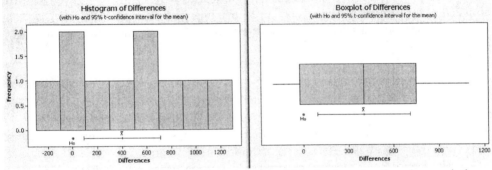

The observed departure from a bell-shape is well within the bounds of random variation, if you consider the small sample size.

If you prefer to examine a normal probability plot of the differences, or a dotplot, then you have to compute the differences, place them in say C3 (**Calc>Calculator: Expression:** *C1 – C2*, **Store in:** *C3*), and proceed to examine C3.

Exercise 9.136 - M/S page 506

Open up the MSSTUDY data file. The paired subjects' ages are in C3 and C6, named *MS-AGE* and *NON-AGE* respectively. To test for a difference in mean age, apply **Stat>Basic Statistics>Paired-t: First sample:** *MS-AGE*, **Second sample:** *NON-AGE*, with **Options: Mean:** *0*, **Alternative:** *Not Equal*. Also, click **Graphs** and select some. We get:

Paired T-Test and CI: MS-AGE, NON-AGE

```
Paired T for MS-AGE - NON-AGE

              N     Mean    StDev   SE Mean
MS-AGE       10  38.7000   6.0194    1.9035
NON-AGE      10  37.9000   5.0431    1.5948
Difference   10  0.800000  3.881580  1.227464


95% CI for mean difference: (-1.976715, 3.576715)
T-Test of mean difference = 0 (vs not = 0): T-Value = 0.65  P-Value = 0.531
```

99

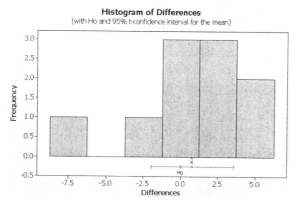

The P-value = 0.531, is large, so there is no significant difference in age between the two groups (i.e., the researchers did a good job of matching up the subjects age-wise).

You can repeat now this analysis for the other two variables, to check for successful matching.

Comparing Two Proportions

To compare proportions computed from two independent samples, enter the sample sizes and counts into **Stat>Basic Statistics>2 Proportions**. This produces both a CI and a test of significance for the difference between the true proportions, using a normal approximation for the sampling distribution of the estimated difference (assuming sufficiently large sample sizes).

Example 9.7 - M/S page 472

To assess whether the proportion of adults who smoke has decreased between 1995 and 2005, enter the observed sample counts of 555 and 578, along with the corresponding sample sizes of 1500 and 1750 respectively, into **Stat>Basic Statistics>2 Proportions**

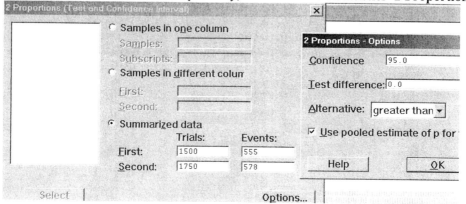

In the **Options** box, the null hypothesis is H_0: $p_1-p_2=0$. We choose the **Alternative** 'greater than' since we are only asking if there is evidence of a decrease over time in the

100

proportion who smoke (i.e. $p_1 > p_2$ or $p_1 - p_2 > 0$). We also tick off **Use pooled estimate of p for test**, since this is the approach used in your textbook. We get

Test and CI for Two Proportions

```
Sample    X     N   Sample p
1        555  1500   0.370000
2        578  1750   0.330286


Difference = p (1) - p (2)
Estimate for difference:  0.0397143
95% lower bound for difference:  0.0121024
Test for difference = 0 (vs > 0):   Z = 2.37   P-Value = 0.009
```

The P-value is highly significant (and $< \alpha = .05$), indicating clear evidence of a decreased proportion of smokers.

In this example, we have *summarized* data, but if instead, we had the raw data, i.e. two columns (one for each year) of length 1500 and 1750, with categorical (smoker or non-smoker) observations in each, we would then select in the dialog box **Samples in Different Columns**. Or, if for each of the 3250 subjects, smoker status were in C1, and year of observation in C2, we would use **Samples in One Column** (C2 = subscripts).

This procedure is based on the normal approximation to the distribution of a sample proportion, and is ***not valid for small sample sizes***. Minitab tries to help, and where sample sizes warrant, will print out the following warning:

```
* NOTE * The normal approximation may be inaccurate for small samples.
```

accompanied by a more accurate P-value calculated from an exact test called *Fisher's exact test*. You can interpret this P-value the usual way.

Exercise 9.61 - M/S page 477

To compare the two remission proportions, use **Stat>Basic Statistics>2 Proportions: Summarized Data: First:** *98* **trials,** *14* **events, Second:** *102* **trials,** *5* **events,** with **Options: Test Difference:** *0*, **Alternative:** *greater than*, **Use pooled estimate of p for test**. The alternative is 'greater than' since we only want to know if the treatment (sample 1) improves chances of remission, i.e. if $p_1 > p_2$ or $p_1 - p_2 > 0$.

Test and CI for Two Proportions

```
Sample    X    N   Sample p
1        14   98   0.142857
2         5  102   0.049020

Difference = p (1) - p (2)
Estimate for difference:  0.0938375
95% lower bound for difference:  0.0258888
Test for difference = 0 (vs > 0):   Z = 2.26   P-Value = 0.012
```

The two proportions are computed and printed out above: 0.142 for those using St. John's wort, and .049 for the placebo group. The P-value is .012 so we have ample evidence of a St. John's wort effect at the 10% level, but just miss significance at the more stringent 1% level.

Comparing Two Variances (optional)

Two procedures for comparing variances are produced by **Stat>Basic Statistics> 2 Variances** - the F-test discussed in the text, and another (non-parametric) test called Levene's test, which is more accurate if the normality assumption is questionable. A two-sided alternative is assumed.

Example 9.11 - M/S page 487

Open up the MICEWTS data file. The data are stacked with Weights in C2 and Supplier in C1. To compare the variability in weight of the supplied mice, use **Stat>Basic Statistics> 2 Variances** as follows:

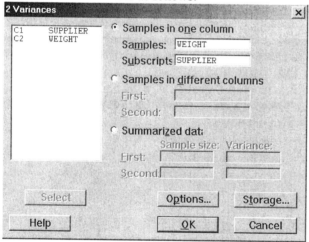

Results pop up in both a graphical window and in the session window. In the latter:

Test for Equal Variances: WEIGHT versus SUPPLIER

```
95% Bonferroni confidence intervals for standard deviations

SUPPLIER   N    Lower     StDev     Upper
       1  13  0.138579  0.202133  0.361467
       2  18  0.070860  0.098193  0.156836

F-Test (normal distribution)
Test statistic = 4.24, p-value = 0.007

Levene's Test (any continuous distribution)
Test statistic = 4.31, p-value = 0.047
```

We see that the F-test statistic = 4.24 (ratio of bigger sample variance over smaller sample variance), and its p-value = 0.007, assuming a two-sided alternative, which is the case here. The p-value is less than the specified α = .10, so we have adequate evidence

102

that the variability in weights is not the same for the two suppliers. If normality were questionable, we would have more trust in Levene's test, whose P-value of 0.047 would lead us to the same conclusion here since it is less than the specified α.

What if the alternative had been one-sided? Since .007 is a two-tailed probability, the probability that F > 4.24 is .007/2. Hence, the P-value would be either .007/2, if the data were consistent with the alternative hypothesis, or 1 - .007/2, if the data pointed the other way.

Chapter 10 – Analysis of Variance

In this chapter, we will construct ANOVA tables for various designs, using
Stat>ANOVA>One-way (or **One-way (Unstacked)**)
Stat>ANOVA>Two-way

Plot the relevant means with
Stat>ANOVA>Main Effects Plot
Stat>ANOVA>Interactions Plot

Split one worksheet into two, according to the distinct values in one column, with
Data>Split Worksheet

Enter a simple pattern of numbers into a column with
Calc>Make Patterned Data

One-way ANOVA

If there is one factor present in a completely randomized design, we may test the null
hypothesis that the factor has no effect on the response (i.e. the treatment means are all
equal), using **Stat>ANOVA** and selecting **One-way.**

Example 10.3 - M/S page 520

To randomly assign the 15 customers entered in C1 below to the three brands, you can
just use **Calc>Random Data>Sample From Columns** as below, in effect randomly
scrambling the 15 customers in C1, producing the second C1 shown below   Then, the
first 5 in the list (# 4, 13, 9, 10, 3) get brand A, the next 5 get brand B, and the last 5 get
brand C. For clarity, write 5 A's, 5 B's, 5 C's in C2.

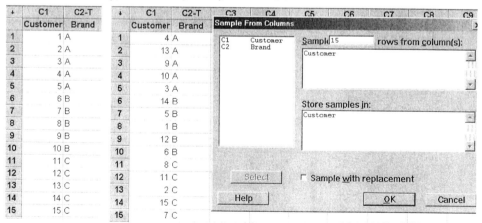

Or, you can start with C1, C2 as on the left, and then scramble C2, instead of C1, to
match each customer in your list with a randomly selected brand.

104

Examples 10.4 & 10.5 - M/S page 525

Open the GOLFCRD data file. The data here are *stacked* with all the distances in C2, and the brands in C1.

↓	C1-T	C2
	Brand	Distance
1	A	251.2
2	A	245.1
3	A	248.0
4	A	251.1

To test the null hypothesis that (mean) distance does not depend on brand, we use **Stat>ANOVA>One-way**. The other selection **ANOVA>One-way (Unstacked)** would be the right choice if the distances were placed in 4 separate columns, one corresponding to each brand, like the (unstacked) layout in Table 10.1 in the text.

STAT>ANOVA>One-way produces the dialog box below:

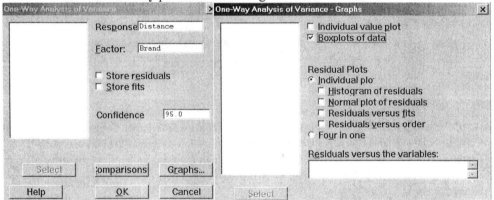

Enter the **Response** and **Factor**, as shown, click on **Graphs**, and tick off **Boxplots**. **OK, OK** produces

One-way ANOVA: Distance versus Brand

```
Source   DF      SS      MS      F      P
Brand     3   2794.4   931.5   43.99  0.000
Error    36    762.3    21.2
Total    39   3556.7

S = 4.602   R-Sq = 78.57%   R-Sq(adj) = 76.78%

                         Individual 95% CIs For Mean Based on
                         Pooled StDev
Level   N    Mean   StDev  --------+---------+---------+---------+-
A      10  250.78   4.74   (---*---)
B      10  261.06   3.87               (---*---)
C      10  269.95   4.50                        (----*---)
D      10  249.32   5.20   (---*---)
                           --------+---------+---------+---------+-
                             252.0     259.0     266.0     273.0

Pooled StDev = 4.60
```

Above, we see the standard ANOVA table, and additional output showing sample means, standard deviations, and confidence intervals for each of the four Brand means. The p-value of .000 in the ANOVA table indicates overwhelming evidence that the brands are not all equivalent. But is the p-value (and 95% confidence levels) trustworthy? Check the **Graph** we requested:

105

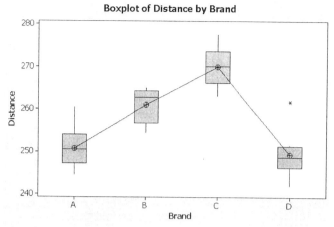

The distributional assumptions are normality and equality of variance. The plots above show little difference in dispersion among the samples, and no severe departures from normality (considering the small sample sizes). The p-value can be trusted.

If you prefer to examine histograms, like those shown in the text, use **Graph>Histogram-Simple: Graph variables:** *Distance*, **Multiple Graphs: By Variables: By:** *Brand* (to include a fitted normal curve, also select **Data View: Distribution: Fit distribution:** *Normal*). Or you can examine normal probability plots via **Graph>Probability Plot-Single: Graph:** *Distance*, **Multiple Graphs: By Variables: By:** *Brand.*

Exercise 10.35 - M/S page 536

Open up the FACES data file, where dominance ratings are in C1 and facial expressions are in C2. To compare ratings among the different facial expressions, run **Stat>ANOVA>One-Way: Response:** *RATING*, **Factor:** *FACE*, **Graphs:** *Boxplots*. We get

One-way ANOVA: RATING versus FACE

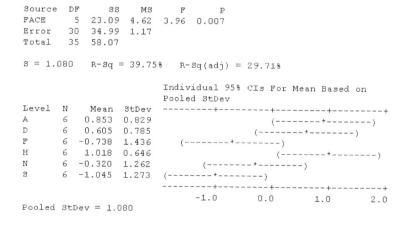

```
Source   DF    SS     MS     F      P
FACE      5  23.09  4.62  3.96  0.007
Error    30  34.99  1.17
Total    35  58.07

S = 1.080    R-Sq = 39.75%    R-Sq(adj) = 29.71%


                          Individual 95% CIs For Mean Based on
                          Pooled StDev
Level  N    Mean  StDev  ---------+---------+---------+---------+
A      6   0.853  0.829                    (--------*--------)
D      6   0.605  0.785                  (--------*--------)
F      6  -0.738  1.436        (--------*--------)
H      6   1.018  0.646                    (--------*--------)
N      6  -0.320  1.262           (--------*--------)
S      6  -1.045  1.273    (--------*--------)
                          ---------+---------+---------+---------+
                               -1.0      0.0      1.0      2.0

Pooled StDev = 1.080
```

106

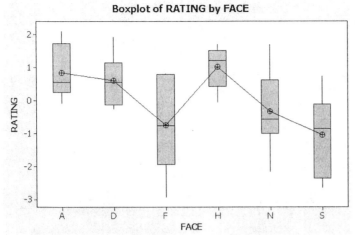

Since the p-value of .007 is less than $\alpha = .10$, we can reject the null hypothesis at the 10% level of significance, and conclude that facial expression does matter.

The graphs allow a check on the assumptions. We see no substantial non-normality (e.g. severe skewness), and while some samples appear more dispersed than others, the samples are very small, so this can easily just be random variation. As well, the sample sizes are equal, improving the procedure's **robustness**. Our conclusions should hold water.

Multiple Comparisons

There are a plethora of multiple comparison procedures that may be used to execute pair-wise comparisons among the means. The text mentions e.g. the Bonferroni and Tukey methods. Tukey's is generally preferable to Bonferroni's, and is included with **ANOVA>One-way** under **Comparisons…**

If you have the *professional version* of Minitab, Bonferroni's method may be applied via **Stats>ANOVA>General Linear Model**: **Responses:** *Response*, **Model:** *Treatments*, **Comparisons:** *Pairwise Comparisons*, **Terms:** *Treatments*, **Method:** *Bonferroni*

Example 10.6 - M/S page 539

Open up the worksheet again. We have already concluded that differences exist among the brands. To apply Tukey's method, select **Stat>ANOVA: One-way,** again, but this time, also click on **Comparisons…**

107

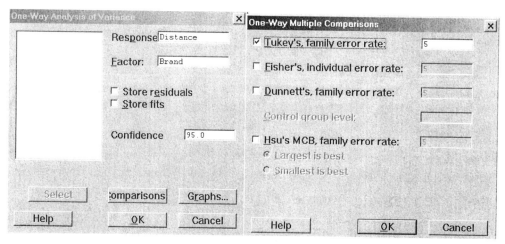

Select **Tukey's** and enter either '5' or '.05' for the family (experiment-wise) error rate, because we want an overall confidence level of 95% for our batch of confidence intervals. **OK, OK**, gives us the same basic ANOVA output as shown earlier in Example 10.4, but additionally, we see the following:

```
Tukey 95% Simultaneous Confidence Intervals
All Pairwise Comparisons among Levels of Brand

Individual confidence level = 98.93%

Brand = A subtracted from:

Brand   Lower   Center   Upper   -------+---------+---------+---------+--
B       4.736   10.280   15.824                       (---*---)
C       13.626  19.170   24.714                             (---*--)
D       -7.004  -1.460    4.084              (---*---)
                                    -------+---------+---------+---------+--
                                         -15        0        15       30


Brand = B subtracted from:

Brand   Lower   Center   Upper   -------+---------+---------+---------+--
C       3.346   8.890    14.434                        (---*---)
D       -17.284 -11.740  -6.196       (---*---)
                                  -------+---------+---------+---------+--
                                       -15        0        15       30


Brand = C subtracted from:

Brand   Lower   Center   Upper   -------+---------+---------+---------+--
D       -26.174 -20.630  -15.086 (--*---)
                                  -------+---------+---------+---------+--
                                       -15        0        15       30
```

Above, we see first brand A compared with B, and with C, and with D, followed by brand B compared with C, and with D, and finally brand C compared with D, so we have all the six possible comparisons. E.g., (4.736, 15.824) is the C.I. for $\mu_B-\mu_A$. Since this interval does not contain '0', we may declare a significant difference between brand A and brand B. Likewise, we see that every comparison above is significant except for the comparison of brand D and brand A (the only interval containing '0').

In order to rank the brands from highest mean distance to lowest, go back to the following portion of the ANOVA output:

108

```
                              Individual 95% CIs For Mean Based on
                              Pooled StDev
Level   N    Mean   StDev   --------+---------+---------+---------+-
A      10  250.78    4.74    (---*---)
B      10  261.06    3.87                     (---*---)
C      10  269.95    4.50                                 (----*---)
D      10  249.32    5.20   (---*---)
                            --------+---------+---------+---------+-
                              252.0     259.0     266.0     273.0
```

Pooled StDev = 4.60

Read off the means, rank and list them. Connect those that do not differ significantly with a solid line. We get:

Mean: 249.32 250.78 261.06 269.95
Brand: D   A   B   C

To construct a confidence interval for any one of the treatment means, read off approximate limits from the output above, or work them out more precisely using:

Sample Mean $\pm$ t$_{\alpha/2}$ s /$\sqrt{\text{n}}$

where **s** is the 'Pooled StDev' given in the output: 4.60 (= $\sqrt{\text{MSE}}$), with 36 df.

Exercise 10.52 - M/S page 545

Open up again the FACES file used in Exercise 10.27. We will compare mean ratings for the 6 facial expressions using Tukey's method with an error rate of 10%, i.e.
Stat>ANOVA>One-way: Response: *Rating*, **Factor:** *Face*, **Comparisons: Tukey's, family error rate:** *10*. We get the one-way ANOVA table followed by the output below:

```
                              Individual 95% CIs For Mean Based on
                              Pooled StDev
Level   N    Mean   StDev   ---------+---------+---------+---------+
A      6   0.853   0.829                  (--------*--------)
D      6   0.605   0.785                (--------*--------)
F      6  -0.738   1.436       (--------*--------)
H      6   1.018   0.646                    (--------*--------)
N      6  -0.320   1.262          (--------*--------)
S      6  -1.045   1.273   (--------*--------)
                            ---------+---------+---------+---------+
                                  -1.0      0.0       1.0       2.0
```

Pooled StDev = 1.080

Tukey 90% Simultaneous Confidence Intervals
All Pairwise Comparisons among Levels of Expression

Individual confidence level = 98.93%

Expression = A subtracted from:

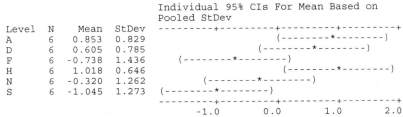

```
Expression  Lower   Center  Upper  ---------+---------+---------+---------+
D          -1.946  -0.248   1.449             (--------*------)
F          -3.289  -1.592   0.106    (-------*--------)
H          -1.532   0.165   1.862           (--------*------)
N          -2.871  -1.173   0.524        (-------*--------)
S          -3.596  -1.898  -0.201   (--------*------)
                                    ---------+---------+---------+---------+
                                          -2.0      0.0       2.0       4.0
```

109

```
Expression = D subtracted from:

Expression   Lower   Center   Upper    ---------+---------+---------+---------+
F           -3.041  -1.343   0.354        (-------*--------)
H           -1.284   0.413   2.111            (-------*-------)
N           -2.622  -0.925   0.772        (-------*--------)
S           -3.347  -1.650   0.047     (--------*-------)
                                         ---------+---------+---------+---------+
                                              -2.0       0.0       2.0       4.0


Expression = F subtracted from:

Expression   Lower   Center   Upper    ---------+---------+---------+---------+
H            0.059   1.757   3.454                (-------*-------)
N           -1.279   0.418   2.116            (-------*--------)
S           -2.004  -0.307   1.391          (-------*--------)
                                         ---------+---------+---------+---------+
                                              -2.0       0.0       2.0       4.0


Expression = H subtracted from:

Expression   Lower   Center   Upper    ---------+---------+---------+---------+
N           -3.036  -1.338   0.359        (-------*--------)
S           -3.761  -2.063  -0.366     (--------*-------)
                                         ---------+---------+---------+---------+
                                              -2.0       0.0       2.0       4.0


Expression = N subtracted from:

Expression   Lower   Center   Upper    ---------+---------+---------+---------+
S           -2.422  -0.725   0.972        (-------*--------)
                                         ---------+---------+---------+---------+
                                              -2.0       0.0       2.0       4.0
```

Rank the facial expressions, using the list of means up above, and then underline or otherwise connect the means that do not differ significantly (intervals containing '0'):

S   F   N   D   A   H
-1.045 -.738 -.320 .605 .853 1.018

In summary, Happy and Angry faces both rate above Sad, and Happy rates above Fearful (find pairs of treatments that are not connected by any bars).

Randomized Block Design

If we group our experimental units into blocks, before assigning treatments, we need to amend the ANOVA to include an extra SS for the block-to-block variation. Since blocks are now a second source of variation in the design, use **ANOVA>Two-way**, to calculate the proper ANOVA table.

   Two-way does not have an option for multiple comparison procedures. You can still request them if you apply **Stat>ANOVA>General Linear Model,** but this procedure is only included with the *professional version of Minitab*.

   To check assumptions, you can use the residual plots listed under **Graphs,** but you should first study the section on residual plots in chapter 12.

Examples 10.8 & 10.9 - M/S page 550

Open up the GOLFRBD data file, containing data from a randomized block design. Note how the data are entered:

↓	C1	C2-T	C3
	Golfer	Brand	Distance
1	1	A	202.4
2	1	B	203.2
3	1	C	223.7
4	1	D	203.6
5	2	A	242.0
6	2	B	248.7

Every row corresponds to one drive of the ball. In column 3, we see the distance. In column 2 we identify the brand used, and in column 1, we identify the golfer.

We are interested in how C3 (distance) depends on C2 (brand) and C1 (golfer). For answers, use the F-ratios printed in the two-way ANOVA output obtained from:

Stat>ANOVA>Two-way, where we enter *Distance* as the **Response,** and *Brand* and *Golfer* in any order next to **Row factor** and **Column factor**

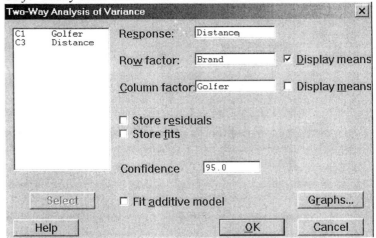

I also ticked off **Display means** for *Brand*, since we will also want to compare these means pairwise, if the global test is significant, to learn more about where the differences lie. We get

Two-way ANOVA: Distance versus Brand, Golfer

```
Source   DF      SS       MS       F       P
Brand     3   3298.7   1099.55   54.31   0.000
Golfer    9  12073.9   1341.54   66.26   0.000
Error    27    546.6     20.25
Total    39  15919.2

S = 4.499    R-Sq = 96.57%    R-Sq(adj) = 95.04%
```

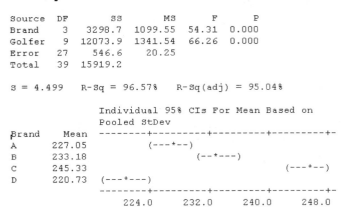

Since each p-value is 0.000, we have strong evidence for differences in mean among brands, and also for differences among golfers (suggesting that blocking was effective).

No multiple comparison procedures are included with **Two-way.** But we can still produce some useful calculations and plots. If you find the *minimum significant difference* between a pair of means, from some multiple comparisons procedure [= 5.7 e.g. with the Bonferroni method with α=.05], you could now apply this criterion to determine which brands differs significantly from which [the means for A and for D differ by more than 5.7, hence this difference is significant]. Unfortunately, you cannot use the 95% CI's shown above to compare the means pair-wise (though non-overlap indicates *some* evidence of a difference).

You can also plot the brand means with **Stat>ANOVA>Main Effects Plot: Response: Distance, Factor: Brand**

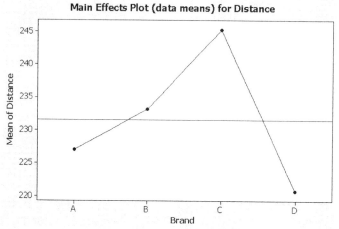

The greatest mean distance is achieved with brand C, and indeed, it differs significantly from all the other brands (see Bonferroni analysis in text).

If you have the professional version of Minitab, you may apply the Tukey or Bonferroni multiple comparisons procedure to the treatment means via **Stat>ANOVA>General Linear Model**, with **Responses:** *Distance*, **Model:** *Brand Golfer.* Select **Comparisons...** and then choose **Pairwise Comparisons** with **Terms:** *Brand.* Select the **Bonferroni** or **Tukey** method. Choose **Test** and/or **Confidence Interval**.

Exercise 10.67 - M/S page 559

Open up the PLANTS data file, where in columns C1 - C3, you will find the student number, plant setting (1 = Live, 2 = Photo, 3 = None) and maximum finger temperature, respectively. Does the type of plant setting affect relaxation level (measured by finger temperature)? Students are the blocks. Run **Stat>ANOVA>Two-way: Response:** *Temperature*, **Row Factor:** *Plant*, **Column Factor:** *Student*

Two-way ANOVA: Temperature versus Plant, Student

```
Source    DF      SS       MS       F      P
Plant      2    0.122   0.06100   0.02   0.981
Student    9   18.415   2.04611   0.63   0.754
Error     18   58.038   3.22433
Total     29   76.575

S = 1.796    R-Sq = 24.21%    R-Sq(adj) = 0.00%
```

The large p-value of .981 for the factor Plant ($> \alpha=.10$) indicates no evidence of a difference in mean relaxation level among the three experimental plant settings.

Factorial Experiments

When the design incorporates two factors, we put the response values in one column, factor 1 settings in a second column, factor 2 settings in a third column (similar to a RBD design), and then use **ANOVA>Two-Way** for the number crunching. With replicates at each factor-level combination, the ANOVA will now include SS's for the *main effect* of factor A, the *main effect* of factor B, and the *interaction* between the factors.

  Two-Way has no option for multiple comparison procedures, but if you have the professional version of Minitab, you may use **Comparisons...** in **Stat>ANOVA>General Linear Model**. Use Minitab's **Help** for more details.

  If you wish to do an initial test of significance on the treatment combinations, you have to enter distinct treatment numbers into a single column and run **ANOVA>One-way**.

Example 10.10 - M/S page 565

Open up the GOLFFAC1 data file. Note that the response Distance is in C3, the factor Club in C1, the factor Brand in C2. Write treatment numbers (1-8) in C4, where 1 denotes the DRIVER/A, 2 denotes DRIVER/B, etc. Name this column 'Treatment'. Or you can use **Calc>Make Patterned Data** (see exercise 10.87 below).

↓	C1-T	C2-T	C3	C4
	Club	Brand	Distance	Treatment
1	DRIVER	A	226.4	1
2	DRIVER	A	232.6	1
3	DRIVER	A	234.0	1
4	DRIVER	A	220.7	1
5	DRIVER	B	238.3	2
6	DRIVER	B	231.7	2

To test for any difference among the 8 treatments, run a **One-Way ANOVA**:

```
One-Way Analysis of Variance                    [X]

C1       Club          Response  Distance
C2       Brand
C3       Distance
C4       Treatment     Factor:   Treatment
```

113

One-way ANOVA: Distance versus Treatment

```
Source      DF      SS       MS       F       P
Treatment    7   33659.8   4808.5   140.35   0.000
Error       24     822.2     34.3
Total       31   34482.0
```

The small p-value tells us that treatments are not all the same. To study the main effects and interaction, proceed with **Stat>ANOVA>Two-Way**

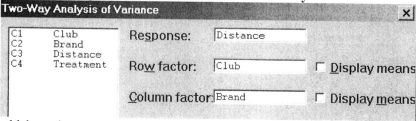

which produces

Two-way ANOVA: Distance versus Club, Brand

```
Source        DF       SS        MS       F       P
Club           1   32093.1   32093.1   936.75   0.000
Brand          3     800.7     266.9     7.79   0.001
Interaction    3     766.0     255.3     7.45   0.001
Error         24     822.2      34.3
Total         31   34482.0
```

The interaction is significant (p =.001 < α=.10), so the relationship between the brands is not the same for drivers and for 5-irons. Maybe the best brand for drivers will not be the best brand for 5-irons. We can examine the nature of the interaction, by plotting the means, using **Stat>ANOVA>Interactions Plot**

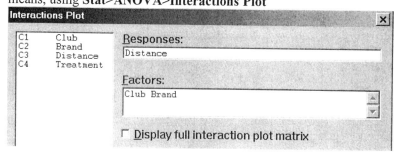

which produces the following plot of means (to put 'Club' on the x-axis, just reverse the order of listing of the factors in the box above):

114

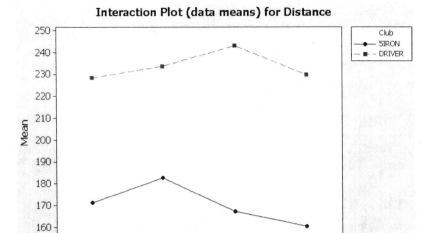

Interaction Plot (data means) for Distance

With 5-irons, brand B appears best, while with drivers, brand C appears best. To confirm these impressions, one simple way to proceed is to split the worksheet according to Club with **Data>Split Worksheet: By:** *Club*. Then, for each type of club (i.e. each new worksheet), run **Stat>ANOVA>One-way** with **Comparisons: Tukey's, family error rate:** *10*. For the Driver, e.g. we get:

```
Tukey 90% Simultaneous Confidence Intervals
All Pairwise Comparisons among Levels of Brand
Individual confidence level = 97.50%

Brand = A subtracted from:

Brand   Lower   Center   Upper   ------+---------+---------+---------+---
B      -5.005    5.300  15.605              (------*-----)
C       4.370   14.675  24.980                 (------*------)
D      -8.980    1.325  11.630           (------*------)
                                 ------+---------+---------+---------+---
                                    -15        0        15        30
```

```
Brand = B subtracted from:

Brand   Lower   Center   Upper   ------+---------+---------+---------+---
C      -0.930    9.375  19.680              (------*------)
D     -14.280   -3.975   6.330        (------*------)
                                 ------+---------+---------+---------+---
                                    -15        0        15        30
```

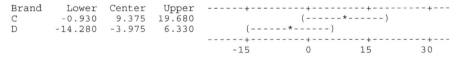

```
Brand = C subtracted from:

Brand   Lower   Center   Upper   ------+---------+---------+---------+---
D     -23.655  -13.350  -3.045   (------*------)
                                 ------+---------+---------+---------+---
                                    -15        0        15        30
```

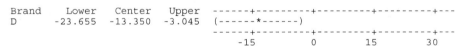

We see that C is significantly better than A and D, but just misses statistical superiority over B (the interval for $\mu_C - \mu_B$ just catches '0'). Repeat for the 5-iron (worksheet), and you will find that brand B is significantly better than the other three brands.

Example 10.11 - M/S page 568

This is similar to Example 10.10. Open up GOLFFAC2, and run a two-way ANOVA:
Stat>ANOVA>Two-Way with the same entries in the dialog box as before (I omit the
test on Treatments). The output is:

Two-way ANOVA: Distance versus Club, Brand

```
Source        DF        SS        MS        F        P
Club           1    46443.9   46443.9   1887.94   0.000
Brand          3     3410.3    1136.8     46.21   0.000
Interaction    3      105.2      35.1      1.42   0.260
Error         24      590.4      24.6
Total         31    50549.8

S = 4.960    R-Sq = 98.83%    R-Sq(adj) = 98.49%
```

There is no evidence of an interaction (p=.260). The main effects of Brand and Club are
both highly significant. The club effect is no surprise, as we already know that drivers
carry the ball further than 5-irons. Distance also depends on brand, and we should
examine this further.

We can plot all the 8 treatment means, like Fig. 10.27 in the text, using
Stat>ANOVA>Interactions Plot: Response: *Distance*, **Factors:** *Club Brand*

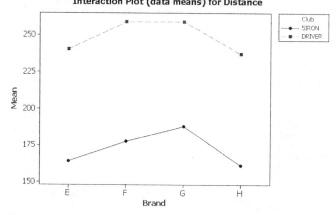

The plot shows little indication of an interaction (i.e. the brand differences look quite
similar for 5-irons and drivers), quite consistent with our test result.

We can also plot the *overall* means for each brand, or for each type of club, using
Stat>ANOVA>Main Effects Plot: Response: *Distance*, **Factors:** *Club Brand*. We get

116

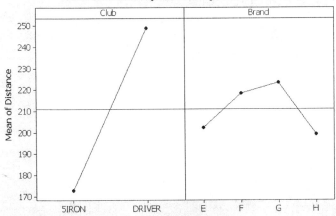

The significant Brand effect appears to be mostly due to the big difference between the high distances achieved by brands F and G, and the low distances achieved by brands E and H. This impression is confirmed in the multiple comparisons analysis shown in the text (Fig 10.26), which can be replicated, using the *professional version of Minitab*, by selecting **Stat>ANOVA>General Linear Model: Response:** *Distance*, **Model:** *Club Brand Club*Brand*, **Comparisons: Pairwise Comparisons, Terms:** *Brand*, **Method:** *Tukey*

Exercise 10.87 - M/S page 576

Open up the VITAMINB data file. *Kidney weight* in C1 is the response. The factor *Rat size* is in C2 and the second factor *Diet* is in C3. To test first for treatment differences, we need to enter some treatment numbers into, say, C4 (name it 'Treatment'). The first 7 rows of data are from treatment 1, the next 7 from treatment 2, etc., so either enter these numbers into C4 directly, or use **Calc>Make Patterned Data>Simple Set of Numbers: Store in:** *Treatment*, **From first value:** *1*, **To last value:** *4*, **In steps of:** *1*, **List each value:** *7* **times**.

We test for treatment differences via **Stat>ANOVA>One-way: Response:** *C1*, **Factor:** *C4*, and also select **Graphs:** *Boxplots,* so that we can do a check on our assumptions. We get

One-way ANOVA: Kidney Weight versus Treatment

```
Source      DF      SS      MS      F       P
Treatment    3   8.1168  2.7056  47.34   0.000
Error       24   1.3715  0.0571
Total       27   9.4883
```

From the low p-value, we are confident that treatment differences exist. You can do a rough check on your assumptions by examining the boxplots shown, or by using other data plots selected from the main menu's **Graph** choices (like normal probability plots), where you must remember to specify **Multiple Graphs> By Variables**

117

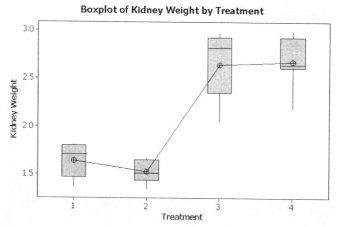

For the more useful factorial part of the analysis, select **Stat>ANOVA>Two-Way: Response:** *C1*, **Row Factor:** *C2*, **Column Factor:** *C3*. We see

Two-way ANOVA: Kidney Weight versus Rat Size, Diet

Source	DF	SS	MS	F	P
Rat Size	1	8.06789	8.06789	141.18	0.000
Diet	1	0.01243	0.01243	0.22	0.645
Interaction	1	0.03643	0.03643	0.64	0.432
Error	24	1.37151	0.05715		
Total	27	9.48827			

There is no evidence of interaction, no evidence of a Diet effect, but strong evidence of an effect from Rat Size.

A picture helps clarify, so use **Stat>ANOVA>Interactions Plot: Response:** *C1*, **Factors:** *C2 C3*. You get the plot on the left below. Repeat, reversing the order of the factors listed, and you get the plot on the right.

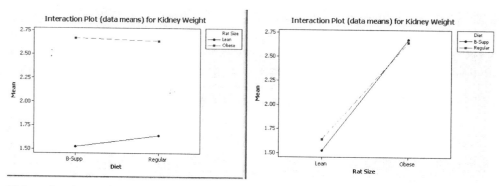

Either plot shows the large effect of rat size, and the comparatively minor (and non-significant) interaction and diet effects.

Chapter 11 – Simple Linear Regression

In this chapter, we will plot the response variable versus the predictor, using
Graph>Scatterplot

Add the least-squares line to the data plot, with
Stat>Regression>Fitted Line Plot

Make inferences and predictions with a regression model, using
Stat>Regression>Regression

Calculate the correlation coefficient, with
Stat>Basic Statistics>Correlation

Fitting by Least-Squares

The method of *least-squares* or *regression* may be used to fit a model showing how a *response variable* depends on another variable called the *predictor*. The simplest model is a straight line, but least squares may also be used to fit a variety of different models, including e.g. quadratic models and models with many predictors (see chapter 12). The fitted line may be calculated and drawn on the plot using **Stat>Regression>Fitted Line Plot**, but **Stat>Regression>Regression** will provide more information and useful options.

Figures 11.3, 11.5, 11.6 - M/S pages 596-601

Take a moment to type in the data into C1 and C2 as shown (or look for the STIMULUS data file):

↓	C1	C2
	Drug	Time
1	1	1
2	2	1
3	3	2
4	4	2
5	5	4

We can produce a scattergram (scatterplot) showing reaction time versus drug amount, using
Graph>Scatterplot - Simple as follows:

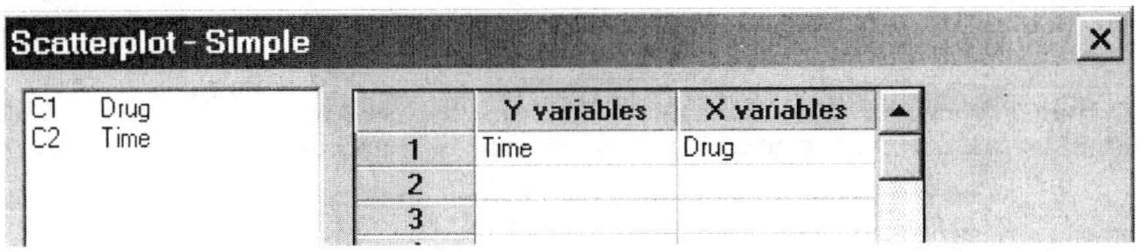

Scatterplot - Simple				×
C1 Drug		**Y variables**	**X variables**	▲
C2 Time	1	Time	Drug	
	2			
	3			

We get

119

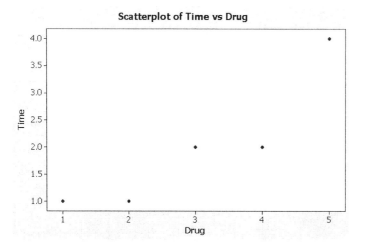

To find the equation of the regression line, and draw it on the plot, we may use
Stat>Regression>Fitted Line Plot

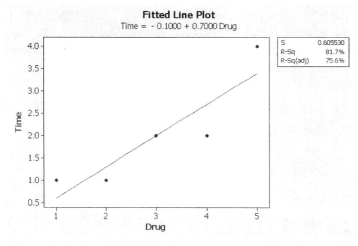

which gives us

Alternatively, to calculate the least-squares equation, along with much additional useful information, to be discussed later, use **Stat>Regression>Regression**

120

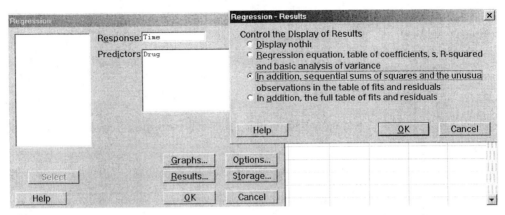

Fill in the **Response** and **Predictors** as above. Click on **Results** and it should appear as above, with the third level of **Display** ticked. You can choose to have less displayed, if you prefer, but I always keep this level, as all of the information displayed will eventually become useful to us. **OK, OK,** and in the session window, we see

Regression Analysis: Time versus Drug

```
The regression equation is
Time = - 0.100 + 0.700 Drug

Predictor      Coef    SE Coef       T       P
Constant    -0.1000     0.6351   -0.16   0.885
Drug         0.7000     0.1915    3.66   0.035

S = 0.605530    R-Sq = 81.7%    R-Sq(adj) = 75.6%

Analysis of Variance

Source           DF       SS       MS       F       P
Regression        1   4.9000   4.9000   13.36   0.035
Residual Error    3   1.1000   0.3667
Total             4   6.0000
```

At the top, the fitted equation is clearly displayed. Additional significant figures for the slope and intercept appear below **Coef**. The **S**um of **S**quared deviations minimized by the regression line appears next to **Residual Error** (1.1000) in the **Analysis of Variance** table. The rest of the information in the table will be discussed later.

Exercise 11.26 - M/S page 610

Open up the OJUICE data file. Note carefully from the question which variable is the predictor and which is the response. Regress with **Stat>Regression>Regression: Response:** *Sweetness*, **Predictors:** *Pectin*. Also, click on **Options** and fill in: **Prediction intervals for new observations:** *300*.

At the top, we get the regression equation:

Regression Analysis: Sweetness versus Pectin

```
The regression equation is
Sweetness = 6.25 - 0.00231 Pectin
```

So, sweetness decreases by .231 for every increase of 100ppm in pectin.

At the bottom, we have:

```
Predicted Values for New Observations
New
Obs     Fit  SE Fit       95% CI            95% PI
  1   5.5589  0.0587  (5.4372, 5.6806)  (5.0967, 6.0211)

Values of Predictors for New Observations
New
Obs   Pectin
  1     300
```

The **Fit** or prediction when the pectin measures 300ppm, is 5.5589 (ignore the other entries on this line, for now). You can check this out by plugging '300' into the regression equation above.

Model Assumptions and Estimating the Variance

It is important to check the assumptions of our model. A scattergram helps, but the best approach is through the use of *residual plots*, which we defer until section 12.13. Our model assumes a constant variance σ^2 for the random errors. The estimate of σ, denoted *s*, is clearly displayed in Minitab's **Stat>Regression>Regression** output or in the **Stat>Regression>Fitted Line Plot** output.

Example 11.2 - M/S page 613

Open up the worksheet (STIMULUS on your CD) used earlier in this chapter, or quickly type in the data on drug amount and reaction time. As before, we use **Stat>Regression>Regression: Response:** *Time*, **Predictor:** *Drug*, and obtain:

Regression Analysis: Time versus Drug

```
The regression equation is  Time = - 0.100 + 0.700 Drug

Predictor      Coef  SE Coef      T      P
Constant    -0.1000   0.6351  -0.16  0.885
Drug         0.7000   0.1915   3.66  0.035

S = 0.605530   R-Sq = 81.7%   R-Sq(adj) = 75.6%

Analysis of Variance
Source           DF      SS      MS      F      P
Regression        1  4.9000  4.9000  13.36  0.035
Residual Error    3  1.1000  0.3667
Total             4  6.0000
```

We see that the estimated standard deviation of the errors is $s = 0.605530$. We may also find the value of s^2 by examining the displayed *Analysis of Variance* table. s^2 is exactly the same as the *Residual Error MS* of *0.3667*, which is just the *Residual Error SS =* *1.1000* divided by the *Residual Error DF* = n − 2 = 3.

Exercise 11.42 - M/S page 616

Open up the CUTTOOL data file, where C1 – C3 contain the data on Cutting Speed, Useful Life (hours) for brand A, and Useful Life (hours) for brand B, respectively. We can examine the data with fitted lines, using **Stat>Regression>Fitted Line Plot: Response:** *HoursA*, **Predictor:** *Speed*, **Type of Regression:** *Linear*, and then repeat changing the **Response** to *HoursB*.

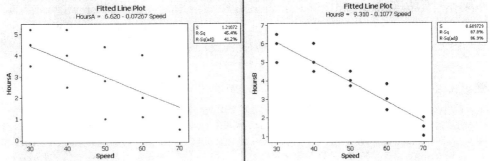

Examining the plots, we see considerably less scatter about the regression line, when predicting brand B lifetimes, which is also evident from comparing the regression standard errors (s), printed in the boxes to the right of each graph: *1.211* for brand A, but only *.6097* for brand B. [The *R-Sq* values printed in the boxes give us similar information, but we will discuss this later]

Inferences about the slope

The **Stat>Regression>Regression** output provides an estimate of the slope (*Coef*), and its standard error (*SE*). Using these, we can then construct a CI or execute a t-test. The *t* value for a test of H: $\beta = 0$ is also displayed in the output with corresponding p-value (two-tailed).

Example 11.3 - M/S page 617

Open up our first simple regression data worksheet again (STIMULUS). Regress reaction time on the amount of drug, i.e. **Stat>Regression>Regression: Response:** *Time*, **Predictor:** *Drug*. Key information about the individual parameters appears right below the regression equation:

123

Regression Analysis: Time versus Drug

```
The regression equation is
Time = - 0.100 + 0.700 Drug

Predictor     Coef    SE Coef      T      P
Constant    -0.1000    0.6351   -0.16   0.885
Drug         0.7000    0.1915    3.66   0.035
```

The slope is the *Coef* of *Drug* in the equation, and equals *.7000*. Its standard error, *SE Coef* equals *.1915*. Hence a 95% C.I. for the true slope would be .7000 ± 3.182 x .1915, where 3.182 is the t value chopping off .025 to the right in a *t* (3df) distribution.

The estimated slope over its standard error, or *Coef/StDev* above, equals *3.66*, and the corresponding p-value is *.035*, from which we may conclude, at the 5% level, that the amount of drug does help to predict reaction time, via a straight-line model. **Note that this p-value assumes a two-sided alternative to the null.** Adjust it appropriately should you encounter a one-sided alternative (divide by 2, if the sample is consistent with the stated alternative).

Exercise 11.115 - M/S page 654

Open up the SEEDGERM data file, showing 'Change in Free Energy' in C1 and 'Temperature' in C2. Obtain a scattergram via **Graph>Scatterplot-Simple: Y:** *Change*, **X:** *Temperature*. To draw the data with fitted line, use **Stat>Regression>Fitted Line Plot: Response(Y):** *Change*, **Predictor(X):** *Temperature*, **Type of Regression:** *Linear*. We get the two plots below.

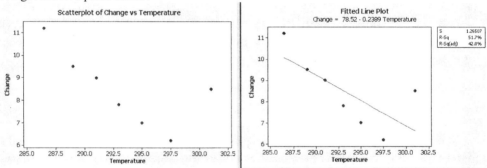

Note how much influence that obvious outlier, approximately (301, 8.5), appears to have on the fitted line, pulling the slope upward.

Now run **Stat>Regression>Regression> Response:** *Change*, **Predictor:** *Temperature*

124

Regression Analysis: Change versus Temperature

The regression equation is Change = 78.5 - 0.239 Temperature

Predictor	Coef	SE Coef	T	P
Constant	78.52	30.31	2.59	0.049
Temperature	-0.2389	0.1033	-2.31	0.069

s = 1.26507   R-Sq = 51.7%   R-Sq(adj) = 42.0%

Analysis of Variance

Source	DF	SS	MS	F	P
Regression	1	8.555	8.555	5.35	0.069
Residual Error	5	8.002	1.600		
Total	6	16.557			

Unusual Observations

Obs	Temperature	Change	Fit	SE Fit	Residual	St Resid
5	301	8.500	6.614	0.929	1.886	2.20R

R denotes an observation with a large standardized residual.

The p-value for the predictor *Temperature*, is *.069* which exceeds $\alpha = .01$, so there is insufficient evidence, at the 1% level, that this model is useful for prediction.

Note that at the very bottom, Minitab **brings to your attention possible outliers**, in a list of *Unusual Observations*, highlighting the point we earlier identified by visual inspection of the plot.

Delete row 5 (the outlier), repeat the steps above, and now we get:

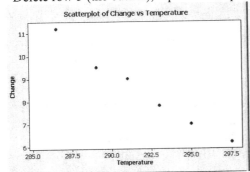 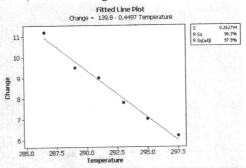

Regression Analysis: Change versus Temperature

The regression equation is Change = 140 - 0.450 Temperature

Predictor	Coef	SE Coef	T	P
Constant	139.759	8.553	16.34	0.000
Temperature	-0.44969	0.02929	-15.35	0.000

s = 0.262794   R-Sq = 98.3%   R-Sq(adj) = 97.9%

Analysis of Variance

Source	DF	SS	MS	F	P
Regression	1	16.279	16.279	235.72	0.000
Residual Error	4	0.276	0.069		
Total	5	16.555			

125

Now, we see the strong linear association (*p = 0.000*), which had been obscured by that outlier in our first analysis.

Exercise 11.110 - M/S page 652

Open up the ORGCHEM data file. We want to plot Maximum Absorption (C2) versus the Hammett Constant (C3), but using different symbols according to the compound. In C1 = Compound, replace the compound variants '1a', '1b', etc, by just '1', and '2a', '2b', etc by just '2'. Now, use **Graph>Scatterplot-With Groups: Y variables:** *Absorption*, **X variables:** *Hammett*, **Categorical variables:** *Compound*, and we see

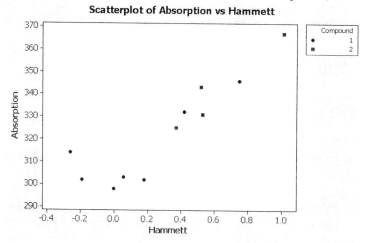

Clearly, there are two different relationships, which should be fitted separately. *For compound 1, though, the relationship appears curved and hence a straight line may not be the most appropriate model choice.*

For (b), (c) and (d), use **Data>Split Worksheet: By variables:** *Compound*, to break up the worksheet into two, according to the compound. Then select **Stat>Regression>Regression** or **Stat>Regression>Fitted Line Plot**, enter *Absorption* and *Hammett* for the response and predictor, respectively. Repeat for each worksheet, to obtain the following two outputs:

Results for: ORGCHEM (Compound = 1)

Regression Analysis: Absorption versus Hammett

```
The regression equation is    Absorption = 308 + 41.7 Hammett

Predictor      Coef  SE Coef       T      P
Constant    308.137    4.896   62.94  0.000
Hammett       41.71    13.82    3.02  0.029

S = 11.9432    R-Sq = 64.6%    R-Sq(adj) = 57.5%

Analysis of Variance
Source          DF      SS      MS      F      P
Regression       1  1299.7  1299.7   9.11  0.029
Residual Error   5   713.2   142.6
Total            6  2012.9
```

126

Results for: ORGCHEM (Compound = 2)

Regression Analysis: Absorption versus Hammett

```
The regression equation is   Absorption = 303 + 64.1 Hammett

Predictor      Coef   SE Coef       T       P
Constant    302.588     8.732   34.65   0.001
Hammett      64.05      13.36    4.79    0.041

S = 6.43656    R-Sq = 92.0%   R-Sq(adj) = 88.0%

Analysis of Variance
Source           DF        SS       MS       F       P
Regression        1    952.14   952.14   22.98   0.041
Residual Error    2     82.86    41.43
Total             3   1035.00
```

The two fitted equations appear to indicate that the absorption rate increases at a much faster rate, as the Hammett constant increases, when using compound 2 (though this disparity may also be attributed to the influence of the two negative Hammett Constant points for compound 1). Both models are statistically significant at the 5% level but not at the 1% level.

The Correlation Coefficient and the Coefficient of Determination

To find the correlation between two variables, r, use **Stat>Basic Statistics>Correlation**. The square of r, called the *coefficient of determination*, is a key measure of the predictive power of the fitted model and is always included in the regression output, written as 'R-Sq'.

Example 11.4 - M/S page 626

Open up the CASINO data file, containing the number of employees in C1 and the crime rate in C2. . To calculate the correlation between the number of employees and the crime rate, we use **Stat>Basic Statistics>Correlation:**

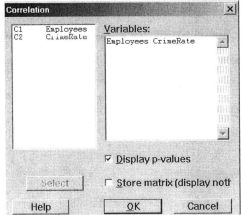

Might as well keep the default selection of **Display p-values**, which give you an assessment of the statistical significance of the observed association (same as testing the slope for a fitted straight line). We get

Correlations: Employees, CrimeRate

```
Pearson correlation of Employees and CrimeRate = 0.987
P-Value = 0.000
```

The very high (and very significant) correlation is no surprise, if we look at the scattergram below (**Graph>Scatterplot-Simple**):

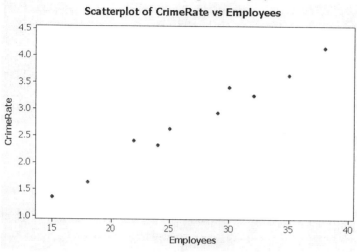

Scatterplot of CrimeRate vs Employees

Example 11.5 - M/S page 630

Open up once again the STIMULUS worksheet. To find r^2, just look for *R-Sq* in the output from: **Stat>Regression>Regression: Response:** *Time*, **Predictor:** *Drug* (or from that little box included in the graph produced by **Stat>Regression>Fitted Line Plot**)

Regression Analysis: Time versus Drug

```
The regression equation is
Time = - 0.100 + 0.700 Drug

Predictor     Coef   SE Coef      T      P
Constant   -0.1000    0.6351  -0.16  0.885
Drug        0.7000    0.1915   3.66  0.035

S = 0.605530   R-Sq = 81.7%   R-Sq(adj) = 75.6%
```

We see that the amount of drug explains (linearly) *81.7%* of the variation among subjects' reaction times, leaving only 18% of this variation unaccounted for (presumably due to other differences among the subjects).

Exercise 11.79 - M/S page 635

Open up the BOXING2 data file, containing blood lactate levels in C1, and perceived recovery in C2. Let's run the regression, to find r^2: **Stat>Regression>Regression: Response:** *Lactate*, **Predictor:** *Recovery*, and at the top of the output, we see:

Regression Analysis: Lactate versus Recovery

```
The regression equation is  Lactate = 2.97 + 0.127 Recovery

Predictor     Coef   SE Coef     T      P
Constant    2.9696    0.7896   3.76  0.002
Recovery   0.12667   0.04878   2.60  0.021

s = 0.950722    R-Sq = 32.5%    R-Sq(adj) = 27.7%
```

So, a modest *32.5%* of the variation on lactate levels is explained by the linear relationship with perceived recovery. If you square root 0.325, you get ±0.57, so *r* must equal +0.57, considering the positive slope of the fitted regression equation above.

Of course, we can find the correlation directly (and square it to determine the coefficient of determination) using **Stat>Basic Statistics>Correlation: Variables:** *Lactate Recovery*

Correlations: Lactate, Recovery

```
Pearson correlation of Lactate and Recovery = 0.570
P-Value = 0.021
```

Confidence and Prediction Intervals

To find a confidence interval for the mean response at some setting of the predictor, or to obtain a prediction interval for a random future observation on the response, use the available **Options** when running your **Regression**. *Both* intervals are printed out when you request **Prediction intervals.** Choose the one that you need.

Examples 11.6 & 11.7 - M/S page 637

Once again, open up the simple data set (STIMULUS) with which we started this chapter. To find either the confidence interval for the mean, or the prediction interval, when we have a 4% drug concentration, run the regression **Stat>Regression>Regression** and use **Options: Prediction intervals for...** Plug in '*4*' as below:

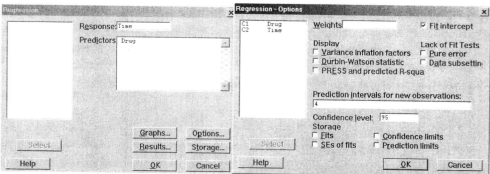

The *default* confidence level of 95% is what we want. Leave **Fit intercept** checked, as un-checking it will fit a line with an intercept of '0'. **OK, OK**, and in the output, in addition to all the stuff that we have seen before, way down at the bottom, you can see:

```
Predicted Values for New Observations
New
Obs    Fit   SE Fit      95% CI              95% PI
 1   2.700    0.332   (1.645, 3.755)   (0.503, 4.897)

Values of Predictors for New Observations
New
Obs  Drug
 1   4.00
```

You can read off the prediction interval, *PI*, or the confidence interval, *CI*, as desired. The *Fit* of 2.700 results from plugging *4.00* for the drug amount in the regression equation. *'SE Fit'* is precisely what your text refers to as the 'estimated standard error of $\hat{y}$', so $2.700 \pm t_{.025}(0.332)$ should produce the exact *CI* displayed above.

Exercise 11.95 - M/S page 644

Open up the CUTTOOL data file again, showing 'Speed', 'HoursA', 'HoursB' in C1 – C3 respectively. For a CI for mean lifetime for brand A when the cutting speed is 45, we use **Stat>Regression>Regression: Response:** *HoursA*, **Predictor:** *Speed*, with **Options: Prediction intervals ...:** *45*, **Confidence level:** *90*, which yields the usual regression output, and, at the bottom:

```
Predicted Values for New Observations
New
Obs    Fit   SE Fit      90% CI              90% PI
 1   3.350    0.332   (2.763, 3.937)   (1.127, 5.573)

Values of Predictors for New Observations
New
Obs  Speed
 1   45.0
```

Repeat, changing the **Response** to '*HoursB*', and we see, at the bottom:

```
Predicted Values for New Observations
New
Obs    Fit   SE Fit      90% CI            90% PI
  1  4.465   0.167   (4.169, 4.761)   (3.345, 5.585)

Values of Predictors for New Observations
New
Obs  Speed
  1   45.0
```

In (a), we are estimating means, and not predicting future random observations, so it is the *CI's* in the outputs above that we use. It is no surprise that the CI for brand A mean lifetime is the wider one, since *s* is bigger there, indicating larger random errors or greater dispersion about the regression line. For (b), use the *PI's* from the outputs.

Section 11.9 - A Complete Example - M/S page 644

Open up the FIREDAM data file. We want to model the relationship between fire damage (C2) and distance from nearest fire station (C1). Start by examining the data and fitted line, using **Stat>Regression>Fitted Line Plot: Response:** *Damage*, **Predictor:** *Distance*, **Type of Regression:** *Linear*. We get

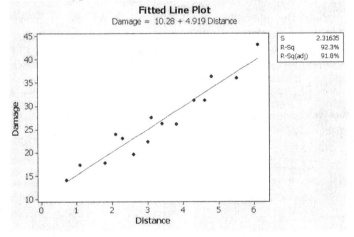

Things looks suitable for regression analysis, i.e. the pattern is linear and the relationship appears to be fairly strong. Also, the scatter is fairly constant about the line, suggesting constant error variance. We also want (see Step 5 in text) a prediction of fire damage at a distance of 3.5 miles. Run **Stat>Regression>Regression: Response:** *Damage*, **Predictor:** *Distance*, with **Options: Prediction intervals for new observations:** *3.5*, and we get:

Regression Analysis: Damage versus Distance

```
The regression equation is    Damage = 10.3 + 4.92 Distance

Predictor    Coef   SE Coef      T      P
Constant   10.278     1.420   7.24  0.000
Distance   4.9193    0.3927  12.53  0.000

S = 2.31635    R-Sq = 92.3%    R-Sq(adj) = 91.8%

Analysis of Variance
Source             DF      SS      MS       F      P
Regression          1  841.77  841.77  156.89  0.000
Residual Error     13   69.75    5.37
Total              14  911.52

Predicted Values for New Observations
New
Obs     Fit   SE Fit       95% CI             95% PI
  1  27.496    0.604  (26.190, 28.801)  (22.324, 32.667)

Values of Predictors for New Observations
New
Obs   Distance
  1       3.50
```

Distance is quite significant as a predictor of Damage, since the *p*-value for the coefficient of Distance is only *0.000* (to 3 decimal places). If you only wish to consider the one-sided alternative that Damage increases with Distance, divide the printed p-value by 2. The model explains *92% (R-Sq)* of the variation in Damages, so this model should be quite useful for prediction. Our prediction of the Damages in a burning house *3.5* miles away from the fire station, is anywhere from *$22,324* to *$32,667*, with *95%* confidence (from the *PI*).

We also estimate that for every extra mile away from the station, average damages increase by *$4,919*, or more precisely by between $4.9193 \pm t_{.025,\, 13DF} (.3927)$ thousands of dollars, i.e. $4070 to $5768, with 95% confidence.

Chapter 12 – Multiple Regression and Model Building

In this chapter, we will, fit a multiple regression model, with
Stat>Regression>Regression

Construct data plots, using
Graph>Scatterplot
Graph>Matrix Plot

Superimpose a fitted quadratic model on a data plot, with
Stat>Regression>Fitted Line Plot

Calculate higher order predictor terms for the model, or log-transform the response, using
Calc>Calculator

Calculate dummy variables with
Calc>Make Indicator Variables

Find p-values for the F distribution, with
Calc>Probability Distributions>F

Run a step-wise regression with
Stat>Regression>Stepwise

Construct a batch of four useful residual plots, with
Stat>Regression>Residual Plots

Check for multicollinearity of predictors, using
Stat>Basic Statistics>Correlation

Fitting the Model

To fit a first order model in k predictors, we again may apply the *least squares criterion*, i.e., choose the plane which minimizes the sum of the squared differences between the observed responses and the predicted or fitted responses. The math is messy. Let Minitab's **Stat>Regression>Regression** do all the hard work.

Example 12.1 - M/S page 665

Open up the REALESTATE data file showing real estate appraisal data for 20 properties.
We can plot *Sale Price*

↓	C1	C2	C3	C4	C5
	Property	SalePrice	LandValue	Improvements	Area
1	1	68900	5960	44967	1873
2	2	48500	9000	27860	928
3	3	55500	9500	31439	1126

133

versus each possible predictor using **Graph>Scatterplot-Simple** with **Multiple Graphs: Multiple Variables:** *In separate panels*, as below:

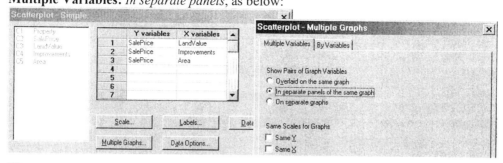

We get

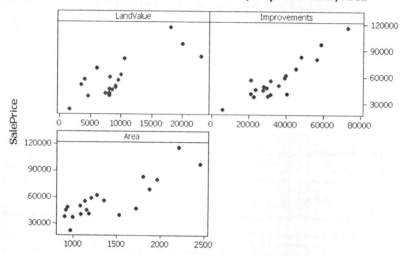

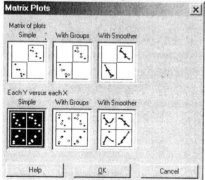

Alternatively, we can produce all three plots on one page using **Graph>Matrix Plot**:

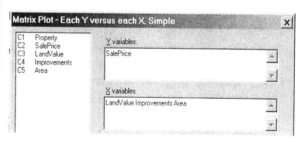

which produces:

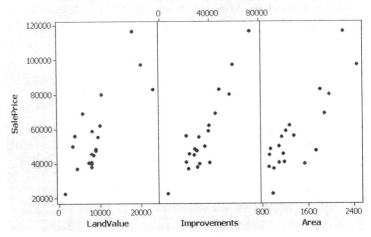

Matrix Plot of SalePrice vs LandValue, Improvements, Area

For each predictor, we see some positive relation with Sale Price, with Improvements Value having the strongest relationship. But perhaps a model with all three predictors will explain Sale Price better than any one predictor alone. To run a multiple regression on all three predictors, simply use **Stat>Regression>Regression**, as before, but now enter all three variables next to **Predictors:**

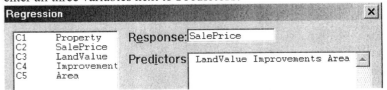

Part of the output is below:

Regression Analysis: SalePrice versus LandValue, Improvements, Area

```
The regression equation is
SalePrice = 1470 + 0.814 LandValue + 0.820 Improvements + 13.5 Area

Predictor        Coef   SE Coef     T      P
Constant         1470      5746   0.26  0.801
LandValue      0.8145    0.5122   1.59  0.131
Improvements   0.8204    0.2112   3.88  0.001
Area           13.529     6.586   2.05  0.057

S = 7919.48   R-Sq = 89.7%   R-Sq(adj) = 87.0%

Analysis of Variance
Source          DF        SS          MS       F      P
Regression       3  8779676741  2926558914  46.66  0.000
Residual Error  16  1003491259    62718204
Total           19  9783168000
```

At the top is the fitted *regression equation* in all three predictors. The parameter estimates also appear under '*Coef*'. The sum of squared errors minimized by this fit is equal to the *Residual Error SS* or *1003491259*, and the estimated standard deviation of the errors = *s* = *7919.48* (or square root the *Residual Error MS*).

Inferences about the parameters and overall model utility

The (Residual) Error Mean Square in the ANOVA table estimates the variance of the errors. The square root of this appears as '*s*'. Next to each *Predictor*, we see the fitted *Coef*ficient and its estimated standard error (*SE*). Their ratio, denoted '*T*' is also displayed, and may be used to test the null hypothesis that the true coefficient (β) equals zero, and the corresponding *p*-value is also shown (assuming a two-sided alternative).

Use the *F* statistic and *p*-value from the *Analysis of Variance* (ANOVA) table to test if the model as a whole is useful for prediction of the response. The coefficient of determination, *R-Sq*, and its adjusted version *R-Sq (adj)*, shown below the ANOVA table, indicate how well the predictors as a whole explain changes in the response.

Examples 12.3 & 12.4 - M/S pages 671 & 676

Open up the GFCLOCKS data file. To see how the auction price depends on both #bidders and age, run **Stat>Regression>Regression: Response:** *Price*, **Predictors:** *Age Bidders*, and we get

Regression Analysis: Price versus Age, Bidders

```
The regression equation is  Price = - 1339 + 12.7 Age + 86.0 Bidders

Predictor      Coef   SE Coef      T      P
Constant    -1339.0     173.8  -7.70  0.000
Age         12.7406    0.9047  14.08  0.000
Bidders      85.953     8.729   9.85  0.000

S = 133.485   R-Sq = 89.2%   R-Sq(adj) = 88.5%

Analysis of Variance
Source          DF       SS       MS       F      P
Regression       2  4283063  2141531  120.19  0.000
Residual Error  29   516727    17818
Total           31  4799790
```

To test the hypothesis that mean auction price increases with the number of bidders, controlling for age, we take the *Coef* of Bidders, and divide by its estimated *SE*, producing the displayed *T* value of 9.85. Assuming a two-sided alternative, Minitab calculates *p* as P(|t|>9.85) = *0.000*. Divide by 2 for the correct p-value here.

To find a 90% CI for the true coefficient of Age, take the *Coef* shown above, and wrap $t_{.05,29df}$(*SE Coef*) around it: *12.7406 ± 1.699(0.9047)*.

We see that this model explains about *89.2% (R-Sq)* of the variation in prices, unadjusted, or *88.5%, adj*usted for sample size and # predictors. The F-value of *120.19* is highly significant with *p* = *.000* (< α = .05), so the model as a whole is useful for predicting prices.

Exercise 12.19 - M/S page 680

Open up the STREETVEN data file, and regress Earnings in C2 on Age in C3 and Hours worked per day in C4, using **Stat>Regression>Regression: Response:** *C2*, **Predictors:** *C3 C4*. We get:

136

Regression Analysis: Earnings versus Age, Hours

```
The regression equation is   Earnings = - 20 + 13.4 Age + 244 Hours

Predictor    Coef   SE Coef      T      P
Constant    -20.4    652.7   -0.03  0.976
Age        13.350    7.672    1.74  0.107
Hours      243.71    63.51    3.84  0.002

S = 547.737    R-Sq = 58.2%    R-Sq(adj) = 51.3%

Analysis of Variance
Source          DF       SS      MS      F      P
Regression       2  5018232 2509116   8.36  0.005
Residual Error  12  3600196  300016
Total           14  8618428
```

The model is highly significant, since $P = 0.005$ for the regression model, and the model explains *58%* of the variation in earnings (*51%*, adjusted). Age is not a significant predictor in the model since the *p*-value for its coefficient is only *.107*, whereas Hours is significant and we can estimate that vendors in any particular age group, who work an extra hour per day, earn on average an additional $243.71 \pm t_{.025,12}(63.51)$ dollars annually.

Using the Model for estimation and prediction

To find a confidence interval for the mean response, or a prediction interval for a future random observation on the response, run the **Regression** procedure with **Options: Prediction Intervals for new observations**, and specify the desired settings for the predictors.

Example 12.5 - M/S page 685

Open up the GFCLOCKS data file again. For a 95% CI for the average auction price for 150 year old clocks with 10 bidders, run **Stat>Regression>Regression** and plug in as follows:

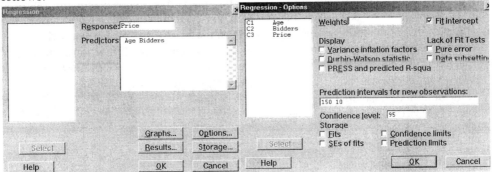

Note that you have to enter the values of the predictors in the exact same order that you entered the predictors in the main dialog box. E.g. if we had entered "10 150", Minitab would assume 10 is the age, with 150 bidders. **OK, OK**, and at the bottom of the session window output, we see:

```
Predicted Values for New Observations
New
Obs     Fit  SE Fit      95% CI             95% PI
  1  1431.7    24.6  (1381.4, 1481.9)  (1154.1, 1709.3)

Values of Predictors for New Observations
New
Obs  Age  Bidders
  1  150    10.0
```

So, with 95% confidence, we can say that the mean auction price for all 150 year old clocks with 10 bidders is between $*1381.40* and $*1481.90*.

To predict the auction price for one 50 year old clock that has 2 bidders, proceed as above, but now plug in "*50 2*" for the **new observations**, and read off the *PI* instead of the *CI* in the output. However, the interval would not be of much use, since this would constitute *extrapolation* well beyond the range of the observed data (always plot the data in various ways before trusting any statistical analysis!). If you proceed naively, Minitab tries to help!! You get the requested output, but with a warning (also note the ridiculous negative intervals):

```
Predicted Values for New Observations
New
Obs    Fit  SE Fit       95% CI            95% PI
  1  -530.0   123.0  (-781.5, -278.5)  (-901.2, -158.8)XX
XX denotes a point that is an extreme outlier in the predictors.

Values of Predictors for New Observations
New
Obs   Age  Bidders
  1  50.0     2.00
```

Exercise 12.35 - M/S page 690

Open up the WATEROIL data file. Since we are predicting Voltage (C2) with only Volume (C3), Salinity (C4), and Surfactant (C7) specified, we cannot use the first order model with all 7 possible predictors (as suggested in the exercise) since we would not know what values to specify for the other 4 predictors. Further analysis though would show that the other 4 predictors do not contribute to the model (e.g. using the approach of section 12.9). So, ignore them and regress just on the 3 indicated predictors, with prediction at the settings of 80 (high volume), 1(low salinity), and 2(low surfactant) respectively. Use **Stat>Regression>Regression: Response:** *Voltage*, **Predictors:** *Volume Salinity Surfactant*, with **Options: Prediction Intervals ...:** *80 1 2*

At the bottom of the output, we have:

```
Predicted Values for New Observations
New
Obs     Fit  SE Fit       95% CI          95% PI
  1  -0.098   0.232  (-0.592, 0.396)  (-1.233, 1.038)

Values of Predictors for New Observations
New
Obs  Volume  Salinity  Surfactant
  1    80.0      1.00        2.00
```

138

So the voltage may be anywhere from 0 to *1.038*, at these settings (95% confidence). However, the large portion of the interval in negative territory (impossible values for voltages) suggests that there may be problems with our model assumptions.

Interaction and Higher Order Models

A first order model in the predictors may not be the correct model. There may be interaction in the way the predictors affect the response, or the predictors may affect the response in a non-linear fashion. Use **Calc>Calculator** to calculate higher order terms such as a product like x_2x_3 or a squared term like x_5^2. Include these among your predictors when you regress, and then examine their contribution.

Example 12.6 - M/S page 691

Open up again the grandfather clock data. Do age and #bidders interact in their effect on clock price? To answer this, enter an interaction term as an additional predictor in the model, and then test its significance. Use **Calc>Calculator**, to calculate the product of *Age* and *Bidders*, and store the result in C4 named '*AgeBid*':

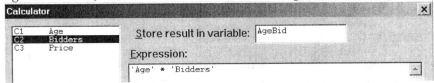

Next, regress on all three predictors, **Stat>Regression>Regression**

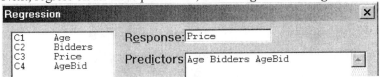

And we get

Regression Analysis: Price versus Age, Bidders, AgeBid

```
The regression equation is  Price = 320 + 0.88 Age - 93.3 Bidders + 1.30 AgeBid

Predictor    Coef   SE Coef      T     P
Constant    320.5     295.1   1.09  0.287
Age         0.878     2.032   0.43  0.669
Bidders    -93.26     29.89  -3.12  0.004
AgeBid     1.2978    0.2123   6.11  0.000

S = 88.9145    R-Sq = 95.4%    R-Sq(adj) = 94.9%

Analysis of Variance
Source           DF       SS       MS       F      P
Regression        3  4578427  1526142  193.04  0.000
Residual Error   28   221362     7906
Total            31  4799790
```

The model as a whole is significant (*p=0.000*). The interaction term *AgeBid* has a *t*-ratio of *6.11* with displayed p-value of *0.000*, but since the alternative tested here is only for a

139

'positive' interaction (i.e. price increases more rapidly with age, when there are more bidders), the true p-value is half the posted value or 0.000/2 = .000. There is strong evidence of interaction.

Example 12.7 - M/S page 698

Open up the ELECTRIC data file. Examine the relationship between electrical usage and home size with **Graph>Scatterplot-Simple: Y:** *Usage*, **X:** *Size*

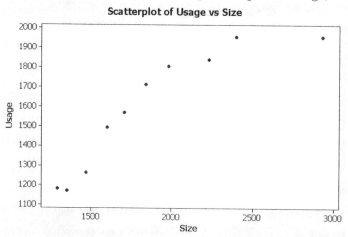

This is quite suggestive of a curved or quadratic type of relation. Let's visually examine first a linear fit, and then a quadratic fit, to the data, using **Stat>Regression>Fitted Line Plot: Type of Regression:** *Linear*

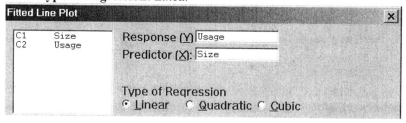

Repeat, but change to **Regression:** *Quadratic*. Here are the two resulting fitted line plots:

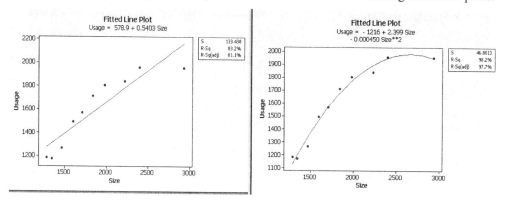

140

What a difference a square makes! The errors or deviations about the line are much smaller, and also more random in nature, for the quadratic fit. Also, note the difference in the *R-Sq* values displayed.

Define a new Size^2 predictor term to include in the model, using **Calc>Calculator**

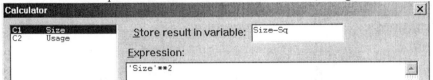

You will see the squares of the home sizes show up in a column named 'Size-Sq'. Now, regress on Size and Size^2, with **Stat>Regression>Regression**

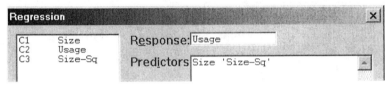

This gives us:

```
The regression equation is  Usage = - 1216 + 2.40 Size - 0.000450 Size-Sq

Predictor          Coef      SE Coef      T      P
Constant        -1216.1        242.8   -5.01  0.002
Size             2.3989       0.2458    9.76  0.000
Size-Sq      -0.00045004   0.00005908  -7.62  0.000


S = 46.8013    R-Sq = 98.2%    R-Sq(adj) = 97.7%

Analysis of Variance
Source          DF      SS      MS       F      P
Regression       2  831070  415535  189.71  0.000
Residual Error   7   15333    2190
Total            9  846402
```

We see that the model as a whole is highly significant in explaining variation in usage, with F = *189.71* and p-value = *0.000*. And the very high R-Sq(adj) value of *97.7 %* indicates a very close fit to the data, indeed.

The Size-Sq coefficient is negative and highly significant in the model with a p-value of *0.000*, indicating strong evidence of a concave downward trend, i.e., the rate of increase in electrical usage slows as home size increases (correct p value for a one sided alternative is .000/2).

Exercise 12.149 - M/S page 774

Open up the PHOSPHOR data file. To investigate the relationship between soil loss and dissolved phosphorus, start with **Graph>Scatterplot-Simple: Y:** *Phosphorus*, **X:** *SoilLoss*

141

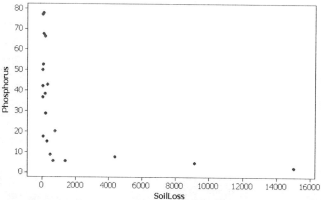

The relationship appears quite non-linear in nature. **Stat>Regression>Fitted Line Plot:
Y:** *Phosphorus*, **X:** *SoilLoss*, **Regression Model:** *Quadratic*, gives us:

Fitted Line Plot
Phosphorus = 42.25 - 0.01140 SoilLoss
+ 0.000001 SoilLoss**2

S	21.8086
R-Sq	34.9%
R-Sq(adj)	27.2%

Again, we see that the relationship is clearly non-linear, but even a quadratic model
appears to be questionable. Form the predictor $SoilLoss^2$, using **Calc>Calculator:
Expression:** *SoilLoss **2*, **Store in:** *SoilLoss-Sq*, and then regress with **Stat>Regression
>Regression: Response:** *Phosphorus*, **Predictors:** *SoilLoss SoilLoss-Sq*, and we get:

```
The regression equation is
Phosphorus = 42.2 - 0.0114 SoilLoss + 0.000001 SoilLoss-Sq

Predictor          Coef      SE Coef       T       P
Constant         42.247        5.712    7.40   0.000
SoilLoss       -0.011404     0.005053   -2.26   0.037
SoilLoss-Sq   0.00000061   0.00000037    1.66   0.115

S = 21.8086    R-Sq = 34.9%    R-Sq(adj) = 27.2%

Analysis of Variance
Source            DF      SS      MS      F       P
Regression         2   4325.4  2162.7   4.55   0.026
Residual Error    17   8085.5   475.6
Total             19  12410.9
```

142

We examine the *SoilLoss-Sq* term, and since its p-value of *0.115* exceeds $\alpha = .05$, we find the evidence for quadratic curvature to be insufficient. While this result appears mysterious, look at the fitted line plot for a straight-line model:

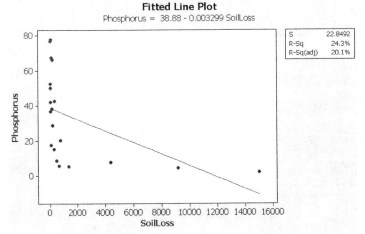

Fitted Line Plot
Phosphorus = 38.88 - 0.003299 SoilLoss

S	22.8492
R-Sq	24.3%
R-Sq(adj)	20.1%

Did the quadratic model reduce the deviations about the fitted line? Not by much! The real problem here is that neither a simple straight line nor quadratic curve fits well. A transformation of one or both of the variables might help, or perhaps the range of values for soil loss is too great for us to fit just one simple model.

Qualitative Predictors – Using dummies

The regression model can only be fit using quantitative predictors, so a special coding is needed to include qualitative variables as predictors. Minitab's **Calc>Indicator Variables** creates k dummy variables for a qualitative variable with k levels, where each dummy distinguishes a particular setting or level of the variable (= '1' for that level, = '0' otherwise). Choose only k-1 of them to use as predictors.

Example 12.9 - M/S page 710

Open the GOLFCRD data file. Notice the stacked arrangement of the data with brand in C1 and distance in C2. *Delete columns C3-C6.* We want to create dummy (indicator) variables to replace the qualitative predictor Brand, in C1.

↓	C1-T	C2	C3
	Brand	Distance	
1	A	251.2	
2	A	245.1	
3	A	248.0	
4	A	251.1	

Use **Calc>Make Indicator Variables:**

143

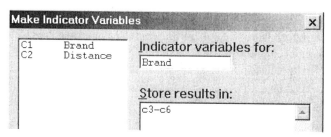

Note that since there are **four brands**, we need to name **four columns** for the indicator variables, with each column being an *indicator* of one of the brands. We get a worksheet now looking like:

↓	C1-T Brand	C2 Distance	C3	C4	C5	C6	C7
1	A	251.2	1	0	0	0	
2	A	245.1	1	0	0	0	
3	A	248.0	1	0	0	0	
4	A	251.1	1	0	0	0	
5	A	260.5	1	0	0	0	
6	A	250.0	1	0	0	0	
7	A	253.9	1	0	0	0	
8	A	244.6	1	0	0	0	
9	A	254.6	1	0	0	0	
10	A	248.8	1	0	0	0	
11	B	263.2	0	1	0	0	
12	B	262.9	0	1	0	0	
13	B	265.0	0	1	0	0	
14	B	254.5	0	1	0	0	
15	B	264.3	0	1	0	0	
16	B	257.0	0	1	0	0	
17	B	262.8	0	1	0	0	
18	B	264.4	0	1	0	0	
19	B	260.6	0	1	0	0	
20	B	255.9	0	1	0	0	
21	C	269.7	0	0	1	0	
22	C	263.2	0	0	1	0	

In C3 is the indicator for Brand A ('1' if Brand A, '0' otherwise), in C4, the indicator for Brand B ('1' if Brand B, '0' otherwise), etc. Now, we regress on the indicators, but we cannot use all of them. Pick any three that you wish, depending on which comparisons you wish to set up via the coefficients of the dummies. To follow the text, we omit the first dummy above, and then **Stat>Regression>Regression: Response:** *C2*, **Predictors:** *C4-C6*, gives us:

144

```
The regression equation is
Distance = 251 + 10.3 C4 + 19.2 C5 - 1.46 C6


Predictor      Coef   SE Coef       T       P
Constant    250.780     1.455  172.34   0.000
C4           10.280     2.058    5.00   0.000
C5           19.170     2.058    9.32   0.000
C6           -1.460     2.058   -0.71   0.483


S = 4.60163    R-Sq = 78.6%    R-Sq(adj) = 76.8%


Analysis of Variance
Source           DF       SS       MS       F       P
Regression        3  2794.39   931.46   43.99   0.000
Residual Error   36   762.30    21.18
Total            39  3556.69
```

The F value = *43.99* and p-value = *0.000* above are *exactly* the same as those obtained when running a one-way ANOVA before. Again we conclude that brand does affect distance.

Exercise 12.74 - M/S page 715

Open up the PEAS data file, with the Yields stacked in C2 and Varieties in C1. Form the dummies for Variety, using **Calc>Make Indicator Variables: Indicator Variables for:** *Variety*, **Store in:** *C3-C7*. Note we needed to specify 5 columns for dummies since there are 5 varieties. Now regress on 4 of the indicators. In the exercise, they suggest using indicators for varieties A, B, C, D, i.e. C3-C6 here, so **Stat>Regression>Regression: Response:** *Yield*, **Predictors:** *C3-C6*, gives us

```
The regression equation is
Yield = 20.4 + 4.75 C3 + 8.60 C4 + 11.3 C5 + 2.10 C6


Predictor      Coef   SE Coef       T       P
Constant    20.3500    0.9445   21.55   0.000
C3           4.750     1.336    3.56   0.003
C4           8.600     1.336    6.44   0.000
C5          11.300     1.336    8.46   0.000
C6           2.100     1.336    1.57   0.137


S = 1.88892    R-Sq = 86.5%    R-Sq(adj) = 82.9%


Analysis of Variance
Source           DF       SS       MS       F       P
Regression        4  342.040   85.510   23.97   0.000
Residual Error   15   53.520    3.568
Total            19  395.560
```

The low p-value for the F statistic in the ANOVA table gives good evidence that mean yield does depend on the variety used.

  Each dummy's *Coef* above estimates the difference between one of the varieties and variety E, since the dummy for variety E was omitted from the regression. So, a 95% CI for the mean difference between varieties D and E is the coefficient of the indicator for variety D (in C6), plus or minus a couple of *SE's*: *2.100* $\pm$ $t_{.025,15df}$(*1.336*). Be sure that you select the correct dummy: check your worksheet, or give the dummies more suggestive names like 'A-Indicator', 'B-Indicator', etc.

145

We can also analyze the data using a one-way ANOVA: **Stat>ANOVA>One-way: Response:** *Yield,* **Factor:** *Variety.* We get

One-way ANOVA: Yield versus Variety

```
Source    DF      SS      MS      F      P
Variety    4  342.04   85.51  23.97  0.000
Error     15   53.52    3.57
Total     19  395.56

S = 1.889   R-Sq = 86.47%   R-Sq(adj) = 82.86%
```

which is identical to the regression ANOVA table.

Using Quantitative and Qualitative Variables in the Model

The regression model (also called the *general linear model*) allows you to fit by least-squares one coherent model with quantitative variables, qualitative variables, and higher order terms for curvature and interaction. Just **Calc** the terms desired and **Regress**.

Example 12.11 - M/S page 719

Open up the CASTING data file. We want to fit a model relating productivity (#castings) to incentive and type of plant, allowing for interaction in the effects of these two variables. First, form dummies for the type of plant. Since there are two types of plant, two dummy variables will be created by Minitab, using **Calc>Make Indicator Variables:**

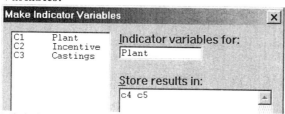

Minitab operates alphabetically, so looking at the worksheet, you can see that C4 is the indicator for Nonunion, and C5 is the indicator for Union. We only need one, so following the text, we will use the indicator for Nonunion, in C4, and let's name C4 'Dummy'.

Next, we need an interaction term, which is formed by taking the product of Incentive in C3 with the dummy for Nonunion: **Calc>Calculator:**

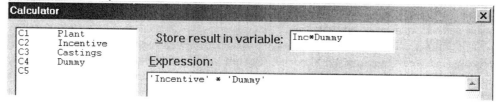

Now, regress on three predictors: **Stat>Regression>Regression:**

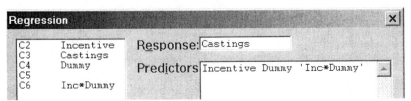

[After double-clicking on C6 to copy it over, its name appears in parentheses. These parentheses are necessary for clarity when using a *column name containing spaces or some special symbols like the '*' that I used*]

We get

```
The regression equation is
Castings = 1366 + 6.22 Incentive + 47.8 Dummy + 0.03 Inc*Dummy

Predictor        Coef   SE Coef       T      P
Constant      1365.83     51.84   26.35  0.000
Incentive       6.217     1.667    3.73  0.002
Dummy           47.78     73.31    0.65  0.525
Inc*Dummy       0.033     2.358    0.01  0.989

S = 40.8387    R-Sq = 71.1%    R-Sq(adj) = 64.9%

Analysis of Variance
Source          DF      SS      MS      F      P
Regression       3   57332   19111  11.46  0.000
Residual Error  14   23349    1668
Total           17   80682
```

The fitted equation appears above. Plug in '1' for Dummy (whence Inc*Dummy = Incentive), to obtain the fitted relation for the non-union plants and plug in '0' for Dummy (whence Inc*Dummy = '0') to obtain the relation for the union plants.

Does the relation between number of castings produced and incentive depend on the type of plant? The coefficient of the product term has a large p-value of *0.989*, so there is no evidence of such a dependence, or interaction.

You can plot the data with superimposed regression lines separately fitted for each group (i.e. plant type) by using: **Scatterplots-With Regression and Groups: Y:** *Castings,* **X:** *Incentives,* **Categorical variables:** *Plant*

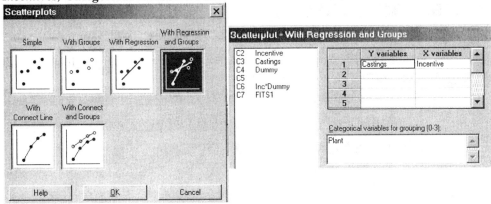

which produces:

147

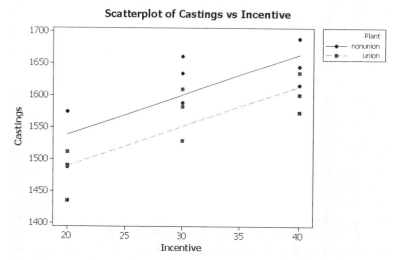

Scatterplot of Castings vs Incentive

Note how nearly parallel the fitted lines are, consistent with our finding of no significant interaction.

[Alternatively, when you run the **Regression**, select **Storage** and save the **Fits** in some column, then **Graph>Scatterplot-With Connect and Groups: Y:** *Fits,* **X:** *Incentive,* **Categorical variables:** *Plant*]

Comparing Nested Models

We can test for significance a portion of the predictors in any model, by running two regressions, one for the *complete model* and one for the *reduced model*, then comparing SSE's. Or you may use the *Sequential SS* shown in the output for the complete model.

Example 12.12 - M/S page 726

Open the CARNATIONS data file. We wish to fit a complete second order model to the data, and assess the contribution of the 2nd order terms. First, calculate the three 2nd order terms (if not already calculated for you in the saved worksheet), using **Calc>Calculator:**

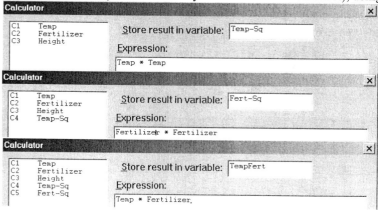

148

Now, regress Height on all predictors with **Stat>Regression>Regression**

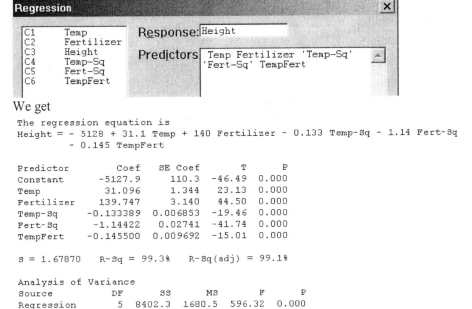

We get

```
The regression equation is
Height = - 5128 + 31.1 Temp + 140 Fertilizer - 0.133 Temp-Sq - 1.14 Fert-Sq
          - 0.145 TempFert


Predictor        Coef   SE Coef        T      P
Constant      -5127.9     110.3   -46.49  0.000
Temp           31.096     1.344    23.13  0.000
Fertilizer    139.747     3.140    44.50  0.000
Temp-Sq     -0.133389  0.006853   -19.46  0.000
Fert-Sq      -1.14422   0.02741   -41.74  0.000
TempFert    -0.145500  0.009692   -15.01  0.000


S = 1.67870    R-Sq = 99.3%    R-Sq(adj) = 99.1%


Analysis of Variance
Source          DF       SS       MS       F      P
Regression       5   8402.3   1680.5  596.32  0.000
Residual Error  21     59.2      2.8
Total           26   8461.4


Source         DF   Seq SS
Temp            1   1510.7
Fertilizer      1    279.3
Temp-Sq         1   1067.6
Fert-Sq         1   4909.7
TempFert        1    635.1
```

The fitted equation appears above. To assess the contribution of the second order terms, there are two ways we may proceed:

(i) Regress again, using only the predictors Temp and Fertilizer (the reduced model):

```
The regression equation is
Height = 106 - 0.916 Temp + 0.788 Fertilizer


Predictor       Coef   SE Coef       T      P
Constant      106.09     55.95    1.90  0.070
Temp         -0.9161    0.3930   -2.33  0.028
Fertilizer    0.7878    0.7860    1.00  0.326


S = 16.6727    R-Sq = 21.2%    R-Sq(adj) = 14.6%


Analysis of Variance
Source          DF       SS      MS      F      P
Regression       2   1789.9   895.0   3.22  0.058
Residual Error  24   6671.5   278.0
Total           26   8461.4
```

Take the difference of the SSE's = $SSE_R - SSE_C$ = *6671.5 - 59.2* = 6612.3, plug into the F statistic, then find the P-value (see further down below).

(ii) **Alternatively** you can find $SSE_R - SSE_C$ by using the Seq SS display in the output for the complete model and adding up the 3 Seq SS's for the 3 second order terms, which equals *1067.6 + 4909.7 + 635.1* = 6612.4, the same as above aside from rounding error. This approach works **only** if we list the predictors being tested last in the *Seq SS* table (so list them last in the Regression dialog box list of predictors).

Once we have found $SSE_R - SSE_C$ by either route, proceed to calculate the F statistic = 6612.3/3 ÷ (*59.2/21*) = 781.6, and then the P-value, either approximately from your F-tables, or more precisely using **Calc>Probability Distributions> F:** *Cumulative Probability*, **Numerator DF:** *3*, **Denominator DF:** *21*, **Input Constant:** *781.6*

Cumulative Distribution Function

```
F distribution with 3 DF in numerator and 21 DF in denominator

     x        P( X <= x )
  781.6            1
```

Hence, the p-value is P(F > 781.6) = 1 – 1 = 0 (to 5 decimal places), providing very strong evidence that the second order model fits the data better than a first order model.

The *Seq SS* for a predictor gives the drop in the SSE caused by adding that predictor to the model with all the earlier listed predictors. To assess the combined contribution of several predictors, list them last, add their *Seq SS* values (same as $SSE_R - SSE_C$), divide by their number (= df), then by the MSE.

Exercise 12.97 - M/S page 733

Again, open up the WATEROIL data file, where C2, C3, C4, C7 display Voltage, Volume, Salinity, and Surfactant readings resp. We wish to fit a model to predict voltage using the latter 3 variables and two interaction terms. Calculate the interaction terms using **Calc>Calculator: Expression:** *C3 * C4*, **Store in:** *VolSalin*, and repeat this calculation with **Expression:** *C3 * C7*, **Store in:** *VolSurf*.   When you use new column names, the first two free columns get chosen, C10 and C11 here. Then, regress with **Stat>Regression>Regression: Response:** *C2*, **Predictors:** *C3 C4 C7 C10 C11*. We get

```
The regression equation is
Voltage = 0.906 - 0.0228 Volume + 0.305 Salinity + 0.275 Surfactant
          - 0.00280 VolSalin + 0.00158 VolSurf

Predictor        Coef    SE Coef       T      P
Constant       0.9057     0.2855    3.17  0.007
Volume      -0.022753   0.008318   -2.74  0.017
Salinity       0.3047     0.2366    1.29  0.220
Surfactant     0.2747     0.2270    1.21  0.248
VolSalin    -0.002804   0.003790   -0.74  0.473
VolSurf      0.001579   0.003947    0.40  0.696

S = 0.504650   R-Sq = 67.9%   R-Sq(adj) = 55.6%

Analysis of Variance
Source          DF       SS      MS      F      P
Regression       5   7.0103  1.4021   5.51  0.006
Residual Error  13   3.3107  0.2547
Total           18  10.3210
```

150

R-Sq (adj) is *55.6%* above. This is lower than the R-Sq(adj) for a first order model in all seven of the original variables (62.5%), hence the latter fits the data a little more closely than the particular interaction model we are examining here (no exact test possible, since models are not nested).

To test for the contribution of the interaction terms in the model fitted above, run the reduced model: **Stat>Regression>Regression: Response:** *C2*, **Predictors:** *C3 C4 C7*

```
The regression equation is
Voltage = 0.933 - 0.0243 Volume + 0.142 Salinity + 0.385 Surfactant

Predictor         Coef   SE Coef      T      P
Constant        0.9326    0.2482   3.76  0.002
Volume       -0.024272  0.004900  -4.95  0.000
Salinity       0.14206   0.07573   1.88  0.080
Surfactant     0.38457   0.09801   3.92  0.001

S = 0.479646   R-Sq = 66.6%   R-Sq(adj) = 59.9%

Analysis of Variance
Source          DF      SS      MS      F      P
Regression       3  6.8701  2.2900   9.95  0.001
Residual Error  15  3.4509  0.2301
Total           18 10.3210
```

Now compare the two Error SS's (*3.4509* and *3.3107*), in a proper F-test. It is **not correct** to try to examine the contribution of these two interaction terms by looking at their two p-values in the complete model, tempting though this approach might seem.

Exercise 12.148 - M/S page 774

Open the SHUTDOWN data file, containing *Downtime* in C1, *Machine Age* in C2, *Machine Type* (type 1 or type 2) in C3, and in C4 a dummy variable, *X2*, for Machine Type (= 1 if type 1; = 0 if type 2). We wish to examine how downtime depends on age and machine type. To fit the proposed model, we need to form the square of age using **Calc>Calculator: Store in:** *Age-Sq*, **Expression:** *MachAge * MachAge*. Regress using **Stat>Regression>Regression: Response:** *Downtime*, **Predictors:** *X2, MachAge, Age-Sq,* and we get

```
The regression equation is
Downtime = - 11.8 + 13.2 X2 + 10.3 MachAge - 0.418 Age-Sq

Predictor       Coef  SE Coef      T      P
Constant     -11.769    3.050  -3.86  0.001
X2            13.244    1.503   8.81  0.000
MachAge       10.294    1.438   7.16  0.000
Age-Sq       -0.4180   0.1613  -2.59  0.020

S = 2.83489   R-Sq = 94.9%   R-Sq(adj) = 94.0%

Analysis of Variance
Source          DF       SS      MS      F      P
Regression       3  2396.36  798.79  99.39  0.000
Residual Error  16   128.59    8.04
Total           19  2524.95

Source   DF   Seq SS
X2        1     1.25
MachAge   1  2341.17
Age-Sq    1    53.95
```

151

(b) Since the *p*-value for age-squared equals *0.020* < α=.05, we have evidence that an age-squared term improves the fit of the model.

(c) Since we listed age and age-squared last in the *Seq SS* table, we can assess their combined contribution to the complete model (difference in SSE's) by adding *2341.17 +* *53.95* = 2395.12. Divide by 2 df, then by the MSE = 8.04, and we get F = 148.95. Then **Calc>Probability Distributions> F:** *Cumulative Probability*: **Numerator DF: 2, Denominator DF:** *16*, **Input Constant:** *148.95*, and we get

Cumulative Distribution Function

```
F distribution with 2 DF in numerator and 14 DF in denominator

      x        P( X <= x )
  148.95        1.00000
```

The p-value = P(F > 148.95) = 1 – *1.00000* = 0.00000. Hence, we have strong evidence that downtime does not depend solely on the machine type; age plays a role too.

Alternatively, for (c), we can explicitly fit the reduced model, i.e. regress on just the dummy variable for machine type: **Stat>Regression>Regression: Response:** *Downtime,* **Predictors:** *X2*

```
Analysis of Variance
Source          DF      SS      MS     F      P
Regression       1     1.2     1.2   0.01   0.926
Residual Error  18  2523.7   140.2
Total           19  2525.0
```

to obtain the SSE$_R$ = *2523.7* . Above we found SSE$_C$ = *128.59* . The difference is 2395.1. Then proceed exactly as above to calculate F = 148.95 with p = 0.00000.

It's always a good idea to plot the data. Use **Graph>Scatterplot-With Groups: Y:** *Downtime,* **X:** *MachAge,* **Categorical variables:** *MachType.* We get:

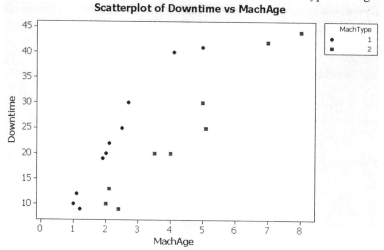

Increased age clearly produces more downtime, but is the effect curved, as indicated by our significant test on age-squared? The general pattern above curves downward, and yet

152

the pattern for machine 2 looks rather straight, and (except for one observation) likewise for machine 1! Indeed, if we fit instead a model with two non-parallel straight lines, R^2 is higher, indicating a closer fit to the data. *Proper specification of models can be a difficult and challenging problem.*

Stepwise Regression

Often we have many candidate predictors available. A common screening procedure to help select a useful subset of the candidate predictors is called stepwise regression. Use Minitab's **Stat>Regression>Stepwise**, and enter all the predictors being considered.

Example 12.13 - M/S page 737

Open the EXECSAL data file. We have 10 candidate predictors X1 – X10 (see text for their descriptions) to consider for a model to be used to predict log executive salary. Run Minitab's stepwise regression procedure: **Stat>Regression>Stepwise**, and fill in the **Predictors** and **Response**. Also, click on **Methods**.

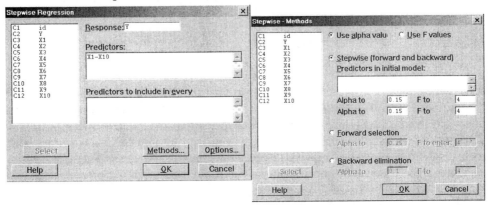

Review the choice of **Methods** and change as desired. The default choices shown above are to enter or remove a variable based on a *15%* significance test. The **Methods** sub-dialog box above actually should read **Alpha to enter**, and **Alpha to remove** (hopefully, these types of errors will be fixed for your version of Minitab). You may choose an **initial** set of predictors to start with, if you wish, or force some **predictors to be included in every model**. Here, we begin screening with no predictors and then add one at a time (with possible removals along the way). The results are:

Stepwise Regression: Y versus X1, X2, X3, X4, X5, X6, X7, X8, X9, X10

```
Alpha-to-Enter: 0.15   Alpha-to-Remove: 0.15

Response is Y on 10 predictors, with N = 100
```

Step	1	2	3	4	5
Constant	11.091	10.968	10.783	10.278	9.962
X1	0.0278	0.0273	0.0273	0.0273	0.0273
T-Value	12.62	15.13	18.80	24.68	26.50
P-Value	0.000	0.000	0.000	0.000	0.000
X3		0.197	0.233	0.232	0.225
T-Value		7.10	10.17	13.30	13.74
P-Value		0.000	0.000	0.000	0.000
X4			0.00048	0.00055	0.00052
T-Value			7.32	10.92	11.06
P-Value			0.000	0.000	0.000
X2				0.0300	0.0291
T-Value				8.38	8.72
P-Value				0.000	0.000
X5					0.00196
T-Value					3.95
P-Value					0.000
S	0.161	0.131	0.106	0.0807	0.0751
R-Sq	61.90	74.92	83.91	90.75	92.06
R-Sq(adj)	61.51	74.40	83.41	90.36	91.64
Mallows C-p	343.9	195.5	93.8	16.8	3.6

There were five steps, and then the process stopped. At **Step 1**, X1 (Experience) was added to the model. The resulting equation (*11.091 + 0.0278 X1*), t-value for the coefficient of X1 (*12.62*), s (*0.161*), and R-squared (*61.90%*), are all shown in the column underneath Step 1. At **Step 2**, X3 was added to the model, and the new coefficients, t-values, s, and R-squared are printed, in the next column. At **Step 5**, X5 was added to the model. No variables can be removed (all are significant at 15% level). The remaining candidates are assessed and none of them contribute significantly. The process stops.

Residual Analysis - Checking the Assumptions

Residual plots help us check our assumptions and gain insight into possible revisions of the model. They can easily be created using the **Graphs** sub-dialog box when running Minitab's **Stat>Regression>Regression** *or* **Fitted Line Plot**.

Example 12.14 - M/S page 742

Open the ELECTRIC data file. Select **Stat>Regression>Regression,** and plug in the response and predictor. To display all the residuals, you need to increase the amount of information displayed, so choose **Results**, and go for the maximal display including *the* *full table of fits and residuals*:

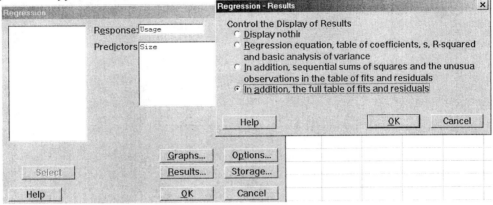

We get:

Regression Analysis: Usage versus Size

```
The regression equation is  Usage = 579 + 0.540 Size

Predictor       Coef   SE Coef       T       P
Constant       578.9     167.0    3.47   0.008
Size         0.54030   0.08593    6.29   0.000

S = 133.438   R-Sq = 83.2%   R-Sq(adj) = 81.1%

Analysis of Variance
Source           DF       SS       MS       F       P
Regression        1   703957   703957   39.54   0.000
Residual Error    8   142445    17806
Total             9   846402

Obs  Size   Usage     Fit  SE Fit  Residual  St Resid
  1  1290  1182.0  1275.9    66.0     -93.9     -0.81
  2  1350  1172.0  1308.3    62.1    -136.3     -1.15
  3  1470  1264.0  1373.2    55.0    -109.2     -0.90
  4  1600  1493.0  1443.4    48.6      49.6      0.40
  5  1710  1571.0  1502.8    44.7      68.2      0.54
  6  1840  1711.0  1573.1    42.3     137.9      1.09
  7  1980  1804.0  1648.7    43.1     155.3      1.23
  8  2230  1840.0  1783.8    51.8      56.2      0.46
  9  2400  1956.0  1875.7    61.5      80.3      0.68
 10  2930  1954.0  2162.0    99.6    -208.0     -2.34R

R denotes an observation with a large standardized residual.
```

Repeat the above, but this time, also select **Graphs** from the **Regression** dialog box above, and fill in:

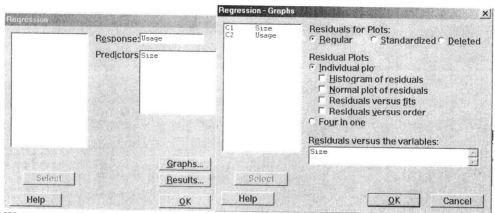

We now get the previous output, along with the following plot of **Residuals versus:** *Size*:

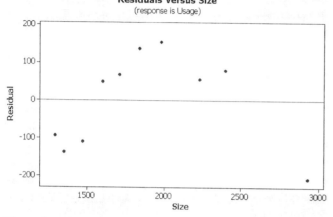

The points show a clear curving pattern, indicating a lack of fit of the straight-line model, and a need to amend it to allow for curvature. You can also easily produce other types of useful residual plots by ticking off your desired choices in the **Regression - Graphs** box.

Next we will try on for size, a quadratic model. We could repeat all of the above, after calculating and adding a 'Size-squared' term to the predictors, or more quickly, let's just select **Stat>Regression>Fitted Line Plot: Type of Regression:** *Quadratic,* with **Graphs: Residuals versus:** *Size,*   and we get

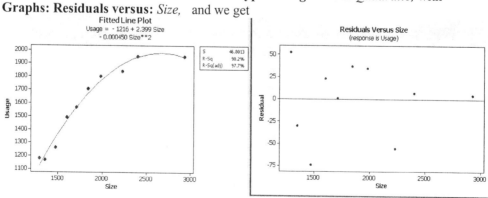

Now, things look better, with no systematic patterns in the residual plot - just a random bobbing about '0'.

Examples 12.15 & 12.16 - M/S pages 745 & 748

Open up the GFCLOCKALT data file. We wish to regress auction price on two predictors and their product, and then perform a residual analysis. First calculate the product term with **Calc>Calculator: Store in:** *AgeBid*, **Expression:** *Age * Bidders* . Then run **Stat>Regression>Regression,** click on **Graphs,** and request the plot of residuals versus Bidders, and two plots that help you check normality of the residuals:

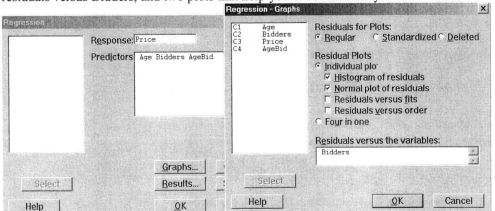

If you want to print out all of the residuals, like shown in Figure 12.36 in the text, use the sub-dialog box for **Results...** and click on *In addition, the full table of fits and residuals.* Usually, though, there is no need to print out all the residuals. **OK, OK** produces:

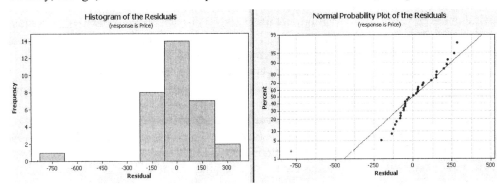

157

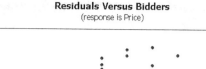

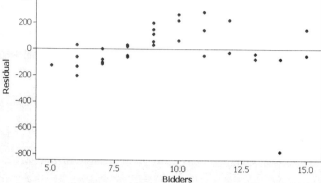

If you use the 3rd or 4th control level of the display of **Results** (see the sub-dialog box displayed above for ex.12.14), Minitab will flag any *unusual observations*, with symbols 'R' or 'X':

```
Unusual Observations
Obs  Age   Price    Fit    SE Fit  Residual  St Resid
 17  170  1131.0  1914.4   116.7    -783.4    -4.80R
 31  194  1356.0  1480.7   133.6    -124.7    -0.83 X

R denotes an observation with a large standardized residual.
X denotes an observation whose X value gives it large influence.
```

'*R*' indicates an observation with a large residual (*St*andardized *Resid*ual bigger than 2 in magnitude). '*X*' indicates an observation with unusual values for the predictors. The latter may have a large influence on the fit, while possessing a small residual.

Minitab has flagged the observation that was altered here, which has an extremely large residual of –783.4, or –4.80 standardized. Standardized residuals are handy for assessing relative size, since standardized values are not scale-dependent. [The standardizing operation does not use 's' as the standard deviation of a residual. It's a bit more complicated.]

The outlier, observation #17, stands out clearly in all the plots (and judging by the curvature in the plot of residuals versus Bidders, we should also consider adding Bidders^2 to the model). Outliers should be investigated and handled with considerable care, as one sole outlier can badly distort the fit and hence your understanding of the true relationship for the general population.

Note that the normal probability plot produced here by Minitab plots '**Percent**' instead of an average normal score on the y-axis. This is just a labeling issue, and does not affect interpretation one bit.

I recommend routine use of: **Stat>Regression>Regression: Graphs: Residual Plots: Four in one**, which produces a nice batch of four useful plots:

158

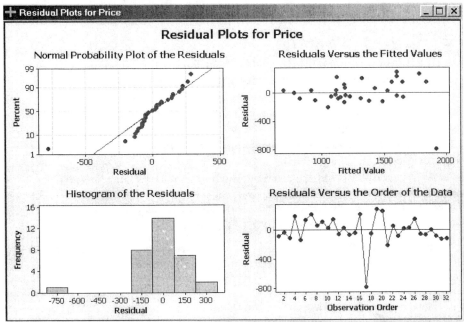

We get a normal probability plot of the residuals, a plot of residuals versus row number (Order of the Data), a histogram of residuals, and residuals versus fits (exact same appearance as residuals versus predictor, except for labeling, when there is only one predictor). In all of these, that one outlier stands out clearly. We can also use these plots to check for various problems with the model (e.g. non-normality, non-constant variance).

If we delete the outlier (row 17), re-run the regression, and use **Graphs: Residual Plots:** *Four in one*, we now get:

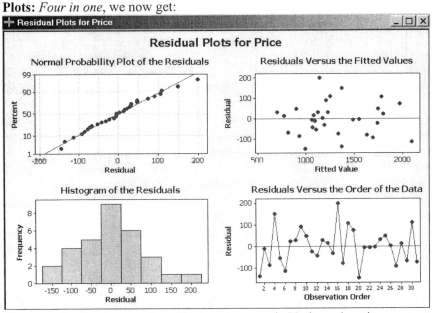

The error distribution now is quite close to normal. Variance is quite constant. You can also verify from the rest of the output that s and R-squared are much improved too.

159

Example 12.17 - M/S page 751

Sometimes, a transformation of the data, like a logarithm, will serve to remedy one or more problems, e.g. stabilizing the variance and/or symmetrizing the distribution of the errors.

Open up the SOCWORK data file, containing data on *salary* in C2 and years of *experience* in C1 for a sample of social workers. Regress salary on experience with **Stat>Regression>Regression: Response:** *Salary*, **Predictors:** *Experience*, and use **Graphs: Residual Plots:** *Four in one*

In the session window, you get the usual output, but the residual plots look like:

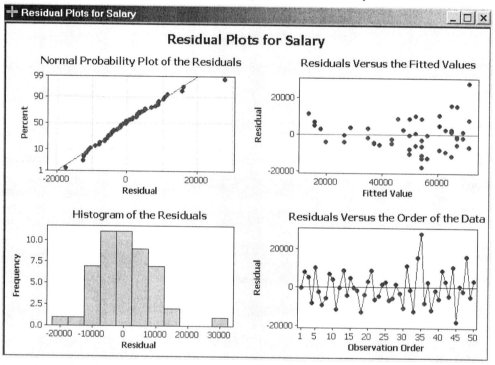

No point in drawing any conclusions from the regression output in the session window, since we have a big problem here: non-constant variance. We see this in the plot of residuals versus fitted values. The good news here is that it is a fairly steady trend, with dispersion increasing gradually as (fitted) salaries increase. Often in this situation, a transformation will remedy the problem. Let's try a log transform of the salary, using **Calc>Calculator.** Under **Functions**, find '*Natural log*' and double click it to copy it (denoted *LOGE,* which stands for 'log base e') under **Expression:**

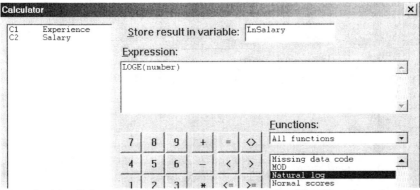

Now double click on '*C2 Salary*' on the left, which will replace '*number*' by '*Salary*' as the argument for the LOGE function. **Store result in:** *LnSalary*. Hit **OK** and the logs appear in a column named *LnSalary*.

Now, regress with the new response variable: **Stat>Regression>Regression: Response:** *LnSalary*, **Predictors:** *Experience*, with **Graphs: Residual Plots:** *Four in one*. **OK, OK**, and we get

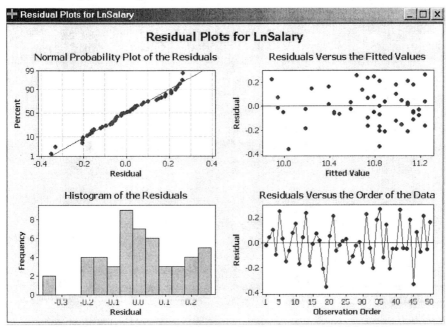

The trend in dispersion is gone! And the distribution is still fairly close to normal. We can now use the regression output in the session window to make reliable inferences with this model.

Exercise 12.151g. - M/S page 775

Open up the MNSALES data. We want to regress the sale prices of apartments (C2) on various characteristics of the apartments (C3-C7), and then perform some diagnostic checks via residual plots. Regress using **Stat>Regression>Regression: Response:** *C2*,

Predictors: *C3-C7*, with **Graphs: Residual Plots:** *Four in one,* and we get the usual session window output, along with the following residual plots:

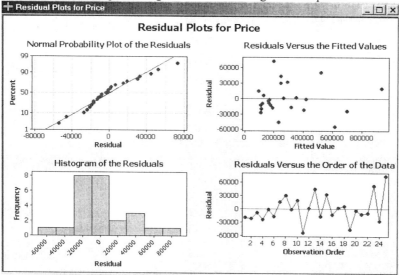

The distribution is close enough to normal, with no clear outliers, and there is no strong pattern when we look at residuals versus fits (well, maybe somewhat smaller dispersion for the very cheap houses). We should also check the regression output for its *Unusual Observations* listing.

We may also wish to plot the residuals versus each individual predictor, which can be done very conveniently using the same sub-dialog box, with **Regression - Graphs: Residual Plots: Residuals versus the variables:** *C3-C7*:

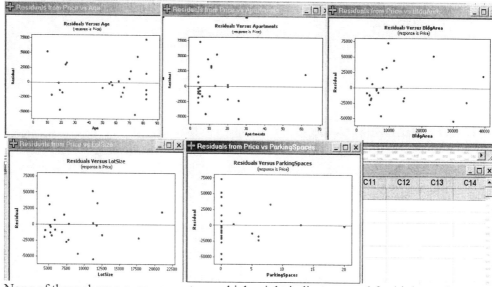

None of these show e.g. any curvature, which might indicate a need for higher order terms in that particular predictor.

Multicollinearity of Predictors

Use **Stat>Basic Statistics>Correlation** to check the correlations among all the predictors being considered for inclusion in the model, so as to spot possible problems of multicollinearity with the fitted model.

Example 12.18 - M/S page 760

Open up the FTC data. We are interested in regressing CO on Tar, Nicotine, and Weight. To look for possible multicollinearity problems, we find the correlations among all three predictors, using **Stat>Basic Statistics>Correlation**:

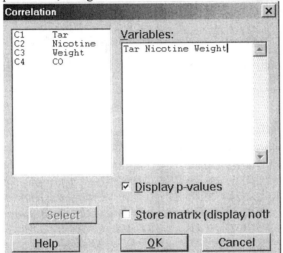

We get a two-way table of correlations:

Correlations: Tar, Nicotine, Weight

```
              Tar   Nicotine
Nicotine    0.977
            0.000

Weight      0.491    0.500
            0.013    0.011

Cell Contents:  Pearson correlation
                P-Value
```

Each entry in the table is a correlation between the variable directly above and the variable directly to the side. Underneath the correlation is its *p-value*. Since all three p-values shown are small, all correlations are highly statistically significant.

We see a very high correlation of *0.977* between Tar and Nicotine. It may be difficult to disentangle the effects of these two predictors in a multiple regression. Weight has a moderate positive correlation with Tar (*0.491*) and with Nicotine (*0.500*).

If CO is regressed on any of these predictors individually, we get a significant positive relationship, as expected, but what happens if we regress CO on all three correlated predictors? Use **Stat>Regression>Regression: Response:** *CO*, **Predictors:** *Tar Nicotine Weight*, and we get:

```
The regression equation is  CO = 3.20 + 0.963 Tar - 2.63 Nicotine - 0.13 Weight

Predictor     Coef   SE Coef     T      P
Constant     3.202     3.462   0.93  0.365
Tar         0.9626    0.2422   3.97  0.001
Nicotine    -2.632     3.901  -0.67  0.507
Weight      -0.130     3.885  -0.03  0.974

S = 1.44573   R-Sq = 91.9%   R-Sq(adj) = 90.7%

Analysis of Variance
Source            DF      SS      MS      F      P
Regression         3  495.26  165.09  78.98  0.000
Residual Error    21   43.89    2.09
Total             24  539.15
```

We see a couple of surprising negative coefficients, for *Nicotine* and *Weight*, but neither coefficient is significant. Only the t-test for *Tar* is significant ($p = 0.001$). There appears to be much redundancy (overlapping information) in these predictors.

Exercise 12.120d. - M/S page 765

Open the BOYS10 data file. There are four variables to be used for predicting oxygen uptake. For their correlations, use **Stat>Basic Statistics>Correlation: Variables:** *Age Height Weight Chest*. We get

Correlations: Age, Height, Weight, Chest

```
            Age   Height   Weight
Height    0.327
          0.356

Weight    0.231    0.790
          0.521    0.007

Chest     0.166    0.791    0.881
          0.647    0.006    0.001

Cell Contents: Pearson correlation
               P-Value
```

It's no surprise that weight, height, and chest depth are all fairly highly correlated with each other (correlations of *0.790, 0.881, 0.791*), and hence, including all of them as predictors for oxygen uptake in the model could easily produce coefficients with small t-values and surprising signs.

Chapter 13 – Categorical Data Analysis

In this chapter, we perform a chi-square test on a one-way table, using
Calc>Calculator
Calc>Column Statistics
Calc>Probability Distributions>Chi-Square

Perform a test of association for two categorical variables, with
Stat>Tables>Cross Tabulation and Chi-Square
Stat>Tables>Chi-Square Test (Table in Worksheet)

And graphically compare the row (or column) distributions, using
Graph>Bar Chart

One categorical variable

We can test a null hypothesis about the distribution of a single categorical variable, by calculating expected counts under the hypothesis, comparing expected counts and observed counts in a chi-square statistic, and then using the chi-square distribution to find the p-value. Unfortunately, there is no shortcut with Minitab, so you will have to work through the calculations step by step.

Example 13.2 - M/S page 788

Enter the four categories of opinion in C1, their respective counts in C2, and the hypothesized proportions in C3.

↓	C1-T	C2	C3
	Opinion	Count	Proportion
1	Legalize	39	0.07
2	Decriminalize	99	0.18
3	ExistingLaws	336	0.65
4	None	26	0.10

We will work out the chi-square statistic, step by step, using **Calc>Calculator**. First, find the expected counts and store them in a C4::

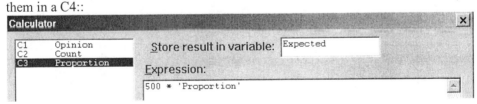

Now, calculate the four individual terms in the chi-square statistic, and store them in C5, named 'Terms':

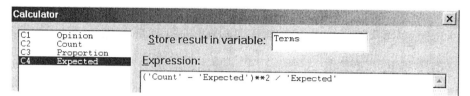

This produces

↓	C1-T	C2	C3	C4	C5
	Opinion	Count	Proportion	Expected	Terms
1	Legalize	39	0.07	35	0.4571
2	Decriminalize	99	0.18	90	0.9000
3	ExistingLaws	336	0.65	325	0.3723
4	None	26	0.10	50	11.5200

Now add up the four terms, to get the value of the chi-square statistic, and store the result in a *stored constant* named 'ChiSq', using **Calc>Column Statistics:**

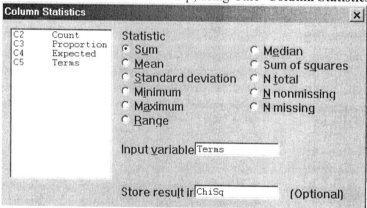

In the session window, we see: Sum of Terms = 13.2495

To evaluate significance, find the p-value or $P(\chi^2 > 13.2495) = 1 - P(\chi^2 < 13.249)$, using **Calc>Probability Distributions>Chi-Square:**

166

```
Chi-Square Distribution                              [X]

 K1    ChiSq          ○ Probability density
                      ⦿ Cumulative probability
                        Noncentrality        [0.0      ]
                      ○ Inverse cumulative probability
                        Noncentrality        [0.0      ]

                        Degrees of freedom: [3        ]

                      ○ Input column:     [          ]
                        Optional storage: [          ]

                      ⦿ Input constant:   [ChiSq     ]
      [  Select  ]      Optional storage: [          ]
```

We get

```
Chi-Square with 3 DF

    x        P( X <= x )
13.2495      0.995873
```

So, the p-value is $1 - .995873 = .004$, and we have strong evidence ($p < \alpha = .01$) that the hypothesized proportions are not correct.

Exercise 13.15 - M/S page 793

To test the hypothesized probabilities, enter the counts into C1 named 'count', and the probabilities into C2 named 'prob'. Now, use **Calc>Calculator** successively to calculate the expected counts **Expression:** *85 * C2*, **Store in : C3**, and then **Expression:** *(C1-C3)**2/C3*, **Store in: C4**. Then **Calc>Column Statistics: Statistic:** *Sum,* **Input:** *C4*, **Store:** *ChiSq* . Then, **Calc>Probability Distributions> Chi-Square:** *Cumulative Probability,* **DF:** *4*, **Input constant:** *ChiSq*, gives

```
Chi-Square with 4 DF

    x   P( X <= x )
17.1631     0.998203
```

So, $p = 1 - 0.998203 = .002$, which gives strong evidence that the placebo percentages do not apply to the patients given the drug.
[For (b) and (c), use **Stat>Basic Statistics>1 Proportion**]

Two way tables- testing for association

To test for an association between two categorical variables, where summarized data (counts) are available, go to **Stat>Tables** and use either **Chi-Square Test (Table in Worksheet)** or **Cross Tabulation and Chi-Square** depending on how the observed counts (frequencies) have been entered in the worksheet. [For raw data, you must select **Cross Tabulation**]

Example 13.3 - M/S page 799

Here we have summary counts from a two-way classification of 500 men with respect to marital status and religion. There are two possible ways to enter such data in a worksheet. One way is to enter the data just as it appears in TABLE 13.8, i.e. as shown below (you can label the columns, but not the rows):

↓	C1	C2	C3	C4	C5
	A	B	C	D	None
1	39	19	12	28	18
2	172	61	44	70	37

For a chi-square analysis, enter the columns containing the table in **Stat>Tables>Chi-Square test (Table in Worksheet):**

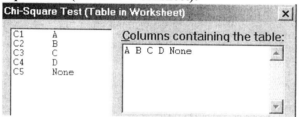

We get

Chi-Square Test: A, B, C, D, None

```
Expected counts are printed below observed counts
Chi-Square contributions are printed below expected counts

               A        B        C        D     None    Total
    1         39       19       12       28       18      116
            48.95    18.56    12.99    22.74    12.76
            2.023    0.010    0.076    1.219    2.152

    2        172       61       44       70       37      384
           162.05    61.44    43.01    75.26    42.24
            0.611    0.003    0.023    0.368    0.650

Total      211       80       56       98       55      500

Chi-Sq = 7.135, DF = 4, P-Value = 0.129
```

Since the *p-value = 0.129* is big (> α), there is insufficient evidence of a relationship between marital status and religion.

A much better way to enter the data, which gives us more flexibility, and better graphing possibilities, is to enter the data as displayed below:

168

↓	C1-T	C2-T	C3
	Religion	MaritalStatus	Counts
1	A	Divorced	39
2	B	Divorced	19
3	C	Divorced	12
4	D	Divorced	28
5	None	Divorced	18
6	A	NeverDivorced	172
7	B	NeverDivorced	61
8	C	NeverDivorced	44
9	D	NeverDivorced	70
10	None	NeverDivorced	37

Note that all the 10 counts are in C3, with religious status in C1, and marital status in C2. For a chi-square analysis, and display of relevant percentages, use

Stat>Tables>Cross Tabulation and Chi-Square

The two categorical variables under analysis must be entered under **Categorical variables** (the first listed variable forms the rows, the second forms the columns), and next to **Frequencies are in,** we indicate the column with the summary counts. If you do not indicate where the summary counts may be found, Minitab will think that each row in the table reflects the outcome of the two variables for just one subject (which would be the case if e.g. your worksheet contained raw un-tallied data). I have also chosen to display the **Counts** and the **Column percents** in the output.

Now, click on **Chi-Square** in the box, and tick:

We get

Tabulated statistics: MaritalStatus, Religion

```
Using frequencies in Counts

Rows: MaritalStatus   Columns: Religion

                 A        B        C        D     None      All

Divorced        39       19       12       28       18      116
             18.48    23.75    21.43    28.57    32.73    23.20
             48.95    18.56    12.99    22.74    12.76   116.00

NeverDivorced  172       61       44       70       37      384
             81.52    76.25    78.57    71.43    67.27    76.80
            162.05    61.44    43.01    75.26    42.24   384.00

All            211       80       56       98       55      500
            100.00   100.00   100.00   100.00   100.00   100.00
            211.00    80.00    56.00    98.00    55.00   500.00

Cell Contents:      Count
                    % of Column
                    Expected count

Pearson Chi-Square = 7.135, DF = 4, P-Value = 0.129
Likelihood Ratio Chi-Square = 6.985, DF = 4, P-Value = 0.137
```

Again, the *p-value of 0.0129* is displayed above (ignore the *p-value=0.137* from the *Likelihood Ratio* test, given at the bottom). In the table, we calculated the relevant percentages, i.e. column rather than row percentages, so that we may see how marital status varies from religion to religion, and not the other way around.

To display, these percentages in a graph, we use **Graph>Bar Chart**

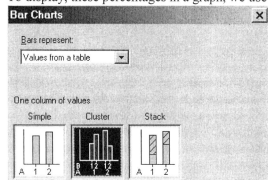

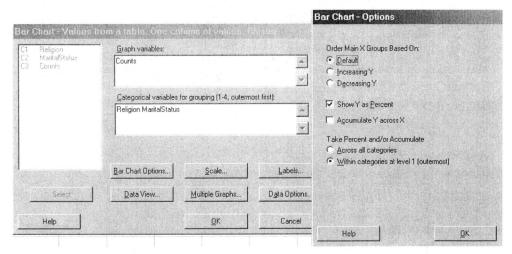

Note we made *Religion* the **outermost** variable in the plot below, and asked to plot **Percent**, rather than the counts, in the display. We get:

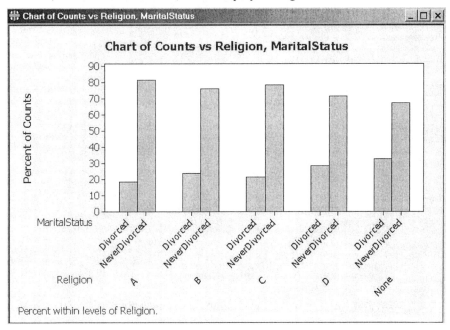

From the above, we can readily compare marital status from religion to religion, e.g. note that religion A has the smallest percent divorced.

Exercise 13.31 - M/S page 808

Open the ETHNIC file, containing Region, Strategy and Count in C1, C2, C3 resp. Run
Stat>Tables>Cross Tabulation and Chi-Square: Categorical Variables: *Region,
Strategy*, **Display:** *Counts, Row Percents*, **Frequencies are in:** *Count*; **Chi-Square:**
Display: *Chi-Square analysis, Expected cell counts*, and we get

Tabulated statistics: Region, Strategy

```
Using frequencies in Count

Rows: Region   Columns: Strategy

            Mobil     None     War      All

Africa         36       39      20       95
            37.89    41.05   21.05   100.00
            38.69    36.62   19.69    95.00

LatinAmer      31       24       7       62
            50.00    38.71   11.29   100.00
            25.25    23.90   12.85    62.00

PostComm       23       32       4       59
            38.98    54.24    6.78   100.00
            24.03    22.74   12.23    59.00

SEAsia         22       11      26       59
            37.29    18.64   44.07   100.00
            24.03    22.74   12.23    59.00

All           112      106      57      275
            40.73    38.55   20.73   100.00
           112.00   106.00   57.00   275.00

Cell Contents:      Count
                    % of Row
                    Expected count

Pearson Chi-Square = 35.412, DF = 6, P-Value = 0.000
Likelihood Ratio Chi-Square = 35.031, DF = 6, P-Value = 0.000
```

Note that the categories for each variable were ordered alphabetically in the display. You
can change this if desired, using **Editor>Column>Value Order,** after clicking anywhere
on the column. Re-ordering the political strategies e.g. here makes sense, since they can
be placed in a natural order from least extreme to most extreme (reordered for the graph
below).

The tiny *p-value* tells us that strategy does vary according to region. For a picture of the
row percentages displayed above, use **Graph>Bar Chart-Cluster:** *Values from a table:*
Graph variables: *Count*, **Categorical variables:** *Region Strategy*, and then select **Bar
Chart Options**: *Show Y as Percent,* **Take percent:** *Within categories at level 1.* We get

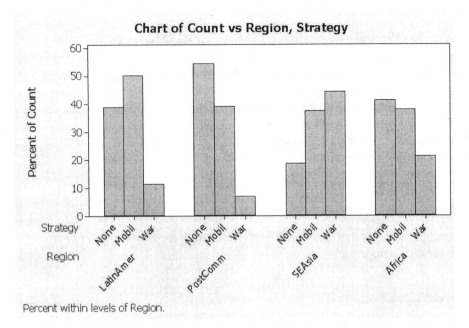

Chart of Count vs Region, Strategy

Percent within levels of Region.

The most striking differences are between South/Southeast/East Asia, and the other three regions. We see terrorism/rebellion/war at comparatively high levels and non-action at comparatively low levels in S./S.E./E. Asia.

Chapter 14 – Nonparametric Statistics

In this chapter, we execute various non-parametric procedures, including
Stat>Nonparametrics>1-Sample Sign
Stat>Nonparametrics>Mann-Whitney (a.k.a. Wilcoxon Rank Sum Test)
Stat>Nonparametrics>1-Sample Wilcoxon (a.k.a. Wilcoxon Signed Rank Test)
Stat>Nonparametrics>Kruskal-Wallace
Stat>Nonparametrics>Friedman

We will rank data, using
Data>Rank

Unstack samples of data using
Data>Unstack Columns

The Sign Test

An alternative to the one-sample *t* test, the sign test is used to test a hypothesis about the *median* of a population (or to find a CI for the median). Look under **Stat>Nonparametrics**.

Figure 14.3 - M/S page 824

Open SUBABUSE or type the data into C1 yourself, from Table 14.1. We wish to test for evidence that the median is less than *1.00*. Go to **Stat>Nonparametrics>1-Sample Sign,** enter your **Test median**, and select the correct **Alternative**:

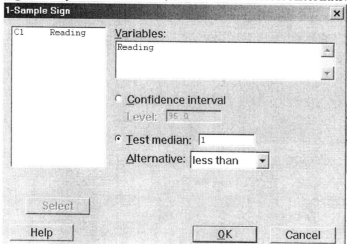

We get

Sign Test for Median: Reading

```
Sign test of median =   1.000  versus  <  1.000

            N   Below   Equal   Above      P   Median
Reading     8       7       0       1  0.0352  0.8350
```

The reported p-value is *0.0352*, less than $\alpha = .05$, so we have sufficient evidence that the median is less than 1.0.

Exercise 14.9 - M/S page 829

Open up the AMMONIA data file, containing the ammonia levels in C1. To test if the median exceeds 1.5, use **Stat>Nonparametrics>1-Sample Sign: Variables:** *C1*, **Test Median:** *1.5*, **Alternative:** *greater than*. We get

Sign Test for Median: Level

```
Sign test of median =   1.500  versus  >  1.500

          N   Below   Equal   Above      P   Median
Level     8       4       1       3  0.7734  1.490
```

Since the sample median is *1.490*, which is less than the hypothesized median of 1.50, we cannot possibly have evidence for a median higher than 1.50. The high p-value just confirms the obvious.

Wilcoxon Rank Sum Test – for 2 independent samples

This procedure is a *rank-transformation* based alternative to the 2-sample *t* test for two independent samples. It does not appear to be on the menu, but Minitab lists another procedure called the Mann-Whitney test, which turns out to be 100% equivalent, though differing in some details of the calculations. So, we will execute the ***Wilcoxon rank sum test*** by applying **Stat>Nonparametrics>Mann-Whitney.** Minitab calculates the p-value using a normal approximation with continuity correction.

Example 14.2 - M/S page 833

Open the DRUGS data file. The reaction times are all stacked in C2 with the drug in C1 (see below). Minitab requires ***unstacked data*** for this procedure. So, first, we have to *unstack* the data using **Data>Unstack Columns**

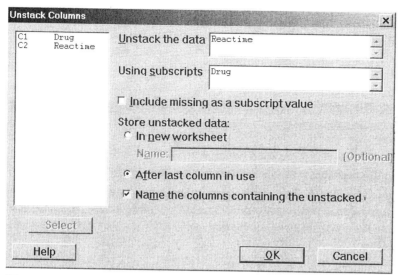

Hit **OK** and now you see the unstacked reaction times in *C3* and *C4*:

↓	C1-T	C2	C3	C4
	Drug	Reactime	Reactime_A	Reactime_B
1	A	1.96	1.96	2.11
2	A	2.24	2.24	2.43
3	A	1.71	1.71	2.07
4	A	2.41	2.41	2.71
5	A	1.62	1.62	2.50
6	A	1.93	1.93	2.84
7	B	2.11		2.88
8	B	2.43		

To test for a difference in reaction time between the two drugs, click on
Stat>Nonparametrics>Mann-Whitney: and indicate where the two samples may be found:

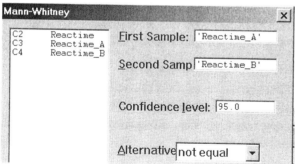

We get:

Mann-Whitney Test and CI: Reactime_A, Reactime_B

```
              N   Median
Reactime_A    6   1.9450
Reactime_B    7   2.5000

Point estimate for ETA1-ETA2 is -0.4950
96.2 Percent CI for ETA1-ETA2 is (-0.9497,-0.1099)
W = 25.0
Test of ETA1 = ETA2 vs ETA1 not = ETA2 is significant at 0.0184
```

The p-value of *0.0184* is less than α=.05, so we have evidence of a difference between the drugs at the specified 5% level of significance. ['*ETA*' in the output is the symbol for a population median]

Minitab estimates the p-value using the normal approximation discussed in the text, rather than doing an exact calculation. The p-value of .0184 here differs slightly from the normal approximation based p-value of .0152 shown in the SAS output in Fig 14.6 in the text, because Minitab includes a *continuity correction* of 0.5 in the top of the z statistic. [The Kruskal-Wallace test, discussed later, may also be applied here, and will give you the same z-value & p-value as displayed in the SAS output]

__Exact Test:__ To perform an *exact test*, take the rank sum from the output, denoted W, and assess significance from a table of exact critical values. 'W' denotes the sum of the ranks for the __First Sample__ in the dialog box. Since the table of critical values in your text assumes that your rank sum is calculated from the smaller sample (fewer observations), make sure that you enter the smaller sample as the __First Sample__ in the dialog box.
   In the example above, the smaller sample was for drug A (6 reaction times). We chose this as our First Sample, so W= *25.0* is the rank sum we need. Compare it with the critical values found in the table in the back of the text: __28 and 56__. Hence, our rank sum is significant at the 5% level.

Exercise 14.25 - M/S page 837

Open up the EYEMOVE data file. We wish to compare visual acuity of deaf and hearing children. Here we find the acuity values for the deaf children in *C1* and the acuity values for the hearing children in *C2*, i.e. *unstacked data*, just what is required by Minitab for this test, so we do not have to waste time unstacking the data as in example 14.2 above.

↓	C1	C2
	DEAF	HEARING
1	2.75	1.15
2	3.14	1.65
3	3.23	1.43

   Run __Stat>Nonparametrics>Mann-Whitney: First Sample:__ *C1,* __Second Sample:__ *C2,* __Alternative:__ *greater than*. We get

Mann-Whitney Test and CI: DEAF, HEARING

```
            N   Median
DEAF       10   2.3750
HEARING    10   1.6450


Point estimate for ETA1-ETA2 is 0.8050
95.5 Percent CI for ETA1-ETA2 is (0.4600,1.2700)
W = 150.5
Test of ETA1 = ETA2 vs ETA1 > ETA2 is significant at 0.0003
The test is significant at 0.0003 (adjusted for ties)
```

(a) $W = 150.5$ is the rank sum for the first sample that you enter in the dialog box (the deaf children). For the exact test, go to the table of critical values in the text, which gives **127** as the upper critical value for a one-sided alternative at the specified $\alpha = .05$. Since W exceeds 127, we have evidence at the 5% level that visual acuity tends to be higher for deaf children.

(b) The very small, normal approximation based, p-value of *0.0003* ($< .05$) displayed in the output supports the conclusion of superior acuity for deaf children. [If you wish to see the numerical value of the z statistic, apply **Stat>Nonparametrics>Kruskal-Wallace** after stacking the data]

Wilcoxon Signed Rank Test – for paired data

This is a rank-based nonparametric alternative to the paired difference t test, which can be applied to paired data. Minitab applies the approximate test based on a normal approximation, but because a continuity correction is also applied, the results will be quite accurate even for rather small samples. You have to form the differences yourself, and then apply the **1-Sample Wilcoxon** test to the single column of differences.

Example 14.3 - M/S page 841

Open up the CRIMEPLAN data file. The data are paired, with each expert's rating of plan A in one column and rating of plan B in another column.

↓	C1	C2	C3
	Expert	A	B
1	1	7	9
2	2	4	5
3	3	8	8
4	4	9	8

Find the difference in ratings for each expert, and place these in a column named *Difference*:

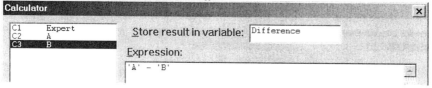

We want to know if there is evidence that plan A has lower ratings, in general, than plan B, or that the 'A – B' differences *tend to fall below zero*, so proceed with **Stat>Nonparametrics>1-Sample Wilcoxon**, as follows:

178

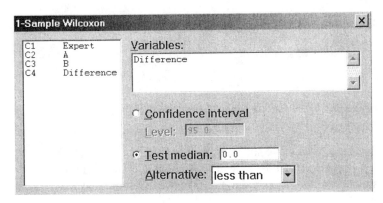

which gives

Wilcoxon Signed Rank Test: Difference

Test of median = 0.000000 versus median < 0.000000

	N	N for Test	Wilcoxon Statistic	P	Estimated Median
Difference	10	9	15.5	0.221	-1.000

The high p-value of *0.221* (> α =.05) is consistent with the null hypothesis of no difference in effectiveness. The exact p-value of .2168 displayed in the text shows how accurate this normal approximation (with continuity correction) is, even with a small sample size.

Exact test: To perform an exact test, use the **Wilcoxon Statistic** displayed in the output. However, Minitab always computes this as the sum of the positive ranks, i.e. T_+. To find T_-, use the fact that $T_+ + T_- = n(n+1)/2$. So, $T_- = (9)(10)/2 - 15.5 = 29.5$. Take the smaller of $\{T_+, T_-\}$ to get $T = 15.5$ which is bigger than the exact critical value of **8**, obtained from the table of critical values in your text. So there is insufficient evidence of a difference in effectiveness, at the 5% level of significance.

Exercise 14.42 - M/S page 846

Open the POWVEP data file containing the 157 day VEP measurements in C2 and the 379 day VEP measurements in C3. Calculate the differences using **Calc>Calculator: Expression:** *C3 – C2*, **Store in:** *Diff*. Test for a significant increase in the VEP scores with **Stat>Nonparametrics>1-Sample Wilcoxon: Variables:** *Diff*, **Test Median:** *0.0*, **Alternative:** *greater than*. We get

Wilcoxon Signed Rank Test: Diff

```
Test of median = 0.000000 versus median > 0.000000

              N
            for    Wilcoxon              Estimated
       N   Test   Statistic      P        Median
Diff  11    11        57.0    0.018       0.9450
```

Since $p = 0.018$ is less than $\alpha=.05$, we have sufficient evidence at the 5% level, to conclude that neurological impairment tends to be greater at the later date. (If you want to check significance using the exact critical value tables in the text, note that $T =$ $(11)(12)/2 - 57.0 = 9$)

Kruskal-Wallace Test - comparing independent samples

This nonparametric alternative to the one-way ANOVA F-test allows you to compare location for any number of independent samples. If you apply it to only two samples, the results are identical to those obtained using the Wilcoxon Rank Sum Test (a.k.a. the Mann-Whitney Test) with normal approximation (no continuity correction).

Figure 14.11 - M/S page 849

Open the HOSPBEDS data file. The data on number of unoccupied beds per hospital are stacked in C2, with the hospital number (#1, 2 or 3) in C1.

↓	C1	C2
	Hospital	Beds
1	1	6
2	1	38
3	1	3
4	1	17

To compare numbers of unoccupied beds at the three hospitals, run **Stat>Nonparametrics>Kruskal-Wallace**:

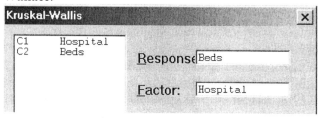

which produces

Kruskal-Wallis Test: Beds versus Hospital

```
Kruskal-Wallis Test on Beds

Hospital    N   Median   Ave Rank       Z
1          10    15.50      12.0     -1.54
2          10    31.50      21.1      2.44
3          10    18.50      13.5     -0.90
Overall    30               15.5

H = 6.10   DF = 2   P = 0.047
H = 6.10   DF = 2   P = 0.047   (adjusted for ties)
```

We have attained significance at the 5% level, since $p = 0.047$, and may conclude that not all three hospitals are equivalent in the distribution of numbers of empty beds.

Exercise 14.93 - M/S page 871

Open up the READ3 data file, containing the student's class in C1 and the increase in reading level in C2. To compare reading level increases for the three classes, use **Stat>Nonparametric>Kruskal-Wallace: Response:** *C2*, **Factor:** *C1*

Kruskal-Wallis Test: INCREAD versus CLASS

```
Kruskal-Wallis Test on INCREAD

                      Ave
CLASS    N   Median   Rank       Z
1        5   0.9000    6.4    -0.98
2        5   1.0000    7.7    -0.18
3        5   1.4000    9.9     1.16
Overall  15            8.0

H = 1.57   DF = 2   P = 0.457
H = 1.58   DF = 2   P = 0.455   (adjusted for ties)
```

The large *p*-value indicates a lack of evidence for differences among methods.

Friedman Test for a Randomized Block Design

If the assumptions necessary to run an ANOVA F-test for a randomized block design are questionable, we may instead choose to apply a nonparametric procedure named after Milton **Friedman**, which again utilizes a rank transformation of the data.

Figure 14.13 - M/S page 856

Open the REACTION2 data file, partially displayed to the side. Note how the data are in *stacked* format, with all the reaction times in one column, C3, drug in another column, C2, and subject number (block) in a third column, C1. This is precisely the format required by Minitab in order to run this procedure (if formatted differently, do some cutting and pasting, or try **Data>Stack**).

↓	C1	C2-T	C3
	SUBJECT	DRUG	TIME
1	1	A	1.21
2	2	A	1.63
3	3	A	1.42
4	4	A	2.43
5	5	A	1.16
6	6	A	1.94
7	1	B	1.48
8	2	B	1.85
9	3	B	2.06

Now run **Stat>Nonparametrics>Friedman**

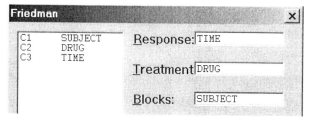

which produces:

Friedman Test: TIME versus DRUG blocked by SUBJECT

S = 8.33   DF = 2   P = 0.016

DRUG	N	Est Median	Sum of Ranks
A	6	1.5283	7.0
B	6	1.7417	12.0
C	6	1.8950	17.0

Grand median = 1.7217

The small *p*-value of *0.016* ($< \alpha$) indicates evidence of a difference in reaction time among the drugs.

Exercise 14.65 - M/S page 858

Open up the PLANTS data file, where in columns C1 - C3, you will find the student number, plant setting (1 = Live, 2 = Photo, 3 = None) and maximum finger temperature, respectively. Does the type of plant setting affect relaxation level (measured by finger temperature)? Students are the blocks. Run **Stat>Nonparametrics>Friedman:**
Response: *Temp,* **Treatment:** *Condition,* **Blocks:** *Student*

Friedman Test: Temperature versus Plant blocked by Student

S = 0.20   DF = 2   P = 0.905

Plant	N	Est Median	Sum of Ranks
1	10	95.600	19.0
2	10	95.683	20.0
3	10	95.817	21.0

Grand median = 95.700

With the large *p*-value, there clearly is no evidence that finger temperature depends on the experimental condition.

Spearman's Rank Correlation

The correlation coefficient discussed in earlier chapters is called *Pearson's (product-moment) correlation coefficient*. An alternative way of examining correlation is to transform the data to ranks, and then calculate Pearson's correlation, as usual, *but on the ranks*. This produces *Spearman's correlation coefficient*.

Example 14.4 - M/S page 862

Open the NEWBORN data file.

To find Spearman's rank correlation between cigarettes smoked (C1) and baby's weight (C2), we first rank the data in each column of data using
Data>Rank

↓	C1	C2
	Cigarettes	**Weight**
1	12	7.7
2	15	8.1
3	35	6.9
4	21	8.2

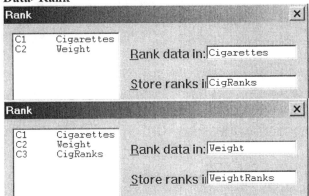

Now, correlate (Pearson) the two columns of ranks with
Stat>Basic Statistics>Correlation:

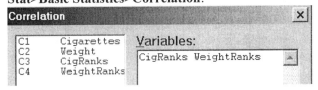

which produces

Correlations: CigRanks, WeightRanks

```
Pearson correlation of CigRanks and WeightRanks = -0.425
P-Value = 0.115
```

So, Spearman's correlation coefficient is *-0.425* here, indicative of a weak negative association, with lighter babies associated with heavier smoking mothers. However, with a *p*-value of *0.115* (divide by 2 for a one-sided alternative), the association fails to attain statistically significance at the 5% level.

Note: The p-value produced this way will be inaccurate if the sample size does not exceed 10, in which case you should compare the reported correlation with a table of critical values (like the one in the back of your text).

183

Exercise 14.75 - M/S page 866

Open the BOXING2 data file, containing Blood *Lactate* Level in C1 and Perceived *Recovery* in C2. To find Spearman's correlation for these two variables, compute and store, in a new column, the ranks of the lactate levels using **Data>Rank: Rank data in:** *Lactate*, **Store ranks in:** *LactateRank*. Similarly compute and store the ranks of the perceived recovery data in *RecoveryRank*.   Apply **Stat>Basic Statistics>Correlation** to the two columns of ranks, and we get

Correlations: LactateRank, RecoveryRank

```
Pearson correlation of LactateRank and RecoveryRank = 0.713
P-Value = 0.002
```

We see a moderately strong positive association, which is also highly significant with a *p*-value of only *0.002* ($< \alpha = .10$). Higher blood lactate levels tend to accompany higher perceived recovery.

Index

TI-83 Manual

Singh Kelly

Pensacola College

Statistics

Tenth Edition

JAMES T. McCLAVE & TERRY SINCICH

Getting Started
The TI-83Plus,   & TI-84

Important Note: If you purchased a **used** calculator your first step is to return the calculator to its original default settings. Follow these steps: Turn **ON** the calculator. Then to reset to original settings as follows:
For the **TI-83 Plus** &   :

Press: **2ⁿᵈ**   +   **4 ENTER**   This will clear all Lists
Press: **2ⁿᵈ**   +   **7   2   2   ENTER** This will reset the calculator

This will help you avoid most unwanted **"ERROR"** messages.

NOTE: In this new addition I have included a special section devoted to **"ERROR"** messages. It is located at the end of this handbook, pp82-89.

After you finish each problem *Press*: **CLEAR**   to return to the home screen. **It may take more than one time**. It is a good idea to begin each new problem from the home screen.☺

Another note here:
▽ Is used for arrow down.   ▷   Is used for arrow right.

△   Is used for arrow up.   ◁   Is used for arrow left.

One feature of the TI-83 Plus & Silver Addition is the use of Scientific Notation when displayed values are of extreme quantities.
(*i.e. 2.33454332E-7 is the calculator's notation for the value 0.000000233454332*) When reading a displayed value take care to read across the entire line. The calculator will always use Scientific Notation for extreme values whether it is set in Scientific MODE or not. Be sure you are familiar with Scientific Notation.

Lastly, the calculator can convert fractions to decimals and decimals to fractions. (*Some decimal approximations represent irrational numbers and of course they cannot be converted into fractions*)

For example: if you have the fraction **7/11** on your screen and you want to convert it to a decimal
press : **MATH   ▽   ENTER   ENTER**   .6363636364 will be displayed.

(This is the decimal approximation for 7/11)

To convert it to a fraction
press: **MATH   ENTER   ENTER**   **7/11** will be displayed.

Contents: McClave/Sincich Text | Calculator Manual

Graphical Methods for Displaying Quantitative Data

To graph a **histogram** using the calculator you first need to enter the data into the list. Let's begin with

Example 2.2   *Percentage of Student Loans in Default.*

To enter the data *press*: **STAT**   **ENTER** you will see these two screens:

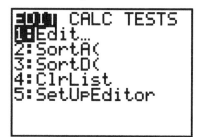 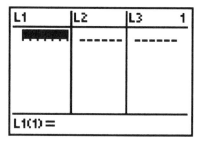

This is the LISTS Menu. You will be using this screen every time you enter data *(The only exception is when entering data for a χ2 test. You will then use the matrix menu)*

Type each number followed by pressing **ENTER**

After typing the first 6 numbers you will see this screen:

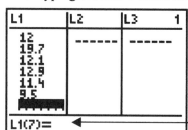

Note: *The bottom of the TI-83 screen indicates which LIST and which entry is active. (LIST1 & it is waiting for the 7th entry)*

Once you have entered all the data *press*:   **2ⁿᵈ**   **Y=**

You will see this screen.
*This is the **Statistical Plots** Menu.*
Whenever you want to plot a graph you will need to adjust the parameters within this menu. You may graph up to 3 things on the same screen by turning on up to 3 Plots.

For now we only want 1 plot (a histogram) of the data we just entered so we will only turn on Plot1. We will enter the parameters section for Plot1 by pressing **ENTER**

After pressing the enter key you will enter the highlighted plot.
(Plot1 in this case)

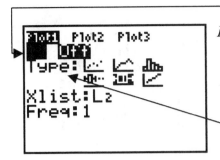

Press: **ENTER** (*This will turn the Plot on*)
You will notice the highlighted switch was moved to "On" as you pressed enter.

Here you have several choices for graph **type.** They are, in order, Scatter-plot, Line-plot, Histogram, Boxplot with outliers, Boxplot without outliers, Normal Probability plot.

For our current problem we will highlight Histogram by following these steps.

Press: ▽   ▷   ▷   *This will highlight the Histogram.*

Press: **ENTER**

Press: ▽   *Beside the "Xlist :" you must enter the list where you put your data. (in my case I used List1)*

Press: **2ⁿᵈ** **1**   *All lists are located as 2nd functions on the corresponding numeral key.(i.e. 2nd 1 gives List1, 2nd 2 gives List2, 2nd 3 gives List3, etc.)*

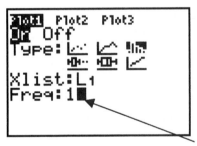

Press: **ALPHA** · 1   **only**
needed if your screen does not show 1 beside **Freq:**.

In some cases the data comes to you in a frequency table, in which case, you will enter the List which contains that frequency here; otherwise, it should be **1.**

ZOOM 9 is a built-in option of the calculator.
By pressing **ZOOM**   9_ you will see this Histogram:

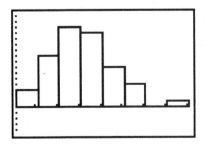

ZOOM 9 *will give you a graph, but that graph may not be appropriate for your problem, as is the case here.*
By pressing **TRACE** *the calculator will display several characteristics of the graph.*

You can see on this next display the min=2.7 and the max<5.1285714.
This is the class width of the first class as chosen by the calculator.

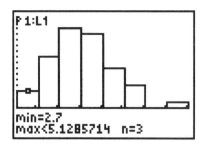

*For our problem we should set our display differently. It is appropriate to set the class width at **1** and the classes should begin at **2.5** and end with **20.5**. Using these setting we will obtain the same display as the SPSS display in the text.*

To adjust the window settings *press*: **WINDOW**

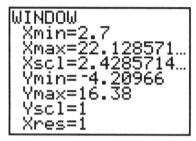

This is the menu where you will adjust the window settings.(It will be necessary for most problems to re-adjust these settings)
For this problem we need to match the settings seen on the next screen. From inspection of the SPSS display found in the text you can discern why we choose these settings.

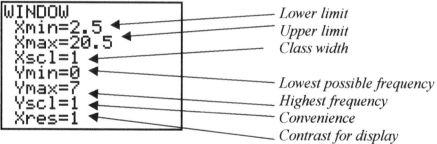

Lower limit
Upper limit
Class width
Lowest possible frequency
Highest frequency
Convenience
Contrast for display

To see the Histogram press: **GRAPH**

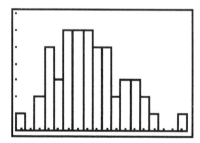

*By pressing **TRACE** you can arrow across to have each class boundary and frequency displayed.*
Note: *the calculator does not display titles units or numerical scale. If you are printing any graph you must add those items yourself using your favorite word processor.*

Now we see the exact display seen in the text. (except for title, numerical scales, and legend) **Remember:** <u>you</u> will need to adjust the window settings <u>for each new problem</u>. The calculator is a wonderful machine but it does not "know" what is appropriate for your problem.☺

Exercise 2.26

To graph a histogram for this problem we must first locate the class mark for
each of the classes. Recall the class mark is the midpoint of the class. We will
enter the class marks into List1 and the corresponding frequency into List2.

Class Mark (L1)	Measurement Class	Relative Frequency (L2)
1.5	.5 – 2.5	.10
3.5	2.5 – 4.5	.15
5.5	4.5 – 6.5	.25
7.5	6.5 – 8.5	.20
9.5	8.5 – 10.5	.05
11.5	10.5 – 12.5	.10
13.5	12.5 – 14.5	.10
15.5	14.5 – 16.5	.05

Begin by typing the data into the Lists.
Press: **STAT ENTER**

To clear the entire list, highlight the
L1 *all the way at the top of the screen.*
And press: **CLEAR ENTER**

To delete any individual entry within a given
list highlight that entry and press: **DEL**

Always enter data for each new problem into an empty list.

After you have entered the class marks into **L1** and the frequencies into **L2**
you should be looking at the upper right screen.

Remember to enter the decimal points as they appear in the text.

We now need to adjust the window and change some settings within the
STAT PLOT MENU.

Press: **WINDOW**

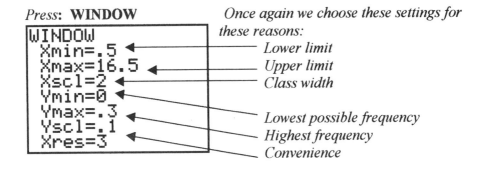

Once again we choose these settings for these reasons:
— *Lower limit*
— *Upper limit*
— *Class width*

— *Lowest possible frequency*
— *Highest frequency*
— *Convenience*

Next *Press*: **2ⁿᵈ** **Y=**

Recall, we left the **STAT PLOT 1** *"on" from the last problem.*

Press: **ENTER** *to enter the menu and make the necessary changes for this problem.* *See screen below.*

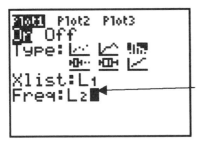

There is only 1 thing to change here for this newest problem. We entered the frequency into **List2** *and so we must tell the calculator the* **Freq:** *is in* **L2**. *Highlight the 1 and press:* **2ⁿᵈ** **2** **L2** *will replace the 1 and now we can plot the graph.*

Press: **GRAPH** you will see the following Histogram:

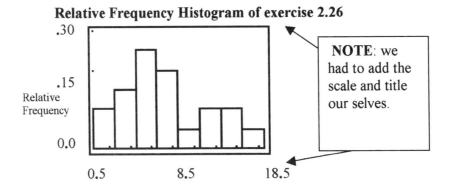

Relative Frequency Histogram of exercise 2.26

NOTE: we had to add the scale and title our selves.

Exercise 2.27

We begin this problem the same as exercise 2.16, That is finding the class marks. We need to enter the **class marks** into **List1** and the corresponding **frequencies** into **List2**.

Press: **STAT   ENTER**   As you can see the old data is still in the lists

L1	L2	L3 2
1.5	.1	------
3.5	.15	
5.5	.25	
7.5	.2	
9.5	.05	
11.5	.1	
13.5	.1	

L2(1)=.1

*Recall: to clear an entire list we highlight the very top: **L1, L2** and press: **CLEAR ENTER***

After you have cleared the two list, L1 & L2 you must enter the new data into the appropriate list.

Class Marks (**L1**)	Magnitude	Number of Earthquakes (**L2**)
0.0	0	1774
0.5	0.1 – 0.9	6
1.5	1.0 – 1.9	580
2.5	2.0 – 2.9	3080
3.5	3.0 – 3.9	4518
4.5	4.0 – 4.9	5548
5.5	5.0 – 5.9	674
6.5	6.0 – 6.9	97
7.5	7.0 – 7.9	9
8.5	8.0 – 8.9	1

Once again we must adjust the window to fit our current data.

Press: **WINDOW**

You will see the following screen:

```
WINDOW
 Xmin=.5
 Xmax=16.5
 Xscl=2
 Ymin=0
 Ymax=.3
 Yscl=.1
 Xres=3
```

Recall: these are the settings from the last problem. The new settings must match our new problem.

The screen below shows the menu <u>after</u> all adjustments have been made.

```
WINDOW
 Xmin=0
 Xmax=9
 Xscl=1
 Ymin=0
 Ymax=6000
 Yscl=1000
 Xres=■
```

Settings after you have made the changes.

Since the last problem used List1 for class marks and List2 for frequency and we graphed a Histogram there are no changes needed within the STAT PLOT Menu.
Press: **GRAPH**

Earthquake data from exercise 2.27

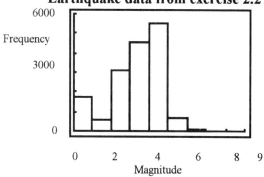

Recall: **you** *must add the title, scales, and any necessary units needed for your problem.*

Example 2.3

To calculate the mean you must first enter the data into the list.
Press: **STAT   ENTER**

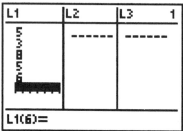

Be sure to start from an empty list. Type all the data into List1. Remember to press **ENTER** *after each number.*

Press: **STAT   ▷**

Since we want **option:1** *all we need to do is Press*: **ENTER** *(see screen below)*

```
1-Var Stats ■
```

You must enter the List where you typed the data.
Press: **2ⁿᵈ** **1** **ENTER**

The screen below left will appear.
Press: ▽ *a few times, and you will see the screen below right.*

```
1-Var Stats
▸x̄=5.4
 Σx=27
 Σx²=159
 Sx=1.816590212
 σx=1.624807681
↓n=5
■
```

```
1-Var Stats
↑n=5
 minX=3
 Q₁=4
 Med=5
 Q₃=7
 maxX=8
■
```

As can be seen from the screen the mean is **5.4**

Exercise 2.53

To calculate the mean and median of the data given we must first enter the data into the calculator. I will use list 1.

```
L1      L2      L3      1
15
12
15
18
16
11
--------
L1(12) =
```

Press: **STAT ENTER**

This will take you to the Lists Menu.

Clear any old data from the list as it may interfere with this new problem.

After you have entered all the data *Press:* **STAT** ▷ **ENTER**

```
1-Var Stats ■
```

At this time you must enter the list designation where you placed the data. Since I put the data in List 1, I will press: **2ⁿᵈ** 1

Next *Press:* **ENTER**

The TI-83 will display a lot of Statistics. Some you will need, some you will not need. I will point out the basic descriptive statistics most often asked for.

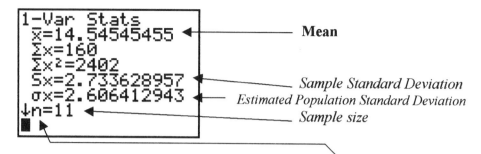

Mean — $\bar{x}=14.54545455$

Sample Standard Deviation — $Sx=2.733628957$

Estimated Population Standard Deviation — $\sigma x=2.606412943$

Sample size — $n=11$

Notice the arrow pointing downward in front of the n=11. This tells us there is more data that can be seen by pressing the arrow down key.

Press: ▽ a few times and you will see this new screen.

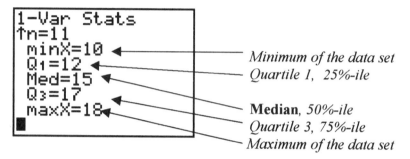

Minimum of the data set — minX=10

Quartile 1, 25%-ile — $Q_1=12$

Median, 50%-ile — Med=15

Quartile 3, 75%-ile — $Q_3=17$

Maximum of the data set — maxX=18

Together these two screens give you most everything wanted when someone asks for "the basic descriptive statistics."

The mode is of course the piece of data with the highest frequency.

For this problem the **mode** = __15__

Example 2.9

To calculate the Variance and Standard Deviation of this data set we will use the same procedures as finding the mean and median.

First enter the data into the calculator.

Press: **STAT ENTER**. (this takes you to the list menu)

After you have entered all the data

Press: **STAT** ▷ **ENTER** (this takes you to the calculations menu)

Once again you must enter the List where you placed the data. Again I used List 1 I *press:* **2ⁿᵈ** 1 **ENTER**

The Standard Deviation can be seen on the resulting screen.

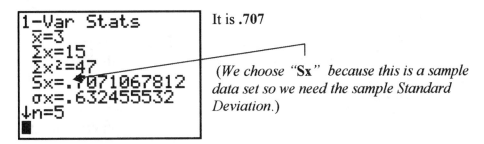

It is .707

*(We choose "**Sx**" because this is a sample data set so we need the sample Standard Deviation.)*

The Variance can be calculated directly from the Standard Deviation as it is the square of the Standard Deviation. So $0.7071067812^2 = 0.5$ Thus we calculate the **Variance = 0.5**

Exercise 2.75

The first step in solving this problem is to enter the data into the calculator.
Press: **STAT ENTER**
Enter all the data remembering to press ENTER after each entry.

After all the data has been entered *press*: **STAT** ▷ **ENTER**
You should now be at the step where you enter the list on which the calculator will perform the 1-Variable Statistics. I *press*: **2nd 1 ENTER** I used List 1.

```
1-Var Stats
  x̄=1.47125
  Σx=11.77
  Σx²=17.3453
  Sx=.0640172968
  σx=.0598826978
↓n=8
■
```

```
1-Var Stats
↑n=8
  minX=1.37
  Q₁=1.415
  Med=1.49
  Q₃=1.52
  maxX=1.55
■
```

a. The range is the maximum – the minimum. For our problem it is $1.55 - 1.37 = \mathbf{0.18}$

b. Recall the variance is the standard deviation squared. For our problem it is $.0640172968^2 = \mathbf{.0041}$

c. The Standard Deviation is seen on the screen to be **.0640**

d. The **morning** drive-times have the **greater** variable ammonia levels as $1.45 > .0641$

Example 2.16

Drawing a box-plot on the TI-83 is the same procedure as drawing a histogram except for 1 step. First you must enter the data into the list. Consider the example problem :

Percentage of Student Loans in Default

To enter the data turn on the calculator and *press*: **STAT ENTER** you will see these two screens as you press each of the buttons:

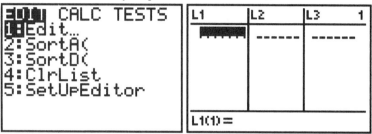

This is the LISTS Menu. You will be using this screen every time you enter data into the calculator. (*The only exception is when entering data for a χ2 test. You will then use the matrix menu*)

Type each number followed by pressing **ENTER**

Once you have entered all the data *press* **2ⁿᵈ**   **Y=**

You will see this screen.

*This is the **Statistical Plots** Menu. Whenever you want to plot a graph you will need to adjust the parameters within this menu. You may graph up to 3 things on the same screen by turning on up to 3 Plots.*

For now we only want 1 plot (a Boxplot) of the data we just entered so we will only turn "on" Plot1. Enter the parameters section by pressing **ENTER**
After pressing the enter key you will enter the highlighted plot.

Press: **ENTER** *This will turn the Plot on. You will notice the highlighted switch was moved to "On" as you pressed enter.*

For our current problem we will highlight Boxplot with outliers, by following these steps.

Press: ▽ ▷ ▷ ▷   *This will highlight the Boxplot.*
Press: **ENTER**
Press: ▽   *Beside the* **"Xlist :"** *you must enter the list where*
Press: **2ⁿᵈ**   **1**   *you put your data. (I used List1)*

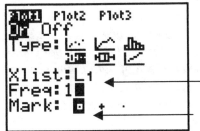

Press: **ALPHA**   1   *This step is* **only**
needed if your screen does not show **1**.
    In some cases the data comes to you
In a frequency table, in which case, you
Will need to enter the List which contains
the frequencies here.

For Mark, I choose the <u>*square*</u>. *Hey, It's Math*

By pressing **ZOOM**   **9**   you will see this Boxplot.

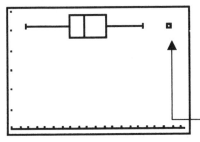

ZOOM   9   *is a built-in function of the TI-83 &*
Plus. It will give you a graph, but that graph
may not be appropriate for your problem. For
any Boxplot, **Zoom**   9   *is usually sufficient.*

This unconnected mark is the possible
outlier. In this case it is also the maximum.

By pressing **TRACE** *the calculator will display several characteristics of the*
graph. For a Boxplot it will display the minimum, Q1, Median, Q3, lower
fence(s), Maximum, one number at a time by flashing the curser on the
appropriate location as you arrow from left to right.

Note: the calculator does not display titles, units or numerical scale. If you are
printing any graph you must add those items yourself using your favorite word
processor. The calculator is a wonderful machine but it does not "know" what is
appropriate for your problem.☺

Exercise 2.126

For this problem we will use two lists and two stat plots.

First enter the data into the appropriate lists. I will put Sample **A** data into **List 1** and Sample **B** data into **list 2**.

Press: **STAT ENTER** (*Clear list 1 and list 2 by highlighting L1 and L2 at the very top of your screen & pressing*: **CLEAR ENTER**)

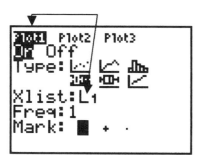

It is good to glance at the two lists to see if they are the same length. If they are not you probably missed a piece of data in the shorter list. Many mistakes occur at data entry. Make sure you have entered the data correctly.
If you make a mistake here nothing else will be correct!

Press: **2ⁿᵈ Y=**

You must turn on each plot one at a time. First plot 1.

Press: **ENTER** Make sure **On** is highlighted.
Press: **ENTER** Arrow over to highlight the boxplot and *press*: **ENTER**

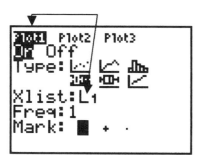

We will use plot 1 for list 1 (A) And plot 2 for list 2 (B).

*Again make sure the **Freq**: is set to **1**. You may choose any Mark you wish.*

Press: **2ⁿᵈ Y=** This time arrow down to **2:Plot2...**
Press: **ENTER** This will take you to plot 2.

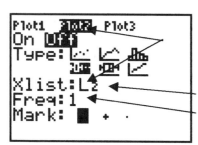

Make sure On is highlighted.
Press: **ENTER**
Arrow over to highlight the boxplot and press: **ENTER**
*Make sure **Xlist**: is **L2** (press*: **2ⁿᵈ 2**)
Freq: *must be* **1**
Any Mark you wish.

Since these are boxplots we can *press*: **ZOOM 9**

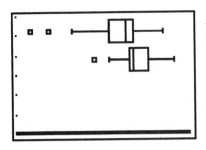

Pressing **TRACE** *will display all important information associated with these boxplots.*

Arrow up/down *to move from plot1/plot2.*

Arrow right/left *to move within a plot.*

The similarity between these two box-plots is they both have possible outliers on the low side. Plot 2 which is the Sample B data has a higher value in every category. (e.g. minimum, Q1, Median, Q3, maximum) Sample A has two possible outliers : **85, 100**. Sample B has 1 possible outlier: **140**.

Exercise 2.139

To construct a Scatter-plot using the TI-83 you must first enter the data into the Lists. The Independent variable (**x**) usually goes in List 1 and the Dependent Variable (**y**) usually goes in List 2. *It is not "carved in stone". It's just a general rule.* For this problem enter Variable #1 into List 1 and Variable #2 into List2.

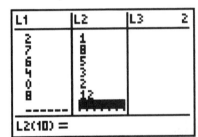

Press: **STAT ENTER**

Note: *For all data given in a frequency table or listed as two or more variables it is necessary to enter the data in the EXACT order it is given.*

Enter each variable into its own list.

Press: **2nd   Y=**

Note: *As we want only one plot make sure Only one plot is turned on.*
  Plot 2 & Plot 3 should be OFF

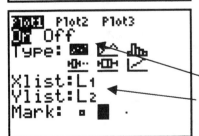

Press: **ENTER**
This will bring you into the Stat Plot 1 menu.

ScatterPlot is the first option of graphs. Make sure your settings match this screen. See page 5 if you need a refresher.

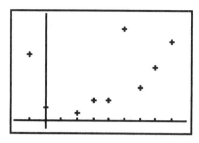

Press: **ZOOM 9**
You will see this screen.
There does appear to be a slight
trend in the data. It appears
Variable #2 tends to increase as Variable #1
increases.

The Counting Principle & Probability

Lets begin by using the factorial symbol, **!**

Example 3.22

Calculate **5 !**

 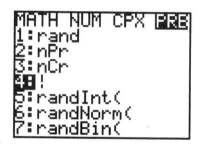

First *press:* 5 , next press **MATH** (see lower left)

Press: ◁ to " **PRB** " Highlight " **4 : !** "

Press: **ENTER** (your screen will now look like this)

Press: **ENTER** 120 will be displayed on the right of the screen.
This is of course the number of permutations of **5** objects, or stated
differently: it is the number of ways you can arrange **5** objects .

Suppose you only want to arrange some of the objects from a group.
There is a different operation on the calculator for just such a desire.

Example 3.28

To permute 5 objects from a group of 5 objects we could actually use **!**
But for our problem lets use the new Permutations rule.

On the calculator it will look slightly different as the calculator uses
computer symbols so P^5_5 will be written as **(5P5)** , {(nPr), *where the*
n ~ N, and r ~ n from the textbook's formula.}

Start from a clear screen. First *press*: **5** then *press*: **MATH**
Press: ◁ to highlight " **PRB** "

Arrow down and Highlight " **2 : nPr** "
Press: **ENTER** (upper right screen will appear)
Press: **5** (Your screen should look like this)

Press: **ENTER**
120 *will be displayed . This is the*
number of permutations of 5 objects
from a group of 5 objects.

Example 3.31

In a <u>combination</u> we are simply interested in the number of ways that
objects can be selected.

For this problem we will also use different notation on the calculator. The
calculator uses computer notation as follows: $\binom{N}{n}$ will be written as nCr
Where n ~ N, and r ~ n from the textbook's formula. So we want (100C5.
Start from a clear screen.

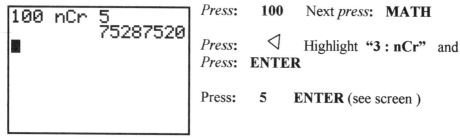

Press: **100** Next *press:* **MATH**

Press: ◁ Highlight **"3 : nCr"** and
Press: **ENTER**

Press: **5** **ENTER** (see screen)

75,287,520 is the number of combinations of 5 objects from 100 objects.
Stated another way; it is the number of ways you can choose 5 objects
from a list of 100 objects.

Example 4.6

To begin this problem we first enter the data into the calculator. The (**x**)'s go
into **list 1** and the p(**x**), [**y**'s] go into **list 2**. *We will use the same procedure*
as we used to calculate the mean and standard deviation from a frequency
table.
Press: **STAT ENTER**

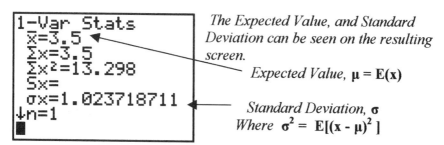

Note: *For all data given in a frequency*
table or listed as two or more variables it is
necessary to enter the data in the EXACT
order it is given.

Enter each variable into its own list.
After you have entered all the data.

Press: **STAT ▷ ENTER** (this takes you to the calculations menu)

Once again you must enter the Lists where you placed the data. I used List 1
and List 2. I *press:* **2nd 1 , 2nd 2 ENTER**

The Expected Value, and Standard
Deviation can be seen on the resulting
screen.

Expected Value, μ = E(x)

Standard Deviation, σ
Where σ² = E[(x - μ)²]

You may think of the Expected Value as "the mean of a future event". So it
should not be a big surprise that it uses the same symbol you're used to seeing
on the calculator.

Exercise 4.31

To begin this problem we first enter the data into the calculator. The (**x**)'s go into **list 1** and the p(**x**), [**y**'s] go into **list 2**. *We will use the same procedure as we used to calculate the mean and standard deviation from a frequency table.*

Press: **STAT ENTER**

Note: *For all data given in a frequency*

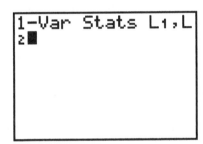

table or listed as two or more variables it is necessary to enter the data in the EXACT order it is given.

Enter each variable into its own list. After you have entered all the data.

Press: **STAT ▷ ENTER** (this takes you to the calculations menu)

Once again you must enter the Lists where you placed the data. I used List 1 and List 2. *I press:* **2ⁿᵈ 1 , 2ⁿᵈ 2 ENTER**

```
1-Var Stats L1,L
2█
```

```
1-Var Stats
 x̄=3.8
 Σx=3.8
 Σx²=25
 Sx=
 σx=3.249615362
↓n=1
█
```

a. $\mu = 3.8$
b. $\sigma^2 = 3.249615362^2 = 10.56$
c. $\sigma = 3.2496$
d. Over the long run the average we would expect is **3.8**
e. Not for this case.
f. Consider the frequency distribution:

$\mu = 2$ is one of the values, so **yes** it can happen

x	1	2	3
P(x)	.25	.50	.25

Example 4.10

We want to calculate the probability of **0, 1, 2, 3,** and **4** successes from a total of **4** independent trials where the probability of success on each trial is **0.1**   Here : **"n" = 4** ,**"x" = 0, 1, 2, 3, 4** and **"p"= 0.10**
We must calculate each "p(x)" one at a time. We will start with x = 0.
Press: **2ⁿᵈ**   **VARS** . (you will see this screen)

```
DISTR DRAW
1:normalpdf(
2:normalcdf(
3:invNorm(
4:tpdf(
5:tcdf(
6:X²pdf(
7↓X²cdf(
```

Arrow down to **"0: binompdf ("**

Press: **ENTER**   (you will see this screen)

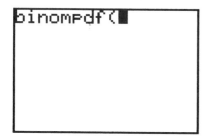

You must enter **n , p , x** *in that order inside the parenthesis.*
For our problem
Press: **4**   ,   **0** .**10**   ,   **0**   **)**

Don't forget the **commas** *or the closed parenthesis!*

Now *press*: **ENTER**   **.6561** will be displayed. This is the probability of x = 0. Now to get the probability of x = 1, you do not have to start over from the beginning. To calculate the probability of **1** success from the same problem you simply press **2ⁿᵈ** **ENTER**

```
binompdf(4,.10,0
)
            .6561
binompdf(4,.10,0
)■
```

(this screen will reappear)

Now **arrow left** *to highlight the* **"0"**
Next press: **1**   *This will replace the 0 with a 1.*

Press: **ENTER** **.2916** *will be displayed.*

Each of the remaining values can be found following the same steps just outlined.

To graph a particular Binomial Distribution follow these steps:
Recall this problem's data where **"n"= 4**, **"p"= 0.10**
Press: **STAT**   "**EDIT**" should be highlighted. *Press*: **ENTER**
Enter the numbers **{0,1,2,3,4}** into **List 1.** These are the **"x"** values.

Now we want to enter a **formula** into **List 2**

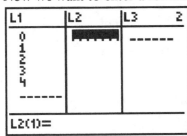

It is the binomial formula with n = 4,
P = .10
If you have never entered a formula into
a machine before take extreme care to
have everything exactly correct!
Even the parentheses, etc.must be exact.

First highlight **L2** *at the very top of the screen*

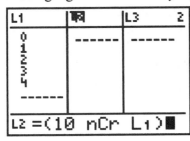

Press: **(**   **4**   **MATH**

Press: ◁   *Highlight* **"3: nCr".**

Press: **ENTER** 2nd **1**   **)**
(this screen will appear)

Press: **(**   **. 10**   **^** 2nd **1**   **)**
You will see the text scroll across the bottom of the screen as you type.

Press: **(**   **. 90**   **^**   **(**   **4**   **-** 2nd **1**   **)**   **)**

You will be able to see all the text by pressing   ◁   or   ▷

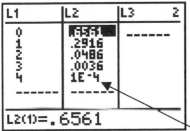

Press: **ENTER**
*The probabilities will appear in **List2**.*

*[The values in **List2** are the probabilities*
*associated with the "**x**" values from **List1***
for this Binomial problem]

Recall: This is given in Scientific Notation

Now to see a plot we must go to the
STAT PLOT so *press:* 2nd **Y=**
(You will see this screen)

"1:Plot1..." *should be highlighted. Press:* **ENTER**

Make sure **On** *is highlighted and*

Press: **ENTER** *(Note: the switch is moved to "On")*

Arrow down and highlight the Histogram.

Press: **ENTER**

Make sure **Xlist: L1 , Ylist: L2**

Now, as always, we need to adjust the **window** to fit our data.

Press: **WINDOW**

These settings are from our new data. You will need to change your settings to match these.

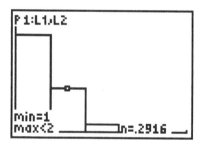

Press: **GRAPH**

By pressing **TRACE** *you can read the probability associated with each class.*

Exercise 4.59

For this problem we see **p = .80, n = 10** We will need to work 3 different problems: 1 for part a, 1 for part b, 1 for part c.

   a.   P(x = 3)

Press: **2ⁿᵈ VARS**

(you will see this screen)

Arrow down to **"0: binompdf ("**
Press: **ENTER**

```
binompdf(█
```

(you will see this screen)

You must enter **n , p , x** *in that order inside the parenthesis.*
For our problem
Press: **10   ,   0 .80   ,   3   )**

```
binompdf(10,.8,3
)
        7.86432E-4
■
```

Don't forget the **commas** *or the closed parenthesis!*

Press: **ENTER**

Note: The answer is given in Scientific Notation.   In common notation it =
0.000786432

b. P(x < 7)   For this part   *Press:* **2ⁿᵈ   VARS**

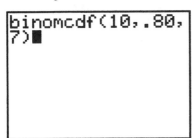

This is not the same as part a.
       !!! Look carefully !!!

Arrow down to **"A: binomcdf ("**
Press: **ENTER**

On the TI-83 the ending "cdf" will calculate a cumulative frequency, starting from 0 and continuing up to and including the number you enter.
Enter **n , p , x** in that order.
For our problem we enter **10 , .80 , 7)**

```
binomcdf(10,.80,
7)■
```

```
binomcdf(10,.80,
7)
        .322200474
■
```

Press: **ENTER**   (see screen above right)
As can be seen on the screen the probability is **.322200474**

c. P(x > 4)   Recall: this is the complement to P(x < 4) which can be done directly on the calculator.

We will work the complementary event and subtract that answer from 1 giving P(x > 4)

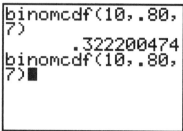

Press: **2ⁿᵈ ENTER**

You will see this screen.

Next, **Arrow left** *and replace the 7 with a* **4**

Press: **ENTER**

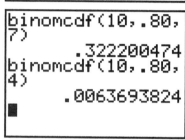

We now need to subtract this complementary event from **1** *to get our answer:*

1 - .0063693824 = **.9936306176**

Example 4.13   Poisson

For this example part a is completely worked in the text and there is no "special way" to do it on the calculator. I will begin with part b.

b.   we want to find the probability of "fewer than 2" which means **P(x < 2)** and we have given information λ = **2.6**

On the calculator we go to the distribution menu

Press: **2ⁿᵈ VARS**   **Arrow down** to "**C : poissoncdf ("**

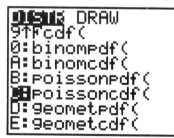

Press: **ENTER**
(you will see screen above right)

Now we must enter λ **, x)** *in that order.*
For our problem enter **2.6**   **,   1   )**

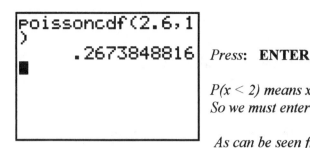

Press: **ENTER**

P(x < 2) means x = 0, or x = 1
So we must enter x = 1

As can be seen from the screen

P(x < 2) is **0.2673848816**

c.   Here we are looking for P(x > 5). λ still equals 2.6.
Once again we must use the complement to find our answer.
The complement to P(x > 5) is P(x ≤ 5). After we calculate the
probability for the complementary event we will subtract that answer
from 1 giving P(x > 5).   Poissoncdf(2.6,5) = .9509628481
So..... P(x > 5) = 1 - .9509628481 = **.0490371519**

d.   For this part we are looking for P(x = 5) so this will be
"**B:Poissonpdf**"   Again λ = 2.6.
On the calculator we go to the distribution menu

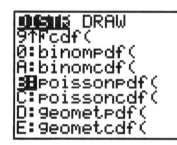

Press: **2ⁿᵈ VARS**

Arrow down to "**B : poissonpdf ("**

Press: **ENTER**

You need only enter λ , **x)**

For our problem we *Press*:   **2.6**   ,   **5**   **)**   **ENTER**

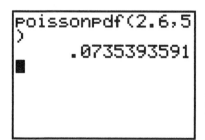

As can be seen from the screen
P(x = 5) is **.0735393591**

Example 5.2

To begin this problem we will go to the Distributions Menu.

Press: **2ⁿᵈ** **VARS**

Highlight option "**2:normalcdf (**"

Press: **ENTER**

To calculate the area under the curve from a **Standard** Normal Distribution you need only enter 2 values: **lower** , **upper**

For our problem we enter **–1.33 , 1.33)**

(*the negative sign is the gray button, <u>not</u> the blue button*)

Press: **-** **1.33** , **1.33** **)**

Press: **ENTER**

```
normalcdf(-1.33,
1.33)
        .8164816037
```

As can be seen from the screen the Probability is **.8164816037**

Example 5.3

For this problem we are still trying to find the area under the curve so again we use **2:normalcdf**

Press: **2ⁿᵈ VARS**

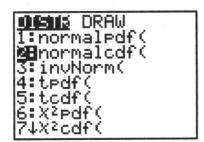

Highlight option "**2:normalcdf (**"

Press: **ENTER**

Recall for **Standard** *Normal Distribution we need only enter 2 values: lower , upper*

For this problem we are given the lower value, 1.64, but we must also enter an upper value.

Based on the theory of the Normal Distribution we choose an upper value that will "cut off" a very small amount of the curve thus giving a very small amount of error in our answer. I choose an upper value of 5 (*This will introduce an error of only .000000287.....*)
Certainly acceptable for most applications.

Press: **1.64 , 5) ENTER**

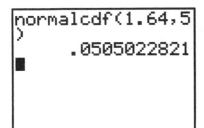

As can be seen from the screen the probability is **.0505022821**

Example 5.4

For this problem we are still trying to find the area under the curve so we use **2:normalcdf** once again.

Press: **2ⁿᵈ VARS**

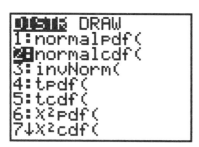

Highlight option "**2:normalcdf (**"

Press: **ENTER**

*Recall for **Standard** Normal Distribution we need only enter 2 values: **lower** , **upper***

For this problem we are given the upper value, .67, but we must first enter a lower value.

Based on the theory of the **Standard** Normal Distribution we choose a lower value that will "cut off" a very small amount of the curve thus giving a very small amount of error in our answer. I choose a lower value of -5
  (*This will introduce an error of only .000000287.....*)
Certainly acceptable for most applications.

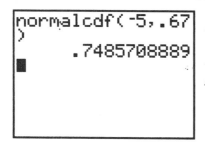

Press: **–5 , .67) ENTER**

As can be seen from the screen the probability is **.7485708889**

Example 5.6

For this problem we are still trying to find the area under the curve so again we use **2:normalcdf**

```
DISTR DRAW
1:normalpdf(
2:normalcdf(
3:invNorm(
4:tpdf(
5:tcdf(
6:X²pdf(
7↓X²cdf(
```

Highlight option "**2:normalcdf (**"

Press: **ENTER**

Since we are using the <u>actual data</u> in this problem and <u>not</u> the z-scores we will need to enter 4 values. The first 2 will be the same (i.e. lower, upper); however, we now must add 2 more values(mean, standard deviation).

```
normalcdf(█
```

For a Normal Distribution we must enter **Lower , upper , mean , std. Dev.** *In that order!*

Press: **8 , 12 , 10 , 1.5)**

```
normalcdf(8,12,1
0,1.5)
        .8175774363
█
```

As can be seen from the screen the probability is **.8175774363**

Note: *there is a slight difference between the text's answer and the TI-83's answer. Error is introduced when rounding the z-value.*

The TI-83 displays the more accurate answer in this case.

Example 5.9

For this problem we will go to the Distributions Menu. But this time we will use **3:invNorm (** because we are not looking for area, we are looking for z.

Press: **2ⁿᵈ VARS**

```
DISTR DRAW
1:normalpdf(
2:normalcdf(
3:invNorm(
4:tpdf(
5:tcdf(
6:X²pdf(
7↓X²cdf(
```

Highlight option "**3:invNorm (**"
Press: **ENTER**

If 95% of the curve is between z and –z, and since the normal distribution is symmetrical, then 47.5% is to the right of the center and 47.5% is to the left of the center. Therefore the rightmost z-value must lie at the 50% + 47.5% = 97.5%-ile.

Press: **.975) ENTER**

```
invNorm(.975)
        1.959963986
■
```

To use this function for this problem you need only enter the percentile and the calculator will return the z-score.

As can be seen from the screen
z = 1.959963986
So −z = **-1.959963986**

Example 5.10

Again we are given area under the curve and asked for the actual value. To begin this problem we will use the Distributions Menu.

Press: **2ⁿᵈ VARS**

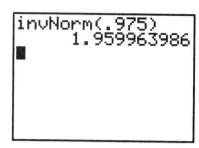

Highlight option "**3:invNorm (**"

To calculate the value of a score given the percentile, mean and standard deviation you need only enter 3 values: **%-ile , mean , standard deviation**

```
invNorm(.90,550,
100)
```

Press: **ENTER**
For our problem we enter **.90 , 550 , 100**

Press: **ENTER**

```
invNorm(.90,550,
100)
        678.1551567
```

As can be seen from the screen the Minimum score on the entrance exam is 678.1551567.

Since you probably can't score this we would say the minimum score is 678.

The Exponential Distribution Function is not a built-in function on the TI-83 so we will use a slightly different procedure for calculating the area under the curve. Since it is not a function that appears on the calculator we must enter the function ourselves.

Example 5.13

For this problem we know $\theta = 2$, and we want $x \geq 5$

So our equation would be : $(1 / 2) e^\wedge (- x / 2)$

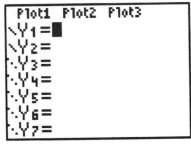

On the calculator you will go to the functions menu.
Press: **Y=**

Now we will need to enter the equation

Press: (1   ÷   2   )   **2ⁿᵈ**

LN   (-)   X,T,0,n   ÷   2   )

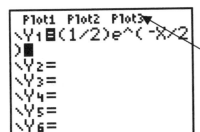

(you will see this text appearing on screen as you type)

Please notice that all 3 STAT PLOTS are OFF. If any of these are highlighted you must turn them off. See p. 4,5 for a refresher.

Now we need to adjust the **WINDOW** to fit our data.

Press: **WINDOW**

$\text{Xmin} = 0$, $\text{Xmax} = 8\,\theta$, $\text{Xscl} = \theta$,
$\text{Ymin} = 0$, $\text{Ymax} = 1/\,\theta$, $\text{Yscl} < 1/\,\theta$

These settings will always give you a nice view of an Exponential graph.

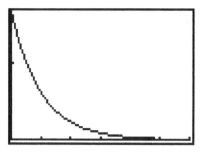

Press: **GRAPH** (you will see this graph)

To find the area under the curve past **x =5** we will need to go to the **Function's Calculations** menu.
Press: **2ⁿᵈ TRACE** (this screen will appear)

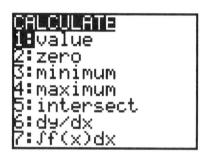

Arrow down to " 7 : { f (x) dx "
[*This choice will integrate the function between two points*]

Press: **ENTER**

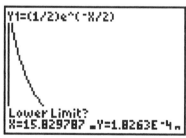

Now we need to enter the lower and upper limits for our problem.
[*Since we wanted* **x ≥ 5 , 5** *will be the lower limit and* **16** *will be our upper limit*]
16 *because our graph window only goes up to* **16**

We could also say [1 – Integral of 0 to 5; which is 1 – the complement]
(1 – the Integral 0 to 5) = (1 – 0.917915) = 0.08208 ≈ .082
You get the same answer either way.

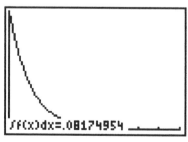

Press: **5 ENTER 16 ENTER**
(this screen will appear)

This screen shows us that the area under the curve past 5 *is* **.08174954**

So the probability that more than 5 hours will pass between emergency arrivals is ≈

0.082

[*To clear the screen after shading go to the* **DRAW** *menu.*]
Press: **2ⁿᵈ PRGM** "1 : ClrDraw " *will be highlighted, press:* **ENTER**
Wait a few seconds then, *Press:* **CLEAR**

Exercise 5.46

To graph a Normal Probability plot for this problem we first enter the data into the calculator. I will use list 1.

Press: **STAT ENTER**

This will take you to the Lists Menu.

Clear any old data from the list as it may interfere with this new problem.

After you have entered all the data *Press*: **2ⁿᵈ   Y=**

"1:Plot1..." *should be highlighted.*

Press: **ENTER**

Make sure **On** *is highlighted*

Press: **ENTER** *(Note: the switch is moved to "On")*

Arrow down and right to highlight the last Plot option.

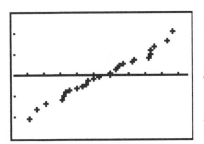

For **Data List** *I entered* **L1** *as that is where I placed my data.*
For **Data Axis** *highlight* **X** *and press*: **ENTER**

Next *Press*: **ZOOM   9**

As you can see from the screen there is no noticeable curve.

It does appear that this data is near Normally Distributed.

In fact you can enter the function y = 0.45x –2.5 and you will see a line graphed through this data. The data follows the line closely ☺

Example 6.8

For Sampling Distribution problems we will use the Normal Distribution
Functions found in the Distributions Menu. The only difference is: instead
of entering the given standard deviation we will adjust the deviation to fit
our sample size. ($\sigma_x = \sigma / \sqrt{n}$)

a. The distribution will be distributed normally with $\mu = 54$ and
$\sigma_x = \sigma / \sqrt{n} = 6 / \sqrt{50} = .85$

b. Find $P(x \leq 52)$ Begin by going to the Distributions Menu.

Press: **2ⁿᵈ** **VARS**

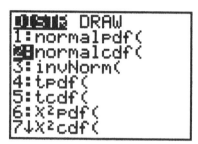

Highlight option "**2:normalcdf (**"

Press: **ENTER**

Since we are using the actual data in this problem and not the z-scores we
will need to enter 4 values. The first 3 will be the same (i.e. lower, upper,
mean); however, we now must adjust the standard deviation to fit our
sample size.

For a Sampling Distribution we must enter
**Lower , upper , mean , adjusted
standard deviation)** *In that order!*

From part a we know $\sigma_x = .85$

For our problem we enter: **0 , 52 , 54 , .85)**
After you have entered all four values *Press*: **ENTER**

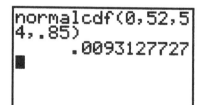

*As can be seen from the screen the
Probability a light bulb last 52 hours or
less is* **.0093**

Exercise 6.31

This is a Sampling Distribution problem so we will use the Normal
Distribution Functions found in the Distributions Menu. The only
difference is: instead of entering the given standard deviation we will
adjust the deviation, as found in part a, to fit our sample size.
We have given information: n = 100, μ = 30, σ = 16.

   a. $\mu_x = 30$, $\sigma_x = 16/\sqrt{100}$

   b.   The shape of the sampling distribution will be approximately
         normal.

   c.   Find $P(x \geq 28)$

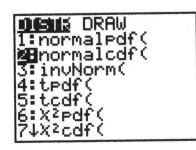

Press: **2nd**   **VARS**

Highlight option "**2:normalcdf (**"

Press: **ENTER**

Since we are using the actual data in this problem and not the z-scores we
will need to enter 4 values. We now must adjust the standard deviation to fit
our sample size. $\sigma_x = 16 / \sqrt{100} = 1.6$

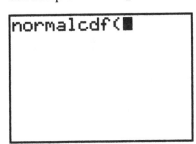

For a Sampling Distribution we must enter
Lower , upper , mean , adjusted
standard deviation) *In that order!*
Don't forget the comma's!!
For our problem we enter:
   28 , 1000 , 30 , 1.6)

*I picked 1000 because it is a very large value for this problem. (> 600
standard deviations from the mean) It will introduce very little error.*

After you have entered all 4 values *Press:***ENTER**

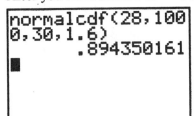

*As can be seen from the screen the
probability is* **.894350161**

d. Find P(22.1 ≤ x ≤ 26.8) *Press:* **2ⁿᵈ VARS**

```
█▓▒▓ DRAW
1:normalpdf(
▓▓normalcdf(
3:invNorm(
4:tpdf(
5:tcdf(
6:X²pdf(
7↓X²cdf(
```

Highlight option "**2:normalcdf (**"

Press: **ENTER**

```
normalcdf(█



```

For a Sampling Distribution we must enter
**Lower , upper , mean , adjusted
standard deviation)** *In that order!*

So we enter: **22.1 , 26.8 , 30 , 1.6)**
Don't forget the comma's!!

After you have entered all four values p*ress:***ENTER**

```
normalcdf(22.1,2
6.8,30,1.6)
         .0227496658
█


```

*As can be seen from the screen the
probability is* **.0227496658**

e. Find P(x ≤ 28.2)

Press: **2ⁿᵈ VARS**

Highlight option "**2:normalcdf (**"
Press: **ENTER**

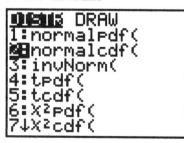

For a Sampling Distribution we must enter:
Lower , upper , mean , adjusted standard deviation) *In that order!*

So we enter: **0 , 28.2 , 30 , 1.6)** *Don't forget the comma's!!*

After you have entered all four values *press:***ENTER**

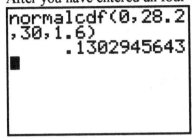

As can be seen from the screen the probability is **.1302945643**

f.   Find $P(x \geq 27.0)$

Press: **2nd**   **VARS**   *Highlight option* "**2:normalcdf (**"
Press: **ENTER**

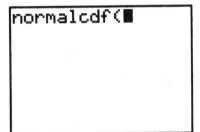

For a Sampling Distribution we must enter:
Lower , upper , mean , adjusted standard deviation) *In that order!*
So for our problem we enter **27.0 , 100 , 30 , 1.6)**

Recall our upper limit must be at least (5σ) beyond the mean so as to introduce very little error. 100 is certainly "out there" for this problem.

After you have entered all 4 values *press:***ENTER**

```
normalcdf(27,100
,30,1.6)
        .9696037028
```

As can be seen from the screen the probability is **.9696037028**

Exercise 6.41

For this problem we will once again use the Normal Distribution Function found in the Distributions Menu. The following information is given in the problem: $\mu = 7$, $\sigma = 2$, $n = 100$, and we are looking for $P(\bar{x} \leq 6.4)$

So $\sigma_{\bar{x}} = 2/\sqrt{100} = .2$

a. *Press:* **2ⁿᵈ** **VARS** *Highlight option* **"2:normalcdf ("**

Press: **ENTER**

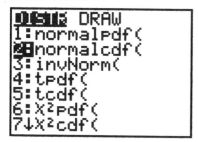 

For a Sampling Distribution we must enter:

Lower , upper , mean , adjusted standard deviation) *In that order!*

So for our problem we enter **0 , 6.4 , 7 , .2)**

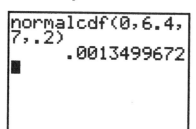

As can be seen from the screen the probability is **.0013499672**

c. We would expect a sample mean of up to 6.4 to occur only .0013 of the time if the average of 7 is correct. Therefore, it is highly probable that the mean number of sick days has been reduced.

Example 7.1

When calculating a confidence interval there are three one-sample choices
on the calculator:
1. **Z-Interval**
2. **T-Interval**
3. **Proportion Interval**

All Intervals are found under the **STAT / TESTS** Menu. For this problem
we want a Z-Interval. We have given information : $\overline{x}$ = 11.6, s = 4.1,
and n = 225 To find the 90% Confidence Interval on the calculator
Press: **STAT** ▷ ▷

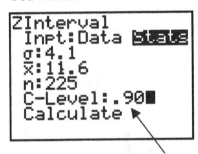

Arrow down *to* 7 : **Z Interval**
Press: **ENTER**

You will see this screen

```
ZInterval
 Inpt:Data Stats
 σ:4.1
 x̄:11.6
 n:225
 C-Level:.90▮
 Calculate ▸
```

For **Inpt** you have 2 choices:
1. *If you choose* **data** *the calculator
will read from a list*
2. *If you choose* **Stats** *then you must
enter the statistics yourself.*

For this problem we choose **Stats** *and
enter the given information.*

Remember to enter the decimal when you enter your confidence level.
Arrow down to **Calculate** and *Press*: **ENTER**

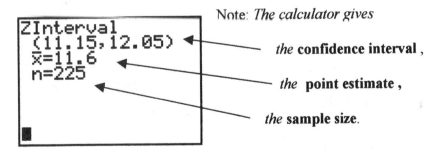

Note: *The calculator gives*

the **confidence interval** ,

the **point estimate** ,

the **sample size**.

As can be seen from the screen the 90% Confidence is: **(11.15 , 12.05)**

Example 7.2

For this problem we want a T-Interval.

a.   To find the 99% confidence on the calculator first enter the data into the calculator. After you have entered all the data into List 1

Press: **STAT**   ▷   ▷

Arrow down *to* **8 : T Interval**
Press: **ENTER**

You will see this screen

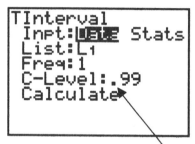

For **Inpt** you have 2 choices:
> 3. *If you choose **data** the calculator will read from a list*
> 4. *If you choose **Stats** then you must enter the statistics yourself.*

*For this problem we choose **data** and enter* **L1** *for the list.*

Remember to enter the decimal when you enter your confidence level.
Arrow down to Calculate and *Press*: **ENTER**

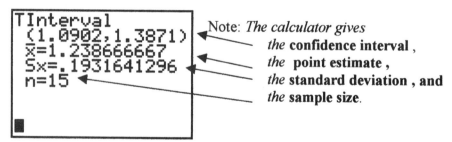

Note: *The calculator gives*
  the **confidence interval** ,
  the **point estimate** ,
  the **standard deviation** , *and*
  the **sample size**.

As can be seen from the screen the 99% Confidence Interval is **(1.09 , 1.39)** Of course for part b. we are assuming that the sample is random and comes from a nearly Normal Distribution.

Example 7.4

When calculating a confidence interval there are three one-sample choices on the calculator: 1. **Z-Interval (known sigma or n ≥ 30)**
                     2. **T-Interval (unknown sigma or n < 30)**
                     3. **Proportions (count data or %-iles)**

All Intervals are found under the **STAT / TESTS** Menu. For this problem we want a Proportion-Interval. To find the 90% Confidence Interval on the calculator

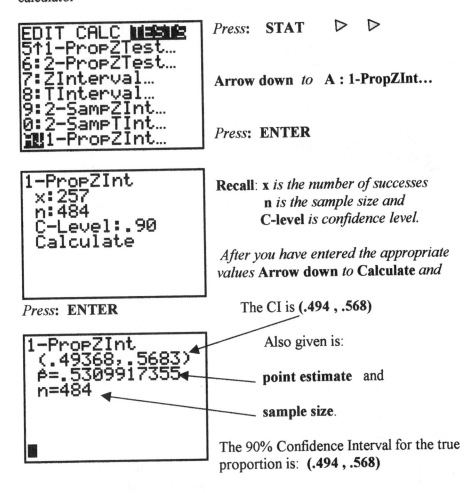

Press: **STAT** ▷ ▷

Arrow down *to* **A : 1-PropZInt...**

Press: **ENTER**

Recall: **x** *is the number of successes*
n *is the sample size and*
C-level *is confidence level.*

After you have entered the appropriate values **Arrow down** *to* **Calculate** *and*

The CI is **(.494 , .568)**

Also given is:

point estimate and

sample size.

The 90% Confidence Interval for the true proportion is: **(.494 , .568)**

Exercise 7.90

For this exercise we want two different Proportion Intervals - one for part a, and one for part b. All Intervals are found under the **STAT / TESTS** Menu.
a. To find this 90% Confidence Interval on the calculator
Press: **STAT** ▷ ▷

Arrow down *to* **A : 1-PropZInt...**

Press: **ENTER**

```
1-PropZInt
 x:11239
 n:43732
 C-Level:.90█
 Calculate
```

Recall: **x** *is the **number** of successes;*
therefore, it must be a whole number!
n *is the sample size and*
C-level *is confidence level.*

After you have entered the appropriate values **Arrow down** *to* **Calculate**
and Press: **ENTER**

As can be seen from the screen the 90% CI is **(.254 , .260)**

```
1-PropZInt
 (.25356,.26043)
 p̂=.2569971645
 n=43732

 █
```

Also given is:
   point estimate and

   sample size.

b.   To find this 90% Confidence Interval on the calculator
               Press: **STAT** ▷   ▷

```
EDIT CALC ▮▮▮▮▮
5↑1-PropZTest…
6:2-PropZTest…
7:ZInterval…
8:TInterval…
9:2-SampZInt…
0:2-SampTInt…
▮▮1-PropZInt…
```

Arrow down   *to*   **A : 1-PropZInt…**

Press: **ENTER**

```
1-PropZInt
 x:10539
 n:43732
 C-Level:.90█
 Calculate
```

Recall: **x** *is the number of successes*
             n *is the sample size and*
             C-level *is confidence level.*

After you have entered the appropriate
values **Arrow down** *to* **Calculate** *and*

```
1-PropZInt
 (.23763,.24435)
 p̂=.240990579
 n=43732
 █
```

Press: **ENTER**
As can be seen from the screen the 90% CI
is **(.238 , .244)**

Also given is:
   point estimate and
   sample size.

Example 8.2

Working a Test of Hypothesis on the calculator is very different and much less complicated than doing it by hand. And while the methods are different the results are ALWAYS the same. For all tests of Hypothesis you will go to the **STAT / TESTS** Menu. For this problem we set up our Hypothesis as :

Ho: $\mu \geq 1.2$

Ha: $\mu < 1.2$ we have given: **n = 100, x = 1.05, s = .5 and α= .01**

Remember ...... the equal sign <u>always</u> goes with   **Ho**

And you <u>always</u>   test   **Ha**

Following the test if the **"p"** value is less than the level of significance (α) we reject **Ho.**

If the **"p"** value is not less than the level of significance then we can not reject **Ho.**   In short: **p < α >>>>>>>>>>>>>> reject Ho,**

p $\geq$ α >>>>>>>> do not reject Ho.

(this is called "the rejection Rule")

Press: **STAT**

Arrow right *to highlight* **TESTS**

Since **Z-Test** *is already highlighted we simply press*: **ENTER**

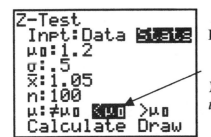

Notice: once again you choose between **Data & Stats**.

There is one new line; the **"Test Line".** *You MUST **match** the inequality in this line to the inequality in* **your Ha.**

Notice I highlighted the "**<μo** " because that is the symbol in Ha.(see above)

The value beside μ_o: under **Inpt** is <u>always</u> the value associated with the Null Hypothesis.   *For this problem it is* **1.2**

After you enter the appropriate values arrow down to **Calculate** and

Press: **ENTER**

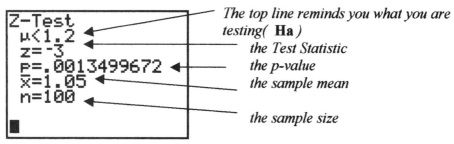

The top line reminds you what you are testing(**Ha**)

the Test Statistic

the p-value

the sample mean

the sample size

On the calculator you will reject Ho if the p-value < α.
There is no other Rule! **It will always work!**

Since **.0013 < .01** we will **reject** Ho.

Exercise 8.43

Everything for this problem is given in the problem. To begin
Press: **STAT**
Arrow right to highlight **TESTS**

Since **Z-Test** *is already highlighted we simply press*: **ENTER**

I choose Stats because I will enter the given information myself.

Notice I choose the last option here because I must match the symbol in **Ha**

After you have entered the appropriate values arrow down to Calculate and
Press: **ENTER**

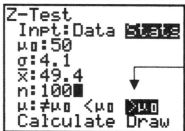

As can be seen from the screen the p-value = .928

If we choose to reject Ho there is a 92.8% chance we are committing an error.

Example 8.5

For this problem we must first enter the data into the calculator.
I will use List 1. We have given information: **α = .01** and we want to test
Ha: μ < 20.

After you have entered all the data into the list *Press*: **STAT**

Arrow right *to* **TESTS**

Arrow down *to* **2 : T-Test...**

Press: **ENTER**

For this problem we choose **Data** *and enter* **L1** *for List*

Notice we still match this line to the inequality in **Ha.**

After you have adjusted your settings to match **arrow down** to **Calculate** and *Press*: **ENTER**

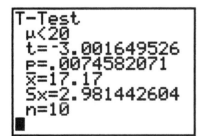

As can be seen from the screen the p-value is .007 which is less than .01 therefore we will **reject Ho.**

So yes, this type of engine does meet the standard that emissions be less than 20 ppm.

Exercise 8.59

a. To begin we note the given information: $\bar{x} = 4.8$, $s = 1.3$, $\alpha = .05$, $n = 5$

   Ho: $\mu \geq 6$

   Ha: $\mu < 6$

   So we will reject Ho if $p < .05$

Press: **STAT**

Arrow left to **TESTS**

Arrow down to **2 : T-Test…**

Press: **ENTER**

Make sure you choose: **Stats**

Notice this inequality must match the symbol in **Ha**.

After you have entered all the values from your problem arrow down to **Calculate**

Press: **ENTER**

As can be seen from the screen $p = .054$ *which is not less than .05 therefore we* **do not reject Ho.**

b. To begin, we note the given information: $x = 4.8$, $s = 1.3$, $\alpha = .05$, $n = 5$

   Ho: $\mu = 6$

   Ha: $\mu \neq 6$

   So we will reject Ho if $p < .05$

Press: **STAT**

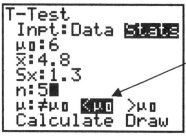

Arrow left to **TESTS**

Arrow down to **2 : T-Test…**

Press: **ENTER**

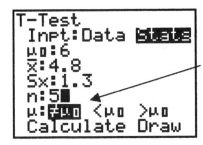

Make sure you choose: **Stats**

Notice this inequality must match the symbol in **Ha.**

After you have entered all the values from your problem arrow down to **Calculate**

Press: **ENTER**

As can be seen from the screen $p = .108$ *which is not less than .05 therefore we* **do not reject Ho.**

b. Since on the calculator we use p-values not Test-Statistics we simply give the p-values as displayed on the screens:
For part (a) $p = .054$, for part (b) $p = .108$

Example 8.7

Again, on the calculator this problem will be worked quite differently than it is by hand yet the results will be the same.

Start by noting the given information. $N = 300$, $x = 10$, $\alpha = .01$,

Ho: $p \geq 5$

Ha: $p < 5$ So we will reject Ho if our p-value $< .01$

Press: **STAT**

Arrow left to **TESTS**

Arrow Down to 5 : **1-PropZTest...**

Press: **ENTER**
You must enter all appropriate values

Once again, this line must match the inequality in your **Ha**

After you have entered the correct values arrow down to **Calculate** and
Press: **ENTER**

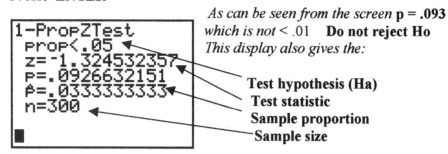

As can be seen from the screen **p = .093**
which is not < .01   **Do not reject Ho**
This display also gives the:

Test hypothesis (Ha)
Test statistic
Sample proportion
Sample size

Example 9.1

On the calculator we will go to the **STAT TESTS** menu once again.
There are 3 different 2-Sample Intervals and 3 different 2-Sample Tests
found within the **TESTS** Menu. Because of the large sample size found in
this problem we will use the 2-Sample **Z** Interval. Begin by entering all data
into List1 and List2.(lowfat diet into list1 and regular diet into list2)

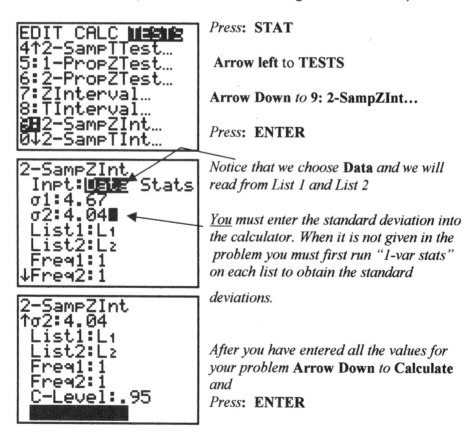

Press: **STAT**

Arrow left to **TESTS**

Arrow Down *to* **9: 2-SampZInt...**

Press: **ENTER**

Notice that we choose **Data** *and we will
read from List 1 and List 2*

*You must enter the standard deviation into
the calculator. When it is not given in the
problem you must first run "1-var stats"
on each list to obtain the standard*

deviations.

*After you have entered all the values for
your problem* **Arrow Down** *to* **Calculate**
and
Press: **ENTER**

As can be seen from the screen the 95% CI of the difference between the means is: **(.68 , 3.12)**

```
2-SampZInt
 (.69072,3.1293)
 x̄₁=9.31
 x̄₂=7.4
 n₁=100
 n₂=100
■
```

Example 9.2

Once again we go to the **STAT TESTS** Menu, and for this problem we will choose a 2-Sample **Z Test** as requested by the problem. We will test the Alternative hypothesis: the two means are not equal, and use alpha = .05 as given in the problem.

Press: **STAT**

```
EDIT CALC TESTS
1:Z-Test…
2:T-Test…
3:2-SampZTest…
4:2-SampTTest…
5:1-PropZTest…
6:2-PropZTest…
7↓ZInterval…
```

Arrow left to **TESTS**

Arrow Down *to* **3: 2-SampZTest…**

Press: **ENTER**

```
2-SampZTest
 Inpt:Data
 σ1:4.732863826…
 σ2:4.037325847…
 x̄1:9.31
 n1:100
 x̄2:7.4
↓n2:100
```

*Once again you must choose **Data** or **Stats**. For our problem we choose **Stats** because we will enter the statistics ourselves.*

Enter the values for your problem.

```
2-SampZTest
↑σ2:4.037325847…
 x̄1:9.31
 n1:100
 x̄2:7.4
 n2:100■
 μ1:≠μ2 <μ2 >μ2
Calculate Draw
```

Don't forget to arrow down and enter all values associated with your problem.

Recall: *you must match this inequality to the one found in your* **Ha**.
        (not equal)

```
2-SampZTest
  μ₁≠μ₂
  z=3.070279301
  p=.0021387241
  x̄₁=9.31
  x̄₂=7.4
↓n₁=100
■
```

After you have set all parameters for this menu **Arrow Down** *to* **Calculate** *and*

Press: **ENTER**

As can be seen from the screen the p-value is **.002** which is less than .05 therefore we **reject Ho.**

Example 9.4

Reading test scores for slow learners.

For this problem we will go back to the **STAT TESTS** Menu, but this time we will choose a 2-Sample **T** Interval as the sample size dictates.

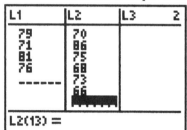

First enter the data.
Each Method into its own List.

I will enter "**New Method**" *into* **List 1** *and* "**Standard Method**" *into* **List 2**.

Press: **ENTER**
Arrow left to **TESTS** and **Arrow Down** to **0: 2-SampTInt...**

```
EDIT CALC TESTS
6↑2-PropZTest…
7:ZInterval…
8:TInterval…
9:2-SampZInt…
0▊2-SampTInt…
A:1-PropZInt…
B↓2-PropZInt…
```

Press: **ENTER**
This time we choose Data as we entered the data into **List 1** and **List 2**.

Notice we choose **Yes,** *since based on the F-test we cannot reject the claim the two sample variances are equal.* **
See note below

> **The F Test :** $Ho : var(pop1) = var(pop2) >>> s_1^2 = s_2^2$ **(equal)**
> $Ha : var(pop1) \neq var(pop2) >> s_1^2 \neq s_2^2$ **(not equal)**
> This test can be performed on the calculator. Go to the tests menu by pressing **STAT** and choose option **"D: 2-SampFTest..."**. You will have the same choices as with any of the other tests. We will highlight **Stats** for this problem since we are entering the statistics ourselves. Adjust your screen to this problem's statistics and at the test line highlight σ1: **"≠σ2"** and press **ENTER.** Arrow down to calculate and press **ENTER** We would reject Ho only if $p < \alpha$ If you reject Ho then choose **NO**, else choose **YES**

After you adjust all the fields in this menu for your problem **Arrow Down** to **Calculate** and *Press*: **ENTER**

```
2-SampTInt
 (-1.399,9.5327)
df=20
x̄₁=76.4
x̄₂=72.33333333
Sx₁=5.83476173
↓Sx₂=6.34369169
■
```

```
2-SampTInt
 (-1.399,9.5327)
↑Sx₁=5.83476173
Sx₂=6.34369169
SxP=6.11991285
n₁=10
n₂=12
■
```

As seen from the screen the 95% CI for the true mean difference is **(-1.4 , 9.5)**

Exercise 9.12

For this problem we will use a 2-Sample T Test and consider the p-value to determine if these two populations produce the same mean.
Press: **STAT**

```
EDIT CALC TESTS
1:Z-Test...
2:T-Test...
3:2-SampZTest...
■4:2-SampTTest...
5:1-PropZTest...
6:2-PropZTest...
7↓ZInterval...
```

Arrow left to **TESTS**

Arrow Down *to* **4: 2-SampTTest...**

Press: **ENTER** For this problem we choose **Stats**.

```
2-SampTTest
↑n1:17
x̄2:7.9
Sx2:4.8
n2:12
μ1:≠μ2 <μ2 >μ2
Pooled:No YES
      Draw
```

For the test line we choose $\neq$ as dictated by the problem.
Recall it must match **Ha**.

Again, based on the results of the F-Test we poole the variances.

After you have adjusted all the fields

Arrow Down *to* **Calculate** *and*
Press: **ENTER**

```
2-SampTTest        2-SampTTest
 μ1≠μ2               μ1≠μ2
 t=-1.645518248    ↑Sx1=3.4
 P=.111456498       Sx2=4.8
 df=27              Sxp=4.0295207
 x̄1=5.4             n1=17
↓x̄2=7.9             n2=12
■
```

As can be seen from the screen the p-value is .11 which is **not less** than .05 therefore we **do not reject Ho**.

b.   Go to **TESTS Arrow Down** to **0 2-SampTInt....**

```
2-SampTInt
 (-5.617,.6173)
 df=27
 x̄1=5.4
 x̄2=7.9
 Sx1=3.4
↓Sx2=4.8
■
```

Press: **ENTER**
Arrow Down *to* **Calculate** *and again*
Press: **ENTER** you will see this screen.

As can be seen from the screen the 95% CI is **(-5.617 , .6173)** *which does include zero.*
Therefore we would **not** reject Ho.

Example 9.5

To perform a Paired Difference Confidence Interval on the calculator first enter the data into List 1 and List 2. I will enter Male into List 1 and Female into List 2. The next step is to create List 3 which will be (List 1 - List 2). Enter your data into List 1 and List 2.

```
L1      L2      ■       3
29300   28800   ------
41500   41600
40400   39800
38500   38500
43500   42600
37800   38000
69500   69200
L3 =
```

For matched pair calculations you **MUST** *enter the data in the* **EXACT** *order it is given.*

After you have entered all the data Arrow up and highlight the **L3** *itself (not the top line of the data)*

Press: (**2ⁿᵈ 1** - **2ⁿᵈ 2**) **ENTER**
You will see the expression scroll across the bottom of your screen.
List 3 will be **created** as you press: ENTER

List 3 is the difference between List 1 and List 2.

Next *Press*: **STAT**

```
EDIT CALC TESTS
2↑T-Test…
3:2-SampZTest…
4:2-SampTTest…
5:1-PropZTest…
6:2-PropZTest…
7:ZInterval…
8▉TInterval…
```

Arrow left *to* **TESTS** *and* **Arrow Down**
to **8: TInterval…**

Press: **ENTER**

```
TInterval
 Inpt:DATA Stats
 List:L₃
 Freq:1▉
 C-Level:.95  ◀
 Calculate
```

Notice we choose **data** *and we are using*
List 3

The problem ask for the 95% CI.

Don't forget to enter it in decimal format.
Arrow Down *to* **Calculate** *and*
    Press: **ENTER**

```
TInterval
 (89.096,710.9)
 x̄=400
 Sx=434.6134937
 n=10

 ▉
```

As can be seen from the screen the 95% CI
is **$(89.10 , 710.90)** rounded to the nearest
penny.

Example 9.6

For the 2-Sample Proportion test we will again use the **STAT TESTS** menu
Note the given information: $n_1 = 1500$, $x_1 = 555$, $n_2 = 1750$, $x_2 = 578$
Ho: $(p_1 - p_2) \leq 0$
Ha: $(p_1 - p_2) > 0$ and we are given $\alpha = .05$
So on the calculator we will reject Ho if the p-value < .05
Press: **STAT**

```
EDIT CALC TESTS
2↑T-Test…
3:2-SampZTest…
4:2-SampTTest…
5:1-PropZTest…
6▉2-PropZTest…
7:ZInterval…
8↓TInterval…
```

Arrow Left to **TESTS** and **Arrow Down**
to **6 : 2-PropZTest…**

Press: **ENTER**
Fill in the appropriate information for
your problem

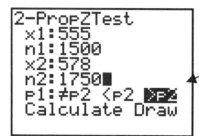

The inequality in this line must match your Ha.

After you have adjusted all fields to fit your problem Arrow Down to Calculate and

Press: **ENTER**

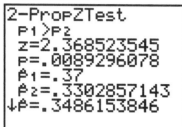

As can be seen from the screen the p-value is .0089 which is less than .05 therefore we reject Ho.

There is sufficient evidence at the $\alpha = .05$ level to conclude that the proportion of adult smokers has decreased.

Example 9.10

To perform an F-Test on the calculator there is one thing you must remember. Data from Supplier 1 goes into list 1 and supplier 2 into list 2.

> No matter what order the text puts the data in, on the calculator the data set that has the **largest variance is called sample 1 !**
> This only applies to F-Test on the calculator.

The Hypothesis for the F-Test are

Ho: s1 = s2

Ha: s1 ≠ s2 , and we will reject Ho if our p-value < .10

We will once again use the **STAT TESTS** Menu.
Press: **STAT**

Arrow Left to **TESTS** and **Arrow up** to **D: 2-SampFTest...**

Press: **ENTER**

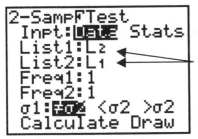

Once again we choose Data as we will enter the statistics ourselves.

Note the order. .The calculator always puts List 2 on top(it must be the sample with the largest variance)
The test line still must match Ha.

After you adjust all fields for your problem **Arrow Down** to **Calculate** and *Press*: **ENTER**

```
2-SampFTest
 σ1≠σ2
 F=4.237545365
 P=.0071180665
 Sx1=.202132858
 Sx2=.098192821
↓x̄1=4.199230769
■
```

As can be seen from the screen the p-value is .007 which is less than .10 therefore we **reject Ho**.

At the $\alpha = .10$ level the two variances are not equal.

Exercise 9.94

For this problem we begin by entering the data into the calculator.

I entered the Sample1 data into List 1 and Sample2 data into List 2.

After you have entered all the data

Press: **STAT**

```
EDIT CALC TESTS
0↑2-SampTInt…
A:1-PropZInt…
B:2-PropZInt…
C:X²-Test…
D:2-SampFTest…
E:LinRegTTest…
F:ANOVA(
```

Arrow Left to **TESTS** *and* **Arrow up** *to* **D: 2-SampFTest…**

Press: **ENTER**

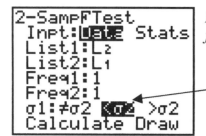

I choose Data so the calculator will read from the lists.

*This line must match your **Ha**.*

After you change all fields for your problem **Arrow Down** to **Calculate** and
Press: **ENTER**

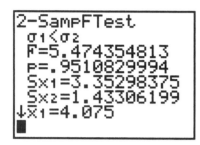

As can be seen from the screen the p-value
is **.951** which is not less than .01 therefore
we **do not reject Ho**.

b.   For this part every step will be the same except one. We will change the
test line since the Alternative Hypothesis for part b is ≠

Press: **STAT**

Arrow Left to **TESTS** and **Arrow up** to
D: 2-SampFTest...

Press: **ENTER**

*Choose **Data** so the calculator will read
from the lists.*

*This line must match your **Ha**!*

After you change all fields for your problem **Arrow Down** to **Calculate** and
Press: **ENTER**

```
2-SampFTest
 σ1≠σ2
 F=5.474354813
 P=.0978340013
 Sx1=3.35298375
 Sx2=1.43306199
↓x̄1=4.075
■
```

*As can be seen from the screen the p-value
is .0978 which is slightly less than .10,
therefore **we reject Ho**.*

*If we were to round our p-value to 2 decimal places it would be .10 which is
not less than itself.
Care must be taken to round according to the instructions given.
As you can see from this problem it could affect your answer.*

Supplemental Exercise 9.112

Here we are looking for a 2-Sample Proportion Interval. We will once again
go to the **STAT TESTS** Menu.
Press: **STAT**

```
EDIT CALC TESTS
5↑1-PropZTest…
6:2-PropZTest…
7:ZInterval…
8:TInterval…
9:2-SampZInt…
0:2-SampTInt…
A:1-PropZInt…
```

Arrow Left to **TESTS** *and* **Arrow
Down** *to* **B: 2-PropZInt…**

Press: **ENTER**

```
2-PropZInt
 x1:211
 n1:588
 x2:26
 n2:123
 C-Level:.90
 Calculate
```

*I put the Mediterranean data in as x1, n1
and the Atlantic data in as x2, n2.*

Remember to enter this in decimal format

Arrow Down to **Calculate** and
Press: **ENTER**

```
2-PropZInt
 (.07872,.2162)
 p̂1=.3588435374
 p̂2=.2113821138
 n1=588
 n2=123
■
```

*As can be seen from the screen the 90% CI
is* **(.079 , .216)**

*There is from a 7.9% increase to a 21.6%
increase in the number of bacteria found
ithin the brill captured from the
Mediterranean Sea.*

Exercise 9.133

For this problem we will use a 2-Sample **Z** Test and consider the p-value
to determine if "…Option III classrooms had significantly more timeout
incidents…" I'll put data for Option II into List 1, and Option III into List 2.

Press: **STAT**

Arrow left to **TESTS**

Arrow Down *to* **3: 2-SampZTest…**

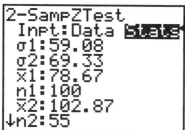

Press: **ENTER**

For this problem we choose **Stats**.

For the test line we choose < *as dictated
by our* **Ha**:

Option III significantly more than **Option II**
  List 2   significantly more than   **List 1**

After you have adjusted all fields to match your problem **Arrow Down** to
Calculate and *Press*: **ENTER**

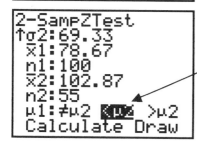

*As can be seen from the screen the
p-value is* ***.014*** *which means if we choose
to reject Ho there is a 1.4% chance we are
committing a Type-I error.*
So at alpha = .014, or anything greater, **yes
Option III has significantly more
Timeout incidents.**

Supplemental Exercise 9.117

For the 2-Sample Proportion test we will use the **STAT TESTS** Menu once again. I let Sun-Shade eggs be group 1.

Note the given information: $n_1 = 70$, $x_1 = 34$, $n_2 = 80$, $x_2 = 31$

Ho: $(p_1 - p_2) \leq 0$

Ha: $(p_1 - p_2) > 0$ and we are given $\alpha = .01$

So we will reject Ho if the p-value < .01

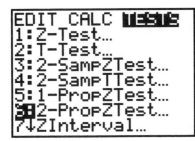

Press: **STAT**

Arrow Left to **TESTS** and **Arrow Down** to **6 : 2-PropZTest…**

Press: **ENTER**

Fill in the appropriate information for your problem

The inequality in this line must match your Ha.

After you have adjusted all fields to fit your problem **Arrow Down** *to* **Calculate** *and*

Press: **ENTER**

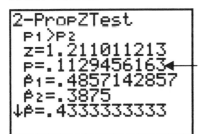

As can be seen from the screen the p-value is **.1129** *which is not less than .01 therefore we* **do not reject Ho**.

Example 10.4

We begin by entering all data into the calculator. Enter each brand into its own list. As with all test of Hypothesis we will test Ha and reject Ho only if our p-value $< \alpha$

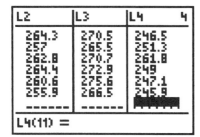

For all ANOVA problems the Hypothesis will be Ho: all means are equal
      Ha: at least one mean is different

After you have entered all the data

Press: **STAT**

Arrow Left to **TESTS** and **Arrow up** to **F : ANOVA(**

Press: **ENTER**

You will see this screen:

For ANOVA you simply enter the list where you placed the data.

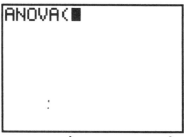

For this problem I will enter L1,L2,L3,L4 as I used said lists.

Press: **2nd 1 , 2nd 2 , 2nd 3 , 2nd 4)**

Press: **ENTER**   *To see all the results you must arrow down.*

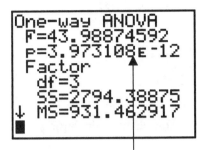

All values requested can be read directly from the screens.

As can be seen the p-value is given in Scientific Notation. In common Notation it is **.000000000003973108** So we feel very confident that all means are not equal!

Exercise 10.25

We begin by entering all data into the appropriate list.

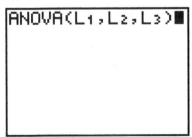

Note: In ANOVA it is not necessary that the list be of equal length.

Press: **STAT**

```
EDIT CALC TESTS
0↑2-SampTInt…
A:1-PropZInt…
B:2-PropZInt…
C:X²-Test…
D:2-SampFTest…
E:LinRegTTest…
F ANOVA(
```

Arrow Left to **TESTS** and **Arrow up** to **F : ANOVA(**

Press: **ENTER**

```
ANOVA(L₁,L₂,L₃)█
```

For ANOVA you simply enter each list where you have the data.

For this problem I will enter L1,L2,L3 as I used said lists.

Press: **2ⁿᵈ** **1** **,** **2ⁿᵈ** **2** **,** **2ⁿᵈ** **3** **)**

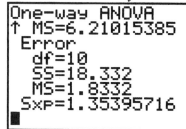

Press: **ENTER**   To see all the results you must arrow down.

All values requested can be read directly from the screens as the calculator uses standard notation.

b.   We see from the first screen the p-value is **.075** which is not less than .01 therefore we **do not reject Ho**.

Example 11.4

Begin by entering the data into the calculator.

Be careful to enter the data in the exact order it is listed.

Notice I place the Crime Rate(dependent) into List 2 and the Number of Casinos (Independent) into List 1.

After you have entered all the data into the calculator

```
EDIT CALC TESTS
2↑2-Var Stats
3:Med-Med
4:LinReg(ax+b)
5:QuadReg
6:CubicReg
7:QuartReg
8↓LinReg(a+bx)
```

Press: **STAT**
Arrow Right *to CALC and* **Arrow Down** *to* **8: LinReg(a+bx)**

Notice option 4 can also be used but it will give the equation in (ax + b). Like algebra (mx+b) form ☺

Press: **ENTER**

```
LinReg(a+bx) ■
```

For regression you need only tell the calculator where you put your data and which place to store the equation.
It must however be in this order:
x's , y's , where

I used List 1 and List 2 for the x's and y's respectively, place results in Y1.

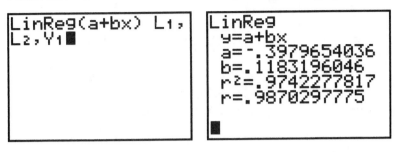

Press: **2ⁿᵈ 1 , 2ⁿᵈ 2 , ᵥARS ▷ ENTER ENTER**
Press: **ENTER**

As can be seen from the screen **r = .987** and **r² = .974**

Example 11.5

To perform a Linear Regression on the calculator first enter the data into the calculator. Enter the Independent variables(x's) into List 1 and the dependent variables(y's) into List 2.

*For all **regressions** both lists MUST be the same length!*

After you have entered all the data you will need to turn on a **STAT PLOT**.
Press: **2ⁿᵈ Y=**

Plot1 *should be highlighted.*

Press: **ENTER**

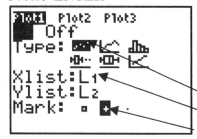

This is the Statistical Plots Menu.

Highlight **On** *and Press*: **ENTER**

Choose the first plot option, Scatterplot.
Enter the list where you put the data.
Choose any Mark you desire.

Press: **ZOOM** 9

This will give you a scatter plot of the data.

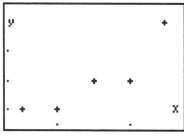

If you want to see the exact picture found
in the text you will need to adjust the
parameters within the **WINDOW** *Menu*
(see p.6 for a refresher)

Now we are ready to run the regression.
Press: **STAT**

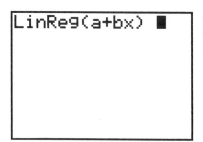

Arrow Right *to* **CALC** *and* **Arrow Down**
to **8: LinReg(a+bx)**

Notice option 4 can also be used but it will
give the equation in form (ax + b).[Like
algebra (mx+b) form ☺]

Press: **ENTER**

```
LinReg(a+bx) ■
```

For regression you need only tell the
calculator where you put your data and
which place to store the equation.
It must however be in this order:
x's , y's , where

I used List 1 and List 2 for the x's and y's respectively and I will place the
results in Y1.

Press: **2nd 1 , 2nd 2 , VARS ▷ ENTER ENTER**

Press: **ENTER**

```
LinReg(a+bx) L₁,
L₂,Y₁█
```

```
LinReg
 y=a+bx
 a=⁻.1
 b=.7
 r²=.8166666667
 r=.9036961141
```

As can be seen from the screen the equation is: **y = -.1 + .7x** , and
r² = .817

If you do not see the r and r² you will need to follow these steps:
Press: **2nd 0 x⁻¹**
Next **Arrow Down** to **DiagnosticsOn**
Press: **ENTER ENTER**
This will turn on the diagnostics and you will see the r and r² on
all subsequent regressions.

Press: **GRAPH**

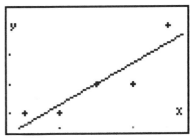

You will see the data and the Least Squares Regression Line.

By pressing **TRACE** *you can move along the line or to each of the data points.*

To **predict a y-value** from a **given x-value** follow these steps:
Press: **2nd TRACE ENTER**

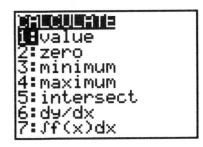

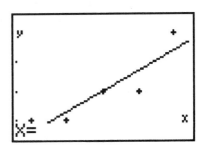

Simply enter the given x-value and *press*: **ENTER**

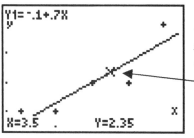

Notice the flashing cursor showing the predicted value's location on the Regression Line

I entered 3.5 and the results, as can be seen from the screen, is **y = 2.35**

Exercise 11.26

We begin by entering the data into the calculator. I will put the x's into List 1 and the y's into List 2.

After you have entered all the data you should be looking at this screen.

Notice I put the pectin level(independent) into list 1 and the sweetness index(dependent) into list 2, since the sweetness index should be based upon the pectin level and not vice/versa.

Press: **STAT**

Arrow Right *to* **CALC** *and* **Arrow Down** *to* **8: LinReg(a+bx)**

Notice option 4 can also be used but it will give the equation inform (ax + b). Like algebra (mx+b) form ☺

Press: **ENTER**

For regression you need only tell the calculator where you put your data and which place to store the equation.
It must however be in this order:
x's , y's , where

I used List 1 and List 2 for the x's and y's respectively and I will place the results in Y1.

Press: **2ⁿᵈ 1 , 2ⁿᵈ 2 , VARS ▷ ENTER ENTER**
Press: **ENTER**

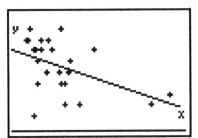

As can be seen from the screen the Least Squares Line is **y = 6.25 - .0023x**
Press: **ZOOM   9**

Once you have the graph displayed you can use the **CALC** *feature of the calculator to predict specific y-values.*

To **predict a y-value** from a **given x-value** follow these steps:
Press: **2ⁿᵈ   TRACE   ENTER**

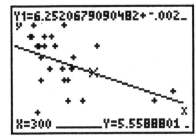

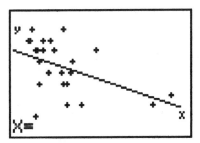

Simply enter the given x-value and *press*: **ENTER**

As can be seen from the screen if x = 300 we predict **y = 5.56**

Exercise 11.115

Begin this problem by entering the data into the calculator. I will put the x's into list 1 and the y's into list 2.

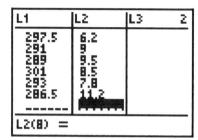

Please note the list are opposite the order given in the textbook.

After you have entered all the data you will need to turn on a **STAT PLOT**.
Press: **2nd   Y=**

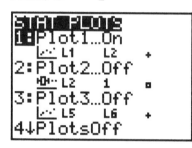

Plot1 *should be highlighted.*

Press: **ENTER**

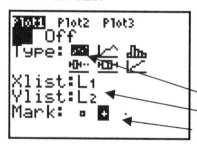

This is the Statistical Plots Menu.

Highlight **On** *and Press*: **ENTER**

Choose the first plot option, Scatterplot.
Enter the list where you put the data.
Choose any Mark you desire.

Press: **ZOOM   9**

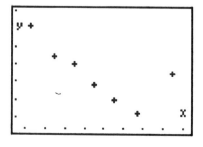

This will give you a scatter plot of the data.

For part **b.** we must run a Linear Regression

Press: **STAT**

Arrow Right *to* **CALC** *and* **Arrow Down** *to* **8: LinReg(a+bx)**
Press: ENTER
I used List 1 and List 2 for the x's and y's respectively and I will place the results in Y1.

Press: **2nd 1 , 2nd 2 , VARS ▷ ENTER ENTER**

Press: **ENTER**

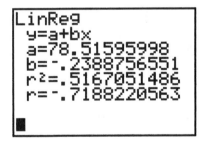

As can be seen from the screen the Equation for the line of best fit is:
$$y = 78.516 - .2389x$$

Press: **GRAPH**

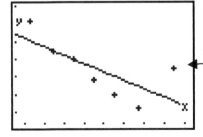

As you can see on the screen there is one point that appears "out of place"

To conduct a test of model adequacy we go to the **STAT TESTS** Menu.

Press: **STAT**

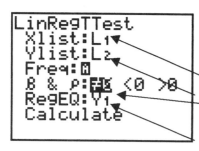

```
EDIT CALC TESTS
0↑2-SampTInt…
A:1-PropZInt…
B:2-PropZInt…
C:χ²-Test…
D:2-SampFTest…
E:LinRegTTest…
F:ANOVA(
```

Recall, the Hypothesis are as follows:

Ho: β = 0, Ha: β ≠ 0
and we reject Ho if $p < \alpha$
(see section 11.5 for details)

Arrow Left *to* **TESTS** *and* **Arrow Up** *to*
E : LinRegTTest…

Press: **ENTER**

```
LinRegTTest
 Xlist:L₁
 Ylist:L₂
 Freq:0
 β & ρ:≠0 <0 >0
 RegEQ:Y₁
 Calculate
```

Enter all fields appropriate to your problem.

I put the x's in List 1
I put the y's in List 2

Ho: $\beta = 0$, **Ha:** $\beta \neq 0$
Put results in Y1

After you have adjusted all fields to match your problem **Arrow Down** to **Calculate** and *Press:* **ENTER**

```
LinRegTTest
 y=a+bx
 β≠0 and ρ≠0
 t=-2.312066453
 p=.0687299155
 df=5
↓a=78.51595998
```

```
LinRegTTest
 y=a+bx
 β≠0 and ρ≠0
↑b=-.2388756551
 s=1.265067737
 r²=.5167051486
 r=-.7188220563
```

As can be seen from the screen the p-value is **.069** which is not less than .01 Therefore for part **d, do not reject Ho**.

To eliminate the one piece of data from the data set you must delete both the x-value and the corresponding y-value. Go to the list.

```
L1      L2      L3     2
295     7       ------
297.5   6.2
291     9
289     8.5
301     9.6
293     7.8
286.5   11.2

L2(5) =8.5
```

*We have to delete both the **301** and the **8.5***

*Highlight **301** and press **DEL***
*Highlight **8.5** and press **DEL***

After you have deleted both values rework the problem exactly as you did above.

Example12.17

For this problem we will begin by looking at a plot of the residuals and then proceed to the data transformation, and finally we will determine the Line of best fit for the transformed data.

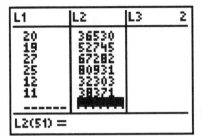

Enter all data into its appropriate list.

After you have entered all the data

Press: **STAT**

Arrow right to **CALC** and **Arrow Down** to **8 : LinReg(a+bx)**

Press: **ENTER**

Enter the location for the X's and Y's

Press: **2nd** **1** , **2nd** **2**

You must run the regression first in order to have the residuals to plot.

```
LinReg
 y=a+bx
 a=11369.41646
 b=2141.309246
 r²=.7869062045
 r=.8870773385
```

Note: The R^2 vaule = 79%

Now we will plot the residuals

Now we will go to the Stat Plot menu and plot the residuals and the original
x values using the scatter plot option.
Press : **2ⁿᵈ Y =**

Press: **ENTER**

Make sure **On** is highlighted and *Press*: **ENTER**

Highlight the scatter plot option and
Press: **ENTER**

Arrow Down to **Ylist**: and *Press*: **CLEAR**

This is where we will place the residuals.

Press: **2ⁿᵈ STAT**

Arrow Down to 7: **RESID** and *Press*:
ENTER

For the **TI-83**

Press: **2ⁿᵈ STAT**

Arrow Down to 4: **RESID** and *Press*:
ENTER

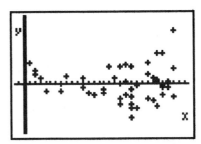

Next Press: **ZOOM   9**

We see the same residual plot as Figure 12.46

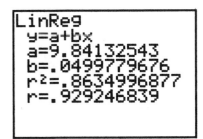

We now want to transform the data to try to reduce the amount of increasing variability.

Press: **STAT   ENTER**

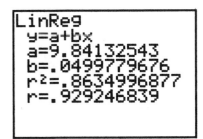

Highlight the L3 all the way at the top of the column and
Press: **LN   2ⁿᵈ   2   ENTER**

The transformed data will appear in List 3.

Now we want to re-plot the residuals using the newly created list 3.

Return to the **STAT CALC** Menu and run the regression using L1 and L3. This will replace the previous residuals with the residuals formed by using the newly transformed data.

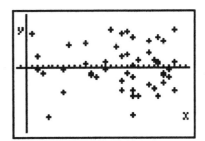

Note: *The R^2 value has also increased.*
To see the new plot of the residuals simply Press: **ZOOM   9**
As you can see from the screen the transformation stabilized the error variances.

Example 12.7

We begin this problem by entering the data into the calculator. After which we will run the Quadratic regression.

Enter size of home into List 1 and Monthly-usage into List 2.

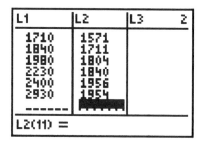

After you have entered all the data
You will want to see a scatter plot of the
data so

Press: **2ⁿᵈ** **Y=**

Plot1 *should be highlighted.*

Press: **ENTER**

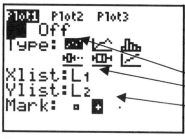

This is the Statistical Plots Menu.

Highlight **On** *and Press:* **ENTER**

Choose the first plot option, Scatterplot.
Enter the list where you put the data.
Choose any Mark you desire.

Press: **ZOOM** **9**

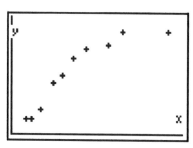

As can be seen from the screen there is a
noticeable curve. A linear regression
would not be appropriate. To run the
Quadratic regression

Press: **STAT** **Arrow Right** to **CALC**
And **Arrow Down** to **5:QuadReg**

At the flashing curser you must enter the lists where you placed the data in this order: x's , y's .

Also, if you want to see the graph you will need to tell the calculator where to send the regression equation.

Make sure your Y1 is clear of any old equations

Press: **2ⁿᵈ**   **1**   ,   **2ⁿᵈ**   **2**   ,   **VARS** ▷   **ENTER**   **ENTER**

Press: **ENTER**

QuadReg L₁,L₂,Y₁

As can be seen from the screen the Quadratic equation that best describes the data is:
Y = −.00045x² + 2.3989x − 1216.143887

To fit the variables given in the text
β_1 = -.00045
β_2 = 2.3989
β_3 = -1216.1439

To see the graph of the regression equation along with the data press: **GRAPH**

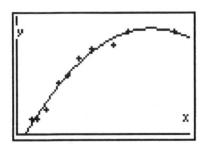

With an R^2 = .9819 we expect the close fit we see on the screen.

Example: Journal of Marketing   Section 13-3

To perform a χ^2 test using the calculator we first enter the observed data into a Matrix (usually A) using the calculator's Matrix Menu. We will then enter the expected data into another Matrix (usually B). From there we can perform the χ^2 test.

Press: **MATRIX**

Arrow Right to **EDIT** and

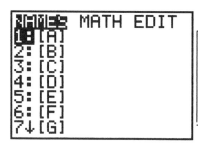

Press: **ENTER**

> For the
> *Press*: **2ⁿᵈ   x⁻¹**
> **Arrow Right** to **EDIT** and
> *Press*: **ENTER**

Enter the dimensions of your matrix. For this problem it is a 2 X 2

Enter each value pressing ENTER after each piece of data.

You have completed Matrix A. Now on to build Matrix B.
Press: **MATRIX**
Arrow Right to **EDIT** then **Arrow Down** and highlight **2:[B]** then *Press*: **ENTER**

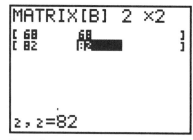

> For the
> *Press*: **2ⁿᵈ   x⁻¹**
> **Arrow Right** to **EDIT** then **Arrow Down** and highlight **2:[B]** then *Press*: **ENTER**

Recall: your Expected Matrix must be the same Dimension as your Observed Matrix.

Enter all values pressing ENTER after each piece of data.

After you have entered all the data into the matrices *press*: **STAT**
Arrow Left to **TESTS** and **Arrow Up** to **C: χ^2–Test....** *Press*: **ENTER**

This is the Chi square test. Recall we placed the Observed values into Matrix A and the Expected values into Matrix B.

```
X²-Test
 Observed: [A]
 Expected: [B]
 Calculate Draw
```

If you do not automatically see **[A]** & **[B]**
Arrow Down *and beside*
Observed: *Press:* **MATRIX   ENTER**

Arrow Down and beside
Expected: *Press:* **MATRIX Arrow Down**
to **[B]** *and Press:* **ENTER**

```
X²-Test
 X²=39.22166428
 P=3.783191E-10
 df=1
```

For the   **Arrow Down** *and*
beside **Observed:** *Press:* **2ⁿᵈ   x⁻¹**
ENTER , **Arrow Down** and beside
Expected: *Press:* **2ⁿᵈ   x⁻¹ Arrow**
Down *to* **[B]** *and Press:* **ENTER**

Next, **Arrow Down** to **Calculate** and
Press: **ENTER**

As can be seen from the screen the χ^2 value is **39.222.** *The associated p-value is* .000000000378, *Which is far below any reasonable α value.*

We therefore conclude that Brand Awareness and Gender are **not** Independent events.

Exercise 13.24

To perform the χ^2 test we first enter the observed data into Matrix A and the expected data into Matrix B. From there we perform the χ^2 test using
$\alpha = .05$

Press: **MATRIX**
Arrow Right to **EDIT** and
Press: **ENTER**

```
NAMES  MATH EDIT
1: [A]
2: [B]
3: [C]
4: [D]
5: [E]
6: [F]
7↓[G]
```

For the
Press: **2ⁿᵈ   x⁻¹**
Arrow Right to **EDIT** and
Press: **ENTER**

Enter the dimensions of your matrix. For this problem it is a 3 X 3

Enter each value pressing ENTER after each piece of data.

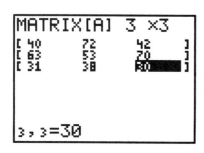

After you have completed Matrix A you must enter the expected values for Matrix B.

Now on to build Matrix B.

Press: **MATRIX**

Arrow Right to **EDIT** then **Arrow Down** and highlight **2:[B]** then

Press: **ENTER**

For the *Press*: **2ⁿᵈ** **x⁻¹** **ENTER**

Arrow Down and beside **Expected**: *Press*: **2ⁿᵈ** **x⁻¹** **Arrow Down** *to*

[B] *and Press*: **ENTER**

Recall: your Expected Matrix must be the same Dimension as your Observed Matrix.

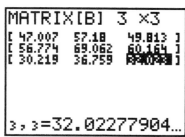

Enter all values pressing ENTER after each piece of data.

After you have entered all the data into the matrices *press*: **STAT**

Arrow Left to **TESTS** and **Arrow Up** to

C: χ^2–Test.... *Press*: **ENTER**

This is the Chi square test. Recall we placed the Observed values into Matrix A and the Expected values into Matrix B.

Now, **Arrow Down** to **Calculate** and *Press*: **ENTER**

```
X²-Test
 Observed:[A]
 Expected:[B]
 Calculate Draw
```

As can be seen from the screen the χ^2 value is **12.327**

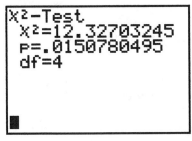

The associated p-value is .015; which is not less than .01 therefore we cannot reject Ho: the events are Independent.

We therefore conclude that **A** and **B** are **Independent events**. ☺

ERROR Messages

DIM MISMATCH is by far the most common **Error** message.

```
ERR:DIM MISMATCH
1▪Quit
```

To clear this message
press [   ] several times

and/or

[   ]   then [   ]

The steps above will return you to the home screen, but they will not remedy the reason for the error! This message is displayed indicating an error in one of two main categories:

A   the Lists are different lengths (dimensions)
B   the designated List may be empty

Error Category A. for DIM MISMATCH

Look at this screen below carefully.

```
L1      L2      L3      2
40      38      2
80      85      -5
25      19      6
50      52      -2
58      58      0
----    51      ----
        ▮▮▮▮▮
L2(11) =
```

The error at the top of this page was given when I tried [   ][   ] using
List1 , List 2
to graph a Histogram.

Look carefully at List 1 and List 2.

They are different lengths.

I thought I was careful in entering the data, yet I received the ERROR message.

First I went back to the Stat Plot.

After checking my Stat Plot and seeing that everything looked correct I decided to check the actual data.
That is when I found the error!
Lists must be the same length!

Error Category **B** for DIM MISMATCH

I was trying to graph a Box & Whisker plot when I received this message:

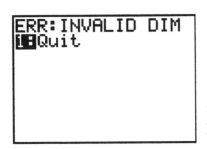

I went to **Stat Plot** and looked at the set-up. Everything seemed to be correct.

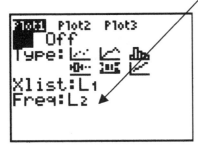

I went to the **Lists** to make sure the Lists were the same length.

They looked the same length.

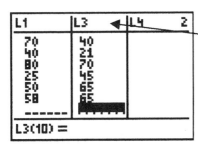

 BUT look closely at the very top of the screen: **L1   L3** ???

Where is **L2** ?
That is where I told the calculator to find the Frequencies.

I had mistakenly deleted List 2 when I was trying to clear it for this problem.

Apparently, I pressed **DEL** instead of **CLEAR** and I deleted List 2!

At this point I could just go to Stat Plot and replace **L2** with **L3** and I will get my Box & Whisker plot, but that will not return my List 2.

To return any or all the lists to the editor follow these steps:

```
EDIT CALC TESTS
1:Edit…
2:SortA(
3:SortD(
4:ClrList
5:SetUpEditor
```

From the home screen *Press:* ☐
Arrow down to "**5:SetUpEditor**"

Press: ☐ ☐

```
SetUpEditor
              Done
```

This screen tells me that I have corrected my deletion and that all Lists are back to where they should be ☺

To verify I *Press:* ☐ ☐
I see that List 2 is back between List 1 and List 3 where it belongs.

```
L1      L2      L3      1
30      ------  25
20              32
60              34
70              40
40              21
80              70
25              45
L1(1)=30
```

Now, to get the Box & Whisker plot I simply go back to Stat Plot and change the frequency list from **L2** to **L3**.

Or

Re-Enter the frequencies into List 2.

You can clear the error message by following the same steps as described on the top right of page 82. ☺

INVALID DIM is another ERROR message that

you may see:

```
ERR:INVALID DIM
1■Quit
```

This error occurs when there is an entry that doesn't "make sense" to the calculator. It most often occurs when you mistakenly enter a stray keystroke.

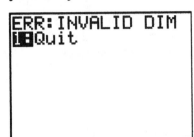

I was trying to graph a Histogram when I mistakenly entered this **"Y"** when I thought I was entering a **"1"** for **Freq:** To enter a **"1"** you must

Press: ☐ **1**

When you arrow down to **Xlist:** *&* **Freq:** *the calculator is already in* **ALPHA MODE** *so by pressing* ☐ *you return the calculator to* **STANDARD MODE**. Yes, I know it seems backwards but it works☺

You can clear the error message by following the same steps as described on the top right of page 82. ☺

SYNTAX is another **ERROR** message that occurs with some regularity.

This error occurs when the calculator is expecting a particular type or number of response and you have neglected to give it what it wants.

This particular screen appeared when I entered a "**-**" sign instead of a "**(-)**" sign when I was adjusting the window.

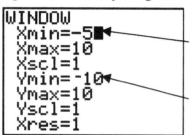

Please note the position and size of the subtraction sign.

It is smaller and placed higher than the negative sign.

When I pressed ENTER I received the above Error message.
You must also be careful when using **any** of the functions from the **Distributions Menu.**

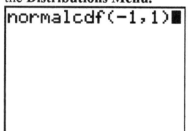

Here I am trying to find the area under the standard normal distribution curve between **negative 1 and positive 1** , but what I've entered is **subtract 1 comma 1.**

Subtract 1 comma 1 makes no sense and the calculator returns the following error message:

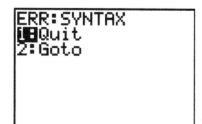

You must take care to use the correct syntax when entering values into your calculator.

You can clear the error message by following the same steps as described on the top right of page 82. ☺

DOMAIN is another **error** message that can occur.

This error occurs when you have entered a value that is outside the expected *domain* (remember algebra)

You can clear the error message by following the same steps as described on the top right of page 82. ☺

I was trying to perform a 1-Proportion Z-Test when I received the error message above.

The error occurred at the very first line. I entered the whole number "30" for the p_0 :

The calculator is expecting a decimal (number between 0 and 1)
Even though I meant 30% the correct entry would be " .30 "
This error can occur when you are using any of the Statistical TESTS or Distributions.

Here is another example of a value outside the expected domain that resulted in this error message.

I was trying to calculate the probability for a problem using the binomial distribution function.

As you can see the second entry is " 25 "
It is the percent that was listed in my book, but recall the proper notation for 25% is " .25 " ; a decimal.

I should have entered " .25 " instead of " 25 "